Advances in Agricultural Extension
Towards Changing the Lives and Livelihoods

Advances in Agricultural Extension

Towards Changing the Lives and Livelihoods

Editors

Prof. S. Venku Reddy

Prof. M. Surya Mani

BSP **BS Publications**

A unit of **BSP Books Pvt. Ltd.**

4-4-309/316, Giriraj Lane, Sultan Bazar,

Hyderabad - 500 095

Phone : 040 - 23445605, 23445688

Advances in Agricultural Extension: Towards Changing the Lives and Livelihoods

Edited by : **Prof. S. Venku Reddy and**

Prof. M. Surya Mani

Published by :

BS Publications

A unit of **BSP Books Pvt. Ltd.**

4-4-309/316, Giriraj Lane, Sultan Bazar,
Hyderabad – 500 095
Phone : 040 – 23445605, 23445688
e-mail : info@bspbooks.net
www.bspbooks.net

Printed at

Aditya Offset Process (I) Pvt. Ltd.
Hyderabad.

Price : Rs. 1995.00

ISBN : 978-93-5230-120-1

A Contribution
From

Sarvareddy Venkureddy
Foundation for Development
(SVFD)

M.S. SWAMINATHAN RESEARCH FOUNDATION

M.S. Swaminathan
Founder Chairman
Ex-Member of Parliament (Rajya Sabha)

Foreword

Agriculture is the back bone of our rural economy. This is why I have often said that if agriculture goes wrong, nothing else will go right. Also the future will belongs to nations with grains and not guns. It is in this context that this present book on agricultural extension is very timely. We are indebted to Prof S Venku Reddy for the trouble he has taken to compile such a useful book. I am particularly happy that the book shows the way to changing lives and livelihoods through knowledge and skill empowerment.

What the country needs is an ever-green revolution leading to increase in productivity in perpetuity without ecological harm. The 2016 marks the beginning of the UN Sustainable Development Decade. Goal 2 of SDGs places emphasis on end hunger, achieve food security and improved nutrition and promote sustainable agriculture. To achieve this goal we need to strengthen our extension services and provide value added information at the right time and place. This book shows the way to achieving this goal. I hope it will be widely read and used for converting scientific know-how into field level do-how.

M S Swaminathan

3rd Cross Road, Taramani Institutional Area, Chennai (Madras) - 600 113, India
Phone: +91-44-2254 2790, 2254 1698 Fax: +91-44-2254 1319
E-mail: founder@mssrf.res.in, swami@mssrf.res.in

PREFACE

Agriculture has been identified as the lead sector for growth in developing countries. In many countries it is a way of life and continues to be the single most important livelihood of the masses. It is also the architect of our food and livelihood security, contributing significantly towards economic growth and social transformation.

Today`s agriculture is facing unique problems of under employment, scarcity of skilled labour and profitable returns to farmers. This poses a great challenge in filling the growing gap between "scientific know how and field level do how".

World over, agricultural extension is in focus due to its progressive metamorphosis in transfer of technology and the sensitivity of various national governments in funding public extension. It is high time that we revisit our extension approaches, skills, methodologies and livelihood strategies to make them adaptable and sustainable.

There is also a need for new direction and retooling the scope of Agricultural Extension. This book is a good collection of extension approaches of many public extension services, private and non government sectors which are worth mentioning for replicating at different agro climatic zones. It is a fine synthesis of issues and strategies authored by veteran extension professionals and need to be widely disseminated. This book will be a valuable addition for reinventing Agricultural Extension and livelihood strategies.

The book is divided into six chapters. The first chapter deals with trends in Agricultural Extension where in the changing roles of agricultural extension for meeting the millennium goals, inclusive growth and total development is projected with a background of extension reforms. The second chapter dealt at length the innovative extension approaches, the need for a broad based extension services, institutional development including the Farmer Producer Associations and public – private partnership in extension. While the third chapter deals with new extension methods and skills, the fourth chapter describes the extension interventions including mass media and Information Communication Technologies.

The potential of Agricultural Extension for changing the lives and livelihoods is presented in chapter five. The final chapter narrated the focused areas for extension intervention, convergence & including climate change to meet the challenges and for ushering evergreen revolution.

The book can be used by students, professionals of Extension research and decision makers. The editors are greatfull to all the authors for their outstanding contributions. Special thanks are due to Dr. V.V. Sadamate, Former Advisor (Agril) Planning Commission GoI for providing insights in to the chapters. We are indepted to Prof. M.S. Swaminadhan for his forward. We acknowledge the efforts of Mrs. A. Chitra, P.S. to Director, EEI and Mrs. C. Maria Victoria, Secretary, PRDIS & SVFD for their assistance in formatting the contents of the book.

- Editors

ABOUT THE EDITOR

Prof. Sarvareddy Venku Reddy, *M.Sc (Ag), Ph.D*, (IARI) Chairman, Sarvareddy Venkureddy Foundation for Development (SVFD), President and Executive Director, of Participatory Rural Development Initiative Society (PRDIS) is an Agricultural Extension and Rural Development Specialist with over four decades of experience in teaching research, training and extension work connected with Agricultural and Rural Development. He has hands on experience in extension management, sustainable agriculture, institutional development, evaluation studies and capacity building programmes. He worked in several capacities in India and abroad as Professor, Advisor, and Consultant to different National and international Institutions including the World Bank, FAO.

As an Agricultural Extension Services Advisor, Ministry of Agriculture, Republic of Uganda, he guided in designing methodologies for implementation of pilot interventions such as Unified Extension Services, Village Level Participatory Approach, Decentralization, Cost – Sharing and out sourcing of Extension and Research services. He also assisted the Ministry in coordinating the projects, consultants, NGOs, media resource centers and other organizations. As a Consultant to World Bank, he participated in appraisal and supervision of IDA funded projects on Agricultural Extension, Research and Irrigation in several districts of India and Africa with Bank Missions.

He worked as Technical Manager to Second Agricultural Extension and Training Project, Somalia and Coordinated with different National and International agencies like FAO, USAID and IFAD for integrating and implementing the extension plans of the sub-project under National Extension System. Dr. Reddy also developed innovative Participatory Extension Methodologies and Communication Techniques for use of field extension and training.

As Professor and Head of Extension Education Institute, ANGR Agriculture University, Hyderabad he was engaged in teaching and guiding post graduate students and organizing number of training programmes for middle and national level functionaries engaged in development. He has several publications, manuals and books at his credit.

As a President and Executive Director of PRDIS and also Chairman of SVFD involved in implementation of development programmes in agriculture and rural development sectors. He had unique opportunity to receive Gold Medal at Ph.D level several honours including Presidential Award and Indira Gandhi Unity Award besides associating with NormanBorlaug, Nobal Lauret and Dr. DanielBenor, Extension Specialist, World Bank. Prof. Reddy served as a working group member of Agriculture Extension of Planning Commission, Knowledge Commission and Expert Member of Agri Biodiversity Committee of GoI. He is at present serving as a Expert member of Biodiversity Management Committees of National Biodiversity Board, Director of J.N.Agri Tech Ltd. of Uganda and also chairperson of 'PARENT' an NGO association.

Email: sarvareddy@yahoo.com

ABOUT THE EDITOR

Prof. Medapati Surya Mani, *M.Sc (Ag), Ph.D,* Director, Extension Education Institute (EEI), Hyderabad has more than 29 years of experience in Teaching, Training, Research, Extension and served as Consultant for International & National Projects funded and supported by FAO (EU Nations), IFAD (Netherlands), AP Bio-technology, WHO, UNDP, GOI, Govt. of Andhra Pradesh and Telangana. She has been constituted as *Planning Commission Working Group Member* for Twelfth Five Year Plan and provided farmer feedback to the group. Guided facilitators and monitored Farmer Field Schools (FFS) on cotton, vegetables and paddy in A.P. and ICM paddy in Chattisgarh CTDP supported by IFAD and sugarcane TOF at RARS, Anakapalle supported by WHO.

She is having expertise in conducting monitoring and evaluation of Rural Development Programmes sponsored by GOI. As an evaluation expert member, Dr. Mani has worked in State level Bt Cotton Impact assessment study funded by DoCD, GoI, Mumbai. She is well experienced in management of programmes connected with Agricultural Extension, Rural Development and worked for formation and strengthening of Self-Help Groups (SHGs). She also handled U.G., P.G and Ph.D. courses besides guiding research scholars and organizing regular training programmes at EEI.

Dr. Mani has number of International and National publications to her credit in addition to, a book on *Biotechnology for sustainable agriculture* and course material for PGDAEM and a chapter in *International Pesticide Policy project-Series No. 9* (FAO, EU IPM programme for cotton Asia) and also manuals for *Adopt a village programme* of CAPART, GOI and WHO supported ToFprogramme on *Sustainable Sugarcane Production*. Her areas of specialization include Farmer Field Schools, Participatory Approaches, Monitoring and Evaluation of Rural Development Programmes, Impact Studies, etc.

e-mail: medapati_mani25@ahoo.co.in

CONTENTS

ABBREVIATIONS

ABI: Agri-Business Incubation
ACABC: Agri-Clinic and Agri-Business Centres
AEO: Agricultural Extension Officer
AES: Agricultural Extension Service
AES: Agro-ecological Systems
AESA: Agro Eco system Analysis
AGB: ATMA Governing Board
AIBP: Accelerated Irrigation Benefit Programme
AICRP: All India Co-ordinated Research Projects
AIS: Agricultural Innovation System
AKIS: Agricultural knowledge information system
ALSS: Agricultural Livelihoods Support Systems
AMC: ATMA Management Committee
AMUL: Anand Milk Union Limeted
ANGRAU: Acharya N.G. Ranga Agricultural University
ANP: Applied Nutrition Programmes
APC: Agricultural Policy Committee
APMC: Agricultural Produce Marketing Committee
APRLP: A.P Rural Livelihood Programme
AR4D: Agricultural Research for Development
ARDF: Agricultural and Rural Development Foundation
ASCI: Agricultural Skill Council of India
ATM: Assistant Technology Managers
ATMA: Agricultural Technology Management Agency

BAP: Block Action Plans
BCI: Better Cotton Initiative
BFAC: Block Farmers Advisory Committee
BTT: Block Technology Team

CAPART: Council for People`s Participation and Rural Technology
CBO: Community Based Organisation
CDAP: Comprehensive District Agriculture Plans
CDP: Community Development Programme
CEDAW: Convention on Elimination of All forms of Discrimination against Women
CGIAR: Consultative Group on International Agricultural Research
CGWB: Central Ground Water Board
CIG: Commodity Interest Group
CIWA: Central Institute of Women in Agriculture
CRS: Community Radio Stations
CSC: Common Service Centres

DAAP: District Agriculture Action Plans
DAC: Department of Agriculture & Cooperation
DAESI: Diploma in Agricultural Extension Services for Input Dealers
DARE: *Department of Agricultural Research and Education*
DEO: District extension officer
DFAC: District Farmers Advisory Committee
DIT: Department of Information Technology
DRCRA: Development of Women and Children in Rural Areas

EEI: Extension Education Institute
ETL: Economic Threshold Level

FAC: Farmer Advisory Committees
FAO: Food and Agriculture Organization
FFS: Farmer Field School
FI: Financial Institutions
FIG: Farmers Interest Groups
FLC: Farmer Life Schools
FO: Farmers Organisation
FPA: Farmers Producers Association
FPC: Farm Producer Companies
FRG: Farmer Research Groups

FSA: Farming systems approach
FTA: Farm-Tele-Advisors
FTC: Farmers Training Centres
FTOC: Farmer Trainer Orientation Course

GATT: General Agreement on Tariffs and Trade
GB: Governing Board
GCARD: Global Conference on Agricultural Research for Development
GDP: Gross Domestic Product
GFAR: Global Forum on Agricultural Research
GI: Grassroots Innovation
GIAN: Grassroots Innovation Augmentation Network
GIS: Geographical Information System
GNH: Global National Happiness
GO: Government Organisation
GOI: Government of India
GPRS: General Packet Radio Service
GPS: Global Positioning System
GRI: Grassroots Innovation

HESA: Human Eco-system Analysis
HRD: Human Resources Development
HYVP: High Yielding Variety Programmes

IADP: Intensive Agricultural District Prgramme
ICAR: Indian Council of Agriculture Research
ICDP: Intensive Cattle Development Programmes
ICM: Integrated Crop Management
ICT: Information and Communication Technology
IDA: International Development Association
IDWG: Inter Departmental Working Group
IFAD: International Fund for Agriculture and Development
IFPRI: International Food Policy Research Institute
INM: Integrated Nutrition Management

IPM: Integrated Pest Management
IPR: Intellectual Property Rights
IRD: Integrated Rural Development
IT: Information Technology
ITD: Innovations in Technology Dissemination
ITK: Indigenous Technical Knowledge
IWDP: Integrated Watershed Development Programme

KCC: Kisan Cal Centre
KKMS: Kisan Knowledge Management System
KVK: Krishi Vigyan Kendra
KVK: Krishi Vigyan Kendra

MAAIF: Ministry of Agriculture, Animal Husbandry and Fisheries
MANAGE: National Institute of Agricultural Extension Management
MARDS: Mahaicony-Abary Rice Development Scheme
MDGs: Millennium Development Goals
MIDH: Mission for Integrated Development of Horticulture
MoA: Ministry of Agriculture
MOOC: Massive Open Online Course
MORD: Ministry of Rural Development
MoU: Memorandum of Understanding
MSRI: Mechanized System of Rice Intensification
MSSRF: M.S. Swaminathan Research Foundation
MVIF: Micro-Venture Innovation Fun
MWCD: Ministry of Women and Child Development

NAADS: National Agricultural Advisory and Development Services
NAARM: National Academy of Agricultural Research Management
NABARD: National Bank for Agriculture and Rural Development
NAEP: National Agricultural Extension Project
NAFED: National Agricultural Cooperative Marketing Federation
NAIP: National Agricultural Innovation Project
NAPCC: National Action Plan for Climate Change

NARO: National Agricultural Research Organization
NARP: National Agriculture Research Project
NARS: National Agriculture Research System
NCA: National Commission on Agriculture
NCDX: National Commodity & Derivatives Exchange of India
NCF: *National Commission* on *Farmers*
NCMSL: National Collateral Management Services Ltd
NDDB: National Dairy Development Board
NDP: National Demonstrations Project
NFDB: National Fisheries Development Board
NFSM: National Food Security Mission
NGO: Non Governmental Organisation
NGRC: National Gender Resource Centre
NHM: National Horticulture Mission
NICRA: National Initiative for Climate Resilience Agriculture
NIF: National Innovation Foundation
NIPHM: National Institute of Plant Health Management
NIRD: National Institute of Rural Development and Panchyati Raj
NMAET: National Mission on Agricultural Extension and Technology
NMOOP: National Mission for Oilseeds and Oil Palms
NREGP:National Rural Employment Guarantee Scheme
NRLM: National Rural Livelihood Mission
NRM: Natural Resources Management
NRSC: National Remote Sensing Centre
NSDC: National Skill Development Corporation
NSDM: National Skill Development Mission
NSSO: National Sample Survey Organization
NTI: National Training Institute
NYKS: Nehru Yuva Kendra Sangathan

PADP: Participatory Action Plan Development
PAR: Participatory Action Research
PIP: Project Implementation Plan
PMKSY: Prime Minister's Krishi Sinchaai Yojana

PPP: Public-Private Partnership
PRDIS: Participatory Rural Development Initiative Society
PRI: Panchayati Raj Institutions
PTD: Participatory Technology Development

R&D: Research and Development
RADP: Rainfed Area Development Programme
RKMP: Rice Knowledge Management Portal
RKVY: Rashtriya KrishiVikasYojana
ROM: Result oriented Management
RUDSETI: Rural Development & Self Employment Training Institutes

SAMETI: State Agricultural Management and Extension Training Institute
SAP: State Agriculture Plan
SAU: State Agricultural University
SDGs: Sustainable Development Goals
SDoA: State Departments of Agriculture
SES: Social Ecological System
SEWP: State Extension Work Plan
SFAC: Small Farmers Agri-business Consortium
SHG: Self Help Group
SIRD: State Institutes of Rural Development
SITE: Satellite Instructional Television Experiment
SLSC: State Level Sanctioning Committee
SMS: Subject matter specialists
SNC: State Nodal Cell
SNO: State Nodal Officer
SOFA: State of Food and Agriculture
SREP: Strategic Research and Extension Plan
SRI: System of rice intensification
SRISTI: Society for Research and Initiatives for Sustainable Technologies and Institutions
SWOT: Strength, Weakness, Threat and Opportunities

T&V: Training and Visit
TANWA: Tamil Nadu Women Association
ToF: Training of Facilitators
TOT: Transfer of Technology

UAES: Unified Agricultural Extension Service
UN: United Nations
UNDP: United Nations Development Programme
UNESCO: United Nations Educational, Scientific and Cultural Organisation
UNFA: Uganda National Farmers Association
USDA: United State Department of Agriculture

VDC: Village Development Committee
VDO: Village Development Officer
VIC: Village Implementation Committee
VLPA: Village Level Participatory Approach
VLWs: Village Level Workers

WHO: World Health Organisation
WTO: World Trade Organisation
WUA: Water User Association
WUE: Water Use Efficiency

ZRS: Zonal Research Stations

Trends in Agricultural Extension

1 Agricultural Extension Retrospect and Prospect

- *Prof. S.V. Reddy & Prof. M. Surya Mani*

Introduction

The focus of Agriculture in India and in developing countries is shifting from Subsistence Agriculture to Agriculture for quality of life through improvement of Livelihood security There is a need for new paradigms for Agricultural growth in research, extension and Institutional development. In this century, more public investment in modern scientific research in agriculture lead to dramatic yield break through to feed the ever-growing population. Today the agriculture sector is faced with several serious challenges: the spiralling demand for food and fibre, declining cultivated area due to population pressure declining agricultural productivity due to natural resource degradation and increasing competition in International markets. It has also become apparent worldwide that intensive agricultural system involving high inputs are not only too costly but are contributing to soil degradation, pest resistance and loss of Bio diversity. One fundamental element in meeting these challenges is the adoption of environmentally friendly, cost effective, and sustainable improved agricultural production and marketing technologies farmers and other rural entrepreneurs.

The sustainable agricultural development and natural resource management, therefore, are important factors for livelihood of millions of small scale, resource poor farmers in India and other developing countries. Sustainable agriculture aims to conserve and save water, regenerate soils by using manures, foregoing deep ploughing to prevent erosion, reclaim unproductive land and minimise use of pesticides and fertilizers and maximize use of Botanical and Bio products. The environment, economics and social factors receives prime consideration under sustainable agricultural system, which can kindle evergreen revolution. Agricultural extension should play a pivot role in ushering such a silent revolution.

Furthermore, the increased dependence on science-based agriculture, bolstered by the rapid agricultural technological advances in the last half century, has placed greater importance on educating farmers on the advanced knowledge and skills to farmers. The future scenario of agriculture with developed pockets and contract farmers will be ranging from placement of seed by aero tractors, monitoring of irrigation, plant protection and other activities by computers to milking of cows by robots and from GPS receivers to Lazar Levelling for precision farming. Biotechnology, Echo technology and information technology will play a decisive role. Farmers tend to be more efficient and analytical with their labour, time and money. The voice of young men and women will be heard demanding for quality services.

Therefore, the transition from a resource based to a technology based system of agriculture; however plays a greater responsibility on the agricultural extension sector, for it is as a vital conduit to transform and educate farmers on the new agricultural information and technologies as well as a conduit to research and policy makers on farmer's problems, needs and concerns. It is now time that agricultural extension should play a major role in ushering management and marketing revolution through quality education of the farmers for not only increasing productivity but also ensuring the quality products to get good prices and thereby the millions of farmers change their lives and livelihoods.

Agriculture Extension as a Profession

Agriculture Extension is a profession like any other discipline. Webster, the lexicographer defines a profession as a calling in which one professes to have acquired some special knowledge used by way of either instructing, guiding or advising others, out of serving them in some act…" Therefore, following attributes qualifies to call agricultural extension as a profession.

- A body of knowledge based on systematic and scientific investigation.
- Continual research and exchange of information among the professionals.
- Established requirement of training and competency for entry into the profession.
- Established technical and ethical standards.
- Personal responsibility by members for self-direction and self-discipline.

There are a large number of institutions all over the world offering degree leading to MSc and PhD in this discipline of agricultural extension. Therefore, it is wrong to assume that anybody can do Agricultural Extension work without a formal training in the discipline.

A great many equate agriculture extension with transfer of technology. To them handing over of improved technology to the clients to make them more productive, is the end but agricultural extension is the means to carry out this 'handing over operation'. I will certainly say that this is a very narrow view of agricultural extension.

Even if accepted for a moment that 'handing over' of technology is the goal, then treating agricultural extension as a means to merely carry out this 'operation', is a gross under-utilization of this powerful means. Several past studies have almost conclusively highlighted that the strategy of development based on transfer of technology has resulted in uneven economic development. The studies have also shown that beneficiaries of the improved technology are those who have higher socio-economic status, higher need to achieve, higher level of aspiration in life, greater efficacy, greater decision making ability and ability to influence etc. This clearly implies that in order to reach the benefits of the technological advancements and innovations to people, it is necessary that they are socio-psychologically enriched by helping them to acquire above mentioned correlates of adoption of the technology. The lead farmers can be used as change agents to uplift the majority of resource poor farmers. Agricultural Extension therefore has to assume this task also by organizing client system besides generating technology. In other words, extension has not only to give them technology, but also to empower them to accept the technology and grow on it. Equality and inclusive growth should be inbuilt into the dynamics of extension.

Agricultural Extension-Concept and Philosophy

Agricultural extension denotes not only extension for crops but also the allied sectors like Veterinary Extension, Animal Husbandry, Horticulture, Home Science, Forestry etc., is one of the policy instruments that the government can use to stimulate agricultural development.

Agricultural extension is an "applied science designed to bring desirable behavioural changes by generation, transmission and application of knowledge and skills in order to improve the enterprises, vocations and living standards of rural people engaged in agriculture".

Extension originated as an informal, out of school education and grown as a distinct discipline. It is also an organization and management system, which requires a definite structure to provide service to the client. It employs people to serve target clients.

The basic tenants of extension are largely educational, service and regulatory in nature. Although very many countries have a mixture of those functions, the trend now is to emphasize the educational aspect of extension.

The philosophy of extension has often been stated as "help the people to help themselves, by starting where people are and with what they have" , thus propelling them to advance as fast as the resources at their disposal can allow . Human behaviour is complex. Sometimes we fail to understand ourselves. But extension discipline is endowed with technologies that can understand the behaviour of clients. Therefore, an extension worker has to study the clients' characteristics, resources and his environment in order to determine the approaches and methods that are needed to effect desired behavioural changes among the clientele.

The History

India has the second largest extension system in the world in terms of professional and technical staff. India is in the process of transforming its agricultural extension systems to become more demand – driven and responsive to farmer needs.

After independence in 1947, the first step of the government towards building an agricultural extension system was expansion of Grow More Food Campaign during the World War II. Administrators and extension workers were exhorted to convince cultivators of the gains in yields that could be obtained through the use of improved seed, compost, farmyard manure, and better cultivation practices, rural agents, often inundated with other assignments, had little or no training for extension work. However, gains in yields were minimal, and Indian leaders came to realize that converting millions of poor farmers to use new technologies was a colossal task.

The community development programme was inaugurated in 1952 to implement a systematic, integrated approach to rural development. The nation was divided into development blocks, each consisting of about 100 villages having population of 60,000 and 70,000. By 1962 the entire country was covered by more than 5,000 such blocks the key person in the programme was the village-level worker, responsible for transmitting to ten villages not only farming technology, but also village upliftment programmes such as cooperation, adult literacy, health, and sanitation. Although each block was staffed with extension workers, the villagers themselves were expected to provide the initiative and much of the needed financial and labour resources, which they were not in a position to do or inclined to do. Although progress had been made by the early 1960s, it was apparent that the programme was spread too thin to bring about the much hoped increase in agricultural production. Criticism of the programme led to more specialized development projects, and some of the functions were taken up by local village bodies. There was only negligible allocation for community development in the sixth plan.

The Intensive Agricultural District Programme, launched in five districts in 1960 by the central government in cooperation with the United States-based Ford Foundation, used a distinctly different approach to boosting farm yields. The programme operated under the premise that concentrating scarce inputs in the potentially most productive districts would increase farm-crop yield faster than would a wider but less concentrated distribution of resources in less productive districts. Under the technical guidance of American cooperative specialists, the programme placed unusual emphasis on saturation approach, organisational structures and administrative arrangements. For the first time, modern technology was systematically introduced to Indian farmers. The intensive Agricultural District Programme was thus a significant influence on the subsequent green revolution.

During the same period, as a prelude to green revolution, wheat and rice varieties were being tested throughout the country, and in 1966 the Government of India imported high-yielding wheat seed from Mexico. Given the urgency of this food crisis, the programme focus of Gram Sevaks of Village Level Workers (VLWs) gave more emphasis to agricultural extension. During this period, the Department of agriculture, along with the other line departments, involved in the distribution and sale of agricultural inputs and services. Although the high yielding wheat and rice varieties had an immediate effect on yields, the lack of attention by both research and extension to the management practices was limited to the overall impact.

The Training and Visit (T&V) extension system was first introduced in India during 1974 through a World Bank Project focusing on crop Management practices. It had an immediate impact on wheat and rice yields. Consequently, this extension approach was adopted throughout the country during the following decade. Implementing T&V extension has largely completed the transformation of the Indian agricultural extension system from community development agency to one concentrating on technology transfer especially for staple food crops. Dr. Daniel Benor, World Bank was the architect of the T & V Extension System.

The extension management system has changed drastically under T&V. At the district level, there was a district extension officer (DEO), 3-4 subject matter specialists (SMS), plus other supporting staff. Agriculture extension officers (AEOs), were assigned at the block level to supervise village extension workers (VEWs). Under T&V projects, most states added large number of VEWs to achieve the recommended ratio of one VEW for approximately 800 farm households who would constitute an extension circle. During 1970s and 1980s, most of these VEWs were secondary school graduates who had in-service training provided under the World Bank financed project. In the past decade, most new VEWs were university graduates. Since SMS positions are filled on the basis of seniority, the technical expertise of this cadre remained weak.

When T&V projects were implemented during 1970s and 1980s, these projects generally financed the salaries of the new staff, especially the expanding VEW cadre, plus travel, and operational costs associated with the T&V approach. Once these projects were completed, then these additional salary costs shifted to the respective state governments. At the point, due to the lack of financial resources, in-services training programmes and the regular schedule of fortnightly visits collapsed in most states. Therefore, during 1990s, extension-operating budget shrank to about 10 percent of recurrent costs, with the programme budget being primarily financed through central government, central projects and schemes. This has led to weakening of Research-Extension-Farmer Linkages, which were built on strong foundation during the operation of T&V Projects.

As a process of Extension Reforms, Agricultural Technology Management Agency (ATMA) was launched in 2005-06 which provides a demand – driven and decentralised plat form; Setting up of an Agri-clincs & Agribusiness Scheme (2002) facilitated promotion of Agribusiness and public Private partnership. The use of ICTs through establishment of kisan call centres (2004) etc., has strengthened the Extension Services.

The ATMA is an autonomous agency aims to make the extension system farmer – driven. An element of convergence of all departments of agril and allied sectors, Bottom up planning, Gender Streamlining multi agency extension strategies, Farm Schools, Block Technology Team (BTTs) Block farmers Advisory Committees, Farmer Interest Groups, Farmer Friend at grass root level were all included in the dynamics of ATMA. However several implementation bottlenecks are being found. The ATMA is being strengthened further through **National Agricultural Extension and Technology Mission** during the 12th plan period. This is the first time government realised the need for a special mission for Agricultural Extension in order to reach the unreached.

Current Scenario of Agriculture Extension

Extension includes all those agencies in the public, private, NGO and community based initiatives that provide a range of agricultural advisory services. While public sector line departments, (mainly the DAC and SDoA and allied departments) were the main agricultural advisory service in the 60's and 70s, the last two decades have witnessed the increasing involvement of private sector, NGOs, community based organizations and media in agricultural extension.

In the public sector, the extension machinery of the allied departments like Veterinary, Horticulture, Diary, Fishery sericulture, women and child welfare, State Departments of Agriculture (SDOA) (all collectively termed as SDoA & ADs) reaches down to the Block and village level. The village extension workers of the SDoA continue to be an important source of information for farmers. ATMA also provides the extension support.

During the last two decades, the number of Krishi Vigyan Kendras (KVKs) established and funded by the ICAR has increased, but the effective reach of these KVKs is marginal. However frontline demonstration and training activities are being carried out amidst the constraint of human resource.

While public sector extension arrangements have weakened, the number and diversity of private extension service providers has increased in the last two decades. These include NGOs, producers associations, input agencies, media and agri-business companies. Many provide better and improved services to farmers, but their effective reach is limited and many of the distant and remote areas and poor producers are neither served by the public nor the private sector. Notwithstanding the important role being played by the private sector extension, there are concerns with regard to wholesomeness of information given, equity and long-term implications.

Considering the changing nature of agriculture, producers currently need a wider range of support, including organisational, marketing, technological, financial and entrepreneurial support. Unfortunately farmers are not able to get this kind of integrated and need-based support from either the public or private sector. This can perhaps be achieved only through forming public-private partnerships.

Broadly speaking three types of technology transfer and advisory service arrangements can be seen in agricultural extension.

(a) ***State and Central Funded Extension***: In this mode, all the staff (field staff, subject matter specialists, training associates and managers) is employed by the government on a permanent basis. Extension by line departments such as SDoA& ADs, SAUs, KVKs are all examples of this type. Extension methods mainly include demonstrations, farm schools, farmer field schools, exposure visits, on farm trials, mass media, ICT, training programme and distribution of inputs. Recent years have witnessed governments establishing additional autonomous organizations such as ATMA (at district level), SAMETI (at state level), MANAGE (at national level), etc. These autonomous organizations (registered under the Indian Societies Act) are expected to work with the typical government organizations indicated earlier.

(b) ***Private Sector Led Agricultural Extension and Advisory Services***: These include extension activities of input agencies and agri-business agencies involved in contract farming/purchases. While inputs agencies focus on demonstrations and trainings, the agri-business companies provide a range of services and this includes, supply of production inputs, field based advisory services and buy-back of produce. The input agencies deploy local input dealers for input-linked extension services. Private consultancy services have also emerged in some of the select crops.

(c) ***Civil Society/ Farmer Organizations Extension Initiatives***: These include efforts by NGOs, farmer association and producer organizations. There is a wide variation in their activities and this is dependent on the focus of NGOs (sustainable agriculture, conservation or organic agriculture, linking poor to markets, micro-finance, gender etc) and the nature of commodities being dealt by the farmer association (e.g. grapes, sugarcane, spices)and producer co-operatives (eg: milk, sugarcane). There are a number of institutional innovations emerging from this sector. One of the recent example is the learning alliance between civil society organizations, para-statal organizations (at district and state level), and sustainable CBOs (below the district level). By and large, this institutional framework is emerging to be very promising which can upscale successful experiences at a much faster speed. Its success however depends upon new type of support systems and incentives which relies less on external subsidies (for individuals) but more on financial support and infrastructure to the groups. The farm producer companies (FPCs) are being formed with support from Govt NABARD SFAC and Private Players.

But it is quite astonishing to note that despite the existence of several public, private and civil society actors having capability to provide all kinds of support (technical, financial, organisational, marketing, etc) the farmers do not get adequate support in addressing their expanding and complex challenges.

Funding

Irrespective of the extension models followed over years, the failures are attributed to not only to the institutional system used by the model and human failure, but also to lack of funding. It appears that without World Bank funding, there is no domestic resource to invest in this most vital sector, which supports livelihoods of more than 700-800 million people. The deployment of fund to agricultural extension by the centre and the states is meager and this financial allocation policy is at the root of collapse of the extension system. Over the recent years, the growth of public investment in agriculture has slipped down. It appears the Constitutional devolution of agriculture also has a role. The states place great importance in maintaining agriculture as its subject while they are not providing much needed funds to this priority sector. The support to agriculture from centre, on the other hand, is not substantial; apparently on the ground that agriculture is not a central subject although centre has been enlarging its strategic control over the sector through international commitments, policies and much legislation. In this context, the recommendation of NCF on shifting agriculture to Concurrent List deserves serious consideration.

New Dimensions

1. **Paradigm Shifts in Agriculture/ Agricultural Extension**

   The following paradigm shifts have taken place in Agriculture / Agricultural Extension which have lot of implications for Agricultural Extension providers to revitalise extension.

   - **Green Revolution to Evergreen Revolution**

     The green revolution has demonstrated India`s Capabilities to balance between Agricultural and Human Growth. However it has also created social economic and Environmental imbalance. Now Agriculture is at the cross roads. In developing countries like India and Africa, agriculture is not just for food production but it is the backbone of the livelihood security of majority of population. Therefore there is a need for effort to produce food, fodder, fibber and other products based on sustainable practices where in large small scale farmers in rain fed areas are involved. Therefore evergreen revolution is important for sustainable food and Nutrition security.

   - **Commodity approach to integrated Farming System Approach**

     India and other developing countries mostly were adopting commodity approach where in a single commodity is grow in a vast tracks of land. The net income from the crops was affected due to internal and external threats owing to pests & diseases, vagaries of Monsoon, market fluctuations as well as GATT & WTO. The small scale farmers in rainfed areas were affected mostly in this process. Hence it is now recommended to follow farming system approach where in intensification, diversification and value addition is done to get more profit out of a unit of land. This could be done through crop mixtures or boarder crops, intercropping etc. and / or in combination of livestock. More directed efforts are required in Farming System Research and Farming System Extension.

- **Mono disciplinary to Interdisciplinary approach**

  The research on Extension has been focusing on mono displinary approach having a team of specialists advising and guiding farmers in different intervals based on their specialisation wasting time and resources of farmers. It is now advised to have interdisciplinary approach where team of experts from different disciplines visit the farmers and advice in a holistic way. Similar trend is followed now in research also.

- **Technology-crop technology Eco-technology and Biotechnology**

  In the past decades crop technology was prominently addressed by Research and Extension. Today, there is a need for an extension worker to know about technologies related to livestock and other farm enterprises as well as about Biotechnology which is rapidly catching up by farmers (Bt cotton which has **revolunterinsed** the cotton industry) Biodiversity and environmental technologies are paramount important to promote sustainable agriculture, climate resilience agriculture, organic farming and Natural farming.

- **Supply driven approach to demand driven approach**

  So far, supply driven approach was followed which has made the farmers to get low prices for the produce due to excess of production and middle men exploitation. Therefore shift is more towards demand driven approach and market led extension where farmers will be advised to grow the crops suitable to the areas and also have demand in domestic and international markets. This calls for providing marketing information and market intelligence. The extension has to be ready to take this role.

- **Farm employment to off-farm employment**

  Since 60 years of independence, extension has been paying more attention for creating on farm employment through increasing the productivity with advocacy to reduce the cost of production.

  However there is a distress among farmers about farming as a viable option for improving livelihoods. This is leading to migration of farmers & youth from rural to urban areas. To arrest these phenomena, extension should also focus on creating viable off-farm employment such as dairy, poultry, sheep & goats, sericulture, mushrooms, fisheries, nursery management, high value crops and other viable enterprises which can provide income and also gainful employment.

- **Self-reliance to self sufficiency**

  The past decades witnessed self reliance society where farmers own abilities, decisions and resources were used to manage the affairs of farm. Today self sufficiency is a buzz word which means today farmers were able to produce supply to their needs and also sell in external market with their own initiative and resources. This needs new direction for extension.

- **Agricultural Development to sustainable Agriculture Development**

  Sustainable agricultural development that meets the needs of the present without compromising the ability of future generation to meet their own needs. The economics, ecology, equity and social factors are important drivers of the development. Earlier agriculture development used to be input intensive with under mining the erosion of natural resources as well as human safety and health.

- **Knowledge and Skills to Empowerment**

  The extension approach to farmers earlier has been to provide knowledge and skills in order to get desirable behavioural changes. This phenomenon has not created a spread effect and motivation to adopt the technologies as well to communicate to other famers. The empowerment on other hand based on the idea that giving farmers knowledge skills, resources, opportunity motivation as well as holding them responsible and accountable for the outcomes of their action. This will contribute to competence and satisfaction so that they contribute also to empower other farmers which will also promote farmer to farm extension and multiplier effect.

- **Agriculture to Agribusiness**

  Agriculture was traditionally used as a livelihood option now it is looked as an Agribusiness. The transforming agriculture as agribusiness means moving from substance farming to modernisation of Agriculture. Agri Business is the key for agricultural production. The farmer need to grow demand driven crops / enterprises, manage the farm to reduce the costs in a value chain way, maintain farm records, increase the production and add the value for the product which can fetch him more income. These trends need to be incorporated in Extension Advisory services.

- **Agricultural Development Approaches**

  - **Productivity to profitability:** There is a need to advocate the small scale farmers about the necessity of reducing the costs specially of input use, increase the production, monitoring the quality, add the value which can give more profitability but also preserve and protect natural resources like land, water and environment.

- **Equity and Sustainability**

  Equity concerns and inclusive growth is important in all extension efforts. The effort should be directed equally to small scale, marginal farmers, rainfed areas, women and other deprived farmers etc. at the same time ecological factors also need to be stressed for enhancing sustainable growth in livelihood context. The gap between have and have-nots need to be bridged.

- **Single Extension Approach to Pluralistic Extension Approach**

  It is well recognised that single extension approach will not be able to solve the farmer's problems. In this context pluralist, demand driven extension system and approaches are essential. However they have to be flexible, supporting and complimenting each other.

- **Public Extension to Private – Public – Partnership**

  In view of inadequacy of resources and man power, extension is not able to reach the unreached farmers and areas. Extension should address concerns of profitability, markets, value chain. The public extension and private extension has its own strengths and weakness. However private – public – partnership in extension can able to plug the loopholes and able to reach more farmers with transparency and accountability. This needs a clear cut division of roles responsibility and resource allocation.

Different Extension Approaches

There are number of extension approaches / models tried by private and public extension service providers. However, some of the approaches projected below are those with which the authors were directly involved in planning and execution.

❖ **Supply Chain Linkage Approach**

A Better Cotton Initiative (BCI) Programme is launched in India aimed at empowerment of farmers on Better Cotton production principles (minimising harmful effects of crop protection practice), improving the soil health, water efficiency, protection of natural habitats & biodiversity, fibre quality)promotion of 'decent work' principles and strong supply chain linkages through Farmers Organisations namely groups, networks/Federation and Producer Companies.

The programme has amply demonstrated as on holistic and futuristic demand driven agricultural extension model, for empowerment and improvement of livelihood of cotton farmers. In addition to improvement there was increase in human, social and economic capital. Majority of farmers were able to get net additional income of about Rs. 10,000 to 15000 / acre (150 to 200 $) through increase in 20 to 25 percent yields, 80 percent reduction in pesticide usage, 20 percent reduction in chemical fertiliser usage and changes in attitude towards environment and application of Decent Work Principles specially in not using banned pesticides, Child Labour and awareness on farmers rights. Besides the programme has demonstrated the implementation of equity and ecological Principles with focus on women and natural resource management. Furthermore, the Linking farmers directly with markets has helped the farmers with better bargaining capacity better prices and elimination of middle men contributing to improvement in livelihoods. This programme and approach is being implemented in several countries.

- **Potential of Farmers Field Schools in promoting Sustainable Sugarcane Production**

In India, production and productivity of sugarcane is severely affected by several biotic and abiotic constraints. The major biotic constrains were due to insect pests, diseases and weeds. For controlling and health hazards. Keeping in view the constraints. Participatory Rural Development Initiative Society (PRDIS). A professional NGO in collaboration with ANGR Agricultural University (ANGRAU) and with the support of World Health Organisation (WHO) has organised a season long Training of Facilitators (ToF) programme during Jan – Dec 2009 on integrated Pest Management (IPM) for sustainable sugarcane production under pesticide free environment through Farmer Field School (FFS) approach. About 20 participants from government and non government organisations have undergone ToF and organized practice FFS as a part of curriculum with small scale resource poor farmers. Simple low cost eco friendly technologies such as innovative planting method to save seed material and to glean higher yields. Non-Pesticides management, soil test based fertilizer management, Agro Eco system Analysis (AESA) based Resource management including water, and intercropping and decent Work principles were used in FFS with IPM strategies. The results indicated that there was significant gain in knowledge and skills (increase in Human Capital) increase in social capital, awareness on health and environmental concerns of the ToF Participants and farmers. Besides, the economic analysis revealed that the yields from FFS fields were about 25 tons / ha higher than the farmer fields with a net additional return of Rs. 35000 per hectare (500 US$). There was also decrease in cost of cultivation, (due to use of less seed material, soil test based inputs and effective management) increase in juice sucrose quality and sugar recovery. The follow-up scaling efforts were undertaken during the years 2010 and 2011. Similar results and higher net gains were observed in FFS and large scale adoption plots. Thus the programme has shown the potential for creating a pest free environment improving the lives and livelihoods of small and Marginal farmers in Irrigation dry and rain fed areas through FFS approach.

- **Community Managed Extension Approach**

This type of extension set-up has been found to be particularly suitable for those successful experiences which are not up scaled due to lack of proper institutional framework at different levels. These experiences may include collective marketing, food security, micro-enterprises, sustainable agriculture, etc. The following specific steps have been taken for upscaling sustainable agriculture and organic farming particularly in A.P Rural Livelihood Programme (APRLP) and IWMP.

1. Identification of successful experiences and motivation of different stakeholders through focused exposure visits.
2. Validation of above experiences on a small scale under the institutional framework of the community (SHGs and their federations).
3. Networking of experienced organizations (under GO, NGO and CBOs sector) which are interested in up-scaling above experiences and willing to provide technical support on regular basis.
4. Linking with Government and other organization for support for (i) community managed extension set-up, capacity building and documentation of experiences (ii) purchase of community equipment's on contributory basis, (iii) purchase of recurring inputs on revolving fund basis, etc.
5. Below the district level, the community managed extension system is operated with the help of SHGs, lead farmers and para workers.

The above experience has shown remarkable synergy among learning alliance partners. It is still a challenge to build up next level of learning alliance in which formal organizations will have the same level of enthusiasm / synergy as being experienced by the present alliance. In this context, the institutional reforms indicated earlier (with autonomous organizations like ATMA, SAMETI, MANAGE, KVKs) may provide a bridge to institutionalize the experiences into the mainstream organizations with convergence.

❖ **Extension through Agri-Clinic and Agri-Business Centres (ACABC)**: The objective of the scheme is to supplement the efforts of public extension and to create gainful employment opportunity for unemployed agricultural graduates. ACABCs are to provide testing, diagnostic and custom hiring services and other consultancies including food and agri-processing besides operating business centers on input supply. These services are to be managed by self-employed graduates / diploma holders in agriculture and allied disciplines, who are specially trained on this aspect. Training is provided by the MANAGE and recognized Training Institutes (NTIs) & SAUs. The programme is being implemented by MANAGE with support from Government of India and NABARD. The business models may include food production, farming, agrochemicals, farm machinery, warehousing, wholesale distribution, and processing, marketing and sale of food products. Bank loan from 2 to 20 lakhs for individuals, with no security up to Rs. 5 lakhs, and up to 1 corer for groups of 5 members with subsidy of 36% for men and 44% for women and special category. Several entrepreneurs have now established agri business and clinics doing private extension services complimenting and supplementing the Government Extension Efforts.

❖ **Private – Public Partnership Approach**

This is a popular approach advocated for reaching the unreached and also for transparency and accountability. This approach was adopted in Andhra Pradesh by Participatory Rural Development Initiative Society (PRDIS) in collaboration with ATMA. During the process of planning a Memorandum of Understanding (MoU) was formulated with clear cut roles and responsibilities of different actors involved in productivity enhancement programme in various crops namely maize, paddy, cotton, red gram, sorghum, castor and groundnut in a given geographical area. The farmers were grouped into commodity groups based on the major commodity that they grown in an area. These groups also select a leader and a facilitator within the group. The extension service is rendered through group meeting, farmer field schools and other mass media interventions. All the extension interventions were jointly planned and executed by both the NGO and ATMA. Joint field visits for monitoring and guiding was undertaken with a schedule drawn jointly. This has helped in increasing the productivity and net income of farmers with a wider coverage. The commodity groups were able to network at village level and sell they produce collectively to get better prices for the quality. This can be replicated in different parts of India.

❖ **Broad based Integrated Extension Services**

The IDH – A sustainable trade initiative organization and sustainable spices initiative in India has been using the broad based Integrated Extension Services giving advice and guidance on range of package of practices namely proactive farming, natural resource management, water management, waste management, community relations, women empowerment, decent work, post harvest, marketing and business concerns through different methods in the process of sustainable spices production. The farmers were able to immensely benefitted by this initiative in terms of reduction of costs through reduction of chemical pesticides of and fertilizers, increasing the yields and the markets tie up for domestic consumption and exports. It will also contribute for livelihood improvement and well being.

❖ **Unified Extension System**

A major change in several government policies on extension in Africa took place in 1992 when Ministry of Agriculture, Animal Husbandry and Fisheries (MAAIF) adopted a unified system of agriculture extension in Uganda and several countries. Under this system, a Field Extension Worker is assigned a geographical area known as a circle. She / he is backstopped by subject matter specialists and follows a farming systems approach, providing guidance to total agricultural resource development including crops, livestock, fisheries, natural resources development including crops, livestock, fisheries, natural resources management etc. Every activity is undertaken following regular scheduled visits to the farmers. The field extension workers are retrained in order to imbibe on different skills in different departments.

Farmers participation in extension programmes was taken a step further by introducing Village Level Participatory Approach (VLPA) to rural development activities. The approach is aimed at reinforcing the "bottom-up" planning process by empowering rural communities to determine their destiny by actively involving them in every step of planning and implementation of development programmes for their villages. It should be noted that VLPA is a multidisciplinary intervention designed to address their problems in a holistic manner and thus can effect sustainable development in rural areas. This can also promote community managed extension services. A similar approach is worth introducing in the ATMA.

Methods and Methodology

❖ **Farmer Field Schools (FFS)**

FFS is a place where farmers undergo a field oriented discovery based training that enables them to become field experts and be able to grow a healthy crop. FFS is used as an effective extension tool for empowering the farming committee, developing self confidence, increase in social and human capital and promote better living. In FFS there will be 25 farmers. Each FFS is facilitated by trained facilitators. The FFS differ from demonstration since it has more emphasis on experimentation, science based learning and skills.

❖ **Information communication technologies (ICT)**

At the juncture when worked is witnessing an information revolution implementation of ICT led media for speedy dissemination and wider accessibility of information to rural countries in vital for accelerating development efforts. Recent past ICT services such as kisan call centres, community radio, T.V. cloud, smart phone and electric farm dires, Green Phone, Mobile phones, video and audio causes are being extensively used.

❖ **Kisan Call Centres (KCC)**

KCC initiative started in January 2004. It takes advantage of telecom network. A call centre based extension services by interdisciplinary experts will be communicating knowledge and information exactly as per the requirement of farming community. The centre will receive calls and provide advice from 6.00 AM to 10.00 PM daily.

❖ **Common Service Centres (CSCs)**

The Department of Information Technology (DIT) GoI has prepared to **roll**over 10,000 CSCs across the country. The objective is to develop a flat form that will enable grants for private and social sector organisations to align their social and commercial goals for the benefit of rural comities in a combination of IT and Non IT based services. This needs a framework to link the extension machinery to CSCs.

- **Portals / Farmers Portals**

  Department of Agriculture and Cooperation has developed many portals applications and websites a farmers portal has been created as to one stop for keeping all farmers needs which can leveraged by extension services.

- **Farmer Life Schools**

  It is an open school that is created for and by farmer. The participants are learning groups of. Farmers, women and farm labour. The main objective is to empowerment of farmers on community development, decent work issues such as discrimination, child labour funding of organisation etc, to enable them to lead a quality and satisfied life with self help, mutual help and cooperation. It is a forum for creating public awareness, discussion and problem solving.

- **Tradition Media**

  The traditional media has played on important role in Indian society in bringing public awareness on community and individual concerns. This can be an effective supplement and complement to ICT & other extension methods. The folk songs, dramas, kalajathaurasetc come under this category.

- **Front line demonstration**

  The KVK is using the front line demonstrations has an effective extension tool to demonstrate the New and Innovative technologies. This can be an effective training ground to impart knowledge and skills as well as to bring attitudinal change.

- **Demonstration**

  The demonstrations since ages have been serving as effective exterior methods to disseminate technologies and also convince farmers for adoption since it involves concept of "seeing in delivering. The demonstrations usually have control plots and method demonstrates are involved in the process.

  In addition, peer reviews, extension talk, teaching of skills, exposure visits, farmers have visits, training; awareness campaigns group dimensions are also used by extension service providers for transfer of technology, changing the behaviour and advocacy. However a single method cannot be said to be effective. A combination of methods based on subjects resources and clientele need to be employed. This pluralistic extension methods & methodologies with proper media mix should be employed to get desirable results.

2. Focused Areas

- Personality Development
- Value chain management
- Producer organisation
- Integrated farming systems

- Livelihoods and well being
- Natural Resource management
- Integrated crop management
- Agri Business
- Empowerment of farmers, farm women and agricultural labours
- Participatory planning and monitoring
- Convergence
- Private – public partnership
- Food and nutrition security
- Gender and equity concerns
- Entrepreneurships skill development
- Extension approaches, methods and methodologies
- Development communication
- Contract farming
- Conservation agriculture
- Organic farming
- Urban Agriculture
- Climate Change
- Biotechnology
- Biodiversity
- Eco-technology
- Horticulture
- Livestock including fisheries
- Food and Nutrition Security
- Livelihoods

Need for New Direction in Scope and Expertise of Extension

There is an increasing realization that dissemination of new knowledge on agricultural technologies alone is not enough to deal with the new set of challenges. Deterioration of natural resources, fragmentation of farm holding, threats and opportunities related to opening up of markets and introduction of new standards for production and marketing have all made agricultural development more complex, while farmers need a range of support - organizational, entrepreneurial, and financial, market. The public sector extension provides is information on technologies, that too often as broad general recommendations, lacking location-specificity. To be successful, farmers require a wide range of knowledge from different sources and support to integrate these bits of knowledge in their production context. They need much more information to charge their lives and livelihoods. The tasks therefore are primarily of three kinds:

- Shifting focus from disseminating technologies to provision of an integrated set of technology packages and services.
- Extension to deal with concerns of Development issues at large namely Natural Resources Management (NRM) Community Relations, Women Empowerment and equity Environmental Entrepreneurship, Watershed Management, Nutrition wellbeing in addition to Agricultural and allied subjects. Farmers need integrated advice for changing life's and livelihoods.
- Strengthening the capacity of the grass root extension system to address location specific issues, and
- Making the system primarily accountable to the community, in larger sense.

Broadening of expertise in extension organizations requires retraining the existing staff and for recruiting staff with diverse skills such as organisational development, market promotion, financial intermediation, Natural Resource Management, value addition (product development), monitoring and learning. However, extension also needs few staff at the district level with some of the above skills and expertise to better design strategies and continuously fine tune the organizational capacity.

The new approach should also focus on organizing farmers into community based organizations (producer companies, cooperatives & federations etc) and transforming these institutions as effective platforms for delivery of an integrated range of technologies and services, cross cutting the value chain.

Public sector extension should also try to learn from the experiences of civil society and the private sector engaged in rendering wide range of farm advisory and support services. The public private partnership seems to want to meet the demands.

Revitalising Training Teaching, Research and Extension Services of Agricultural Extension

1. Training

Extension training as a key input for human resources development can contribute a lot to face the challenges by all concerned. Extension Training in provided by both public and private sector institutions at National, Regional, State, District and grass root levels. Although there are significant efforts made during the past decade to improve the quality and relevance of training, still there is a greater scope for further improvement since the training needs are dynamic.

The following problems that emerged through several studies and experiences need to be redressed in order to revitalize the extension training.

- ❖ Lack of relevance and skill orientation
- ❖ Unfavourable attitudes towards training
- ❖ Duplication of training programmes
- ❖ Inadequate follow up
- ❖ Limited Professional trainers

The following strategies need to be initiated for improving the quality and relevance of the training programmes.

- ❖ Initiating a comprehensive study on training need assessment and manpower development in order to formulate training programmes at various levels based on the study findings.
- ❖ Introduction of innovative training programmes and strategies to make the extension programmes more useful and interesting.
- ❖ Rationalizing and harmonizing the training programmes among the training providers at various levels.
- ❖ Capacity building of training providers to met the present challenges of extension and upgrading skills.
- ❖ Giving due importance and budget provision for intensive follow up of the training programmes.
- ❖ Provision of adequate resources for research and development in several areas of extension training.

With the background cited above, the following innovative extension training programmes are suggested in order to meet the challenges of extension during this decade. I am sure some of them are already offered by different institutions. My appeal to training institutions is to add new programmes based on need assessment of their clientele as we move forward in the journey to reach the sustainable agriculture development.

- ➢ Institutional Development in Agriculture.
- ➢ Gender concerns in agriculture.
- ➢ Result oriented management.
- ➢ Extension methods and strategies for popularizing Eco technology, INM, IPM, NRM etc.
- ➢ Project formulation, implementation and evaluation.
- ➢ ICT for sustainable agricultural development.
- ➢ Community organisation and group dynamics.
- ➢ Participatory Approach for Agricultural and Rural Development.

- ➢ Value chain analysis and management in different crops and enterprises.
- ➢ Planning and management skills.
- ➢ Climate resilience agriculture.
- ➢ Decent work.

2. Teaching

There is need for revitalizing and reorienting the teaching curriculum in Extension Education to meet the challenges of futuristic agricultural development. This calls for substantial changes in the curriculum of undergraduate and postgraduate programmes. While there is need to make modifications in the existing courses, it is also suggested the following new courses could be introduced wherever necessary based on situational analysis.

- ❖ Entrepreneurship development
- ❖ Private-public partnership in extension
- ❖ Value chain management
- ❖ Agri-business Management
- ❖ Environment education
- ❖ Decent work disaster management
- ❖ Institutional development
- ❖ Innovative extension approaches
- ❖ Information communication technologies

3. Research

"You cannot carry an atom bomb in a bullock card and try to win over a war".

The relevance, quality and adequacy of Research in Extension are questionable, while most of the research is carried out through MSc and PhD students. The staff research programmes are very limited. The researches have confined mostly to ex-post facto designs but not experimental. This needs accentuated efforts on the part of extension researchers towards undertaking more experimental and action research programmes. In future, researchers can concentrate their efforts more on the following areas and also think for new areas for experimentation.

- ❖ Experiments on combination of methods and extension strategies to be used in a given situation for a particular crop, technology, subject, socio economic status and ecosystem.
- ❖ Research to develop standards for measurement of extension work (Quality and relevance).
- ❖ Process research of New Extension Approach polices and systems (Farmers Fields Schools, ATMA et).

- ❖ Research for development of indicators for measuring extension concepts such as;
  - Sustainability
  - Extension motive
  - Empowerment
  - Participatory Technology Development (PTD)
  - Systems Research
  - Policy Research

4. Extension Service

There are many extension service providers working parallel at gross root level in advantage areas in one hand and lack of access to extension services in the less endowed areas especially for vulnerable sections of society on the other hand. In light of the above, I suggest the following points for consideration of all concerned.

- ❖ Facilitate prepare participatory strategic action plan at village/mandal/district level with convergence and client consultation using village level participatory approach (VLPA) and implement them with farmer organizations/SHGs, CROs etc.,
- ❖ Introduce result oriented management.
- ❖ Encourage Farmer-to-Farmer extension lead farmer centric extension through Farmer Field Schools demonstration and other media / methods for realizing multiplication effort.
- ❖ Reach the unreached especially vulnerable section of society and less endowed areas.
- ❖ Recognize the role of NGOs, CBOs and private services providers as an asset to compliment and support the Government Extension Service.
- ❖ Employ modern means of communications demonstrations (FFS, community Radio, TV ICT etc) with traditional media to reach and teach the farmers.
- ❖ Retraining the existing extension staff periodically to build their capacity to handle futuristic extension needs is an important activity for the extension services.
- ❖ Strengthening institutional support including mobility with reward systems introduced in Extension.
- ❖ There is need for recognition of the fact that there is need to maintain a sizable ratio of extension workers to farmers to educate and empower them. Besides, it can be good support mechanism to instil confidence and hope of success.
- ❖ The MANAGE, EEIs, SAUs and Professional NGOs should intensify research and development activities to build models which can be replicable by the state department of agriculture and private service providers to keep pace with futuristic extension needs.

The Challenges

While the global forces are shaping future agricultural extension worldwide, national systems are experiencing institutional reforms. The following are enormous challenges that should be addressed by agricultural extension in making the extension services effective and meaningful for meeting the demands of future agricultural development.

- Organising client system (farmers and consumers) into groups, associations, federations, cooperatives and producer companies.
- Creating Public awareness on the need for Evergreen Revolution for sustainable agriculture through multimedia communication strategies.
- Participating in technology development process with focus on R&D besides in this type research in extension.
- Empowering farmers, farm women and farm labour on the sustainable agricultural technologies (Knowledge and skills) with minimum distortion through Farmer Field Schools (FFS), Information Communication Technology (ICT) and other media.
- Educating and motivating client system to adopt the technologies as well as empowering them and to create spread effect through FFS, demonstrations lead farmer centric extension and other methods/media.
- Initiating interactions among the various technology uptake pathways to rationalize and harmonize resources through convergence.
- Participatory monitoring and evaluation with focus to provide feedback to research, clients, markets and policy makers.
- Communicating the policies of the Government to the farmers and facilitating access to credit input and marketing.
- Educating farmers and farm women on their rights, WTO, GATT implications as well as decent work principles through farmer life schools, traditional and other media.
- Human resource development, which includes training of farmers and extension staff to meet the demands of new dimensions in extension.
- Provision of providing advisory sciences in agri-business, farm management and other integrated services.
- Development of personal efficacy and hope of success among farmers.
- Development of achievement motivation entrepreneurship, and desire for competing with standards of excellence among farmers and extension workers.
- Improving on overall well-being and gross domestic happiness by changing the farm families their lives and livelihoods.

Futuristic Extension Model for Changing the Lives and Livelihoods of Farmers

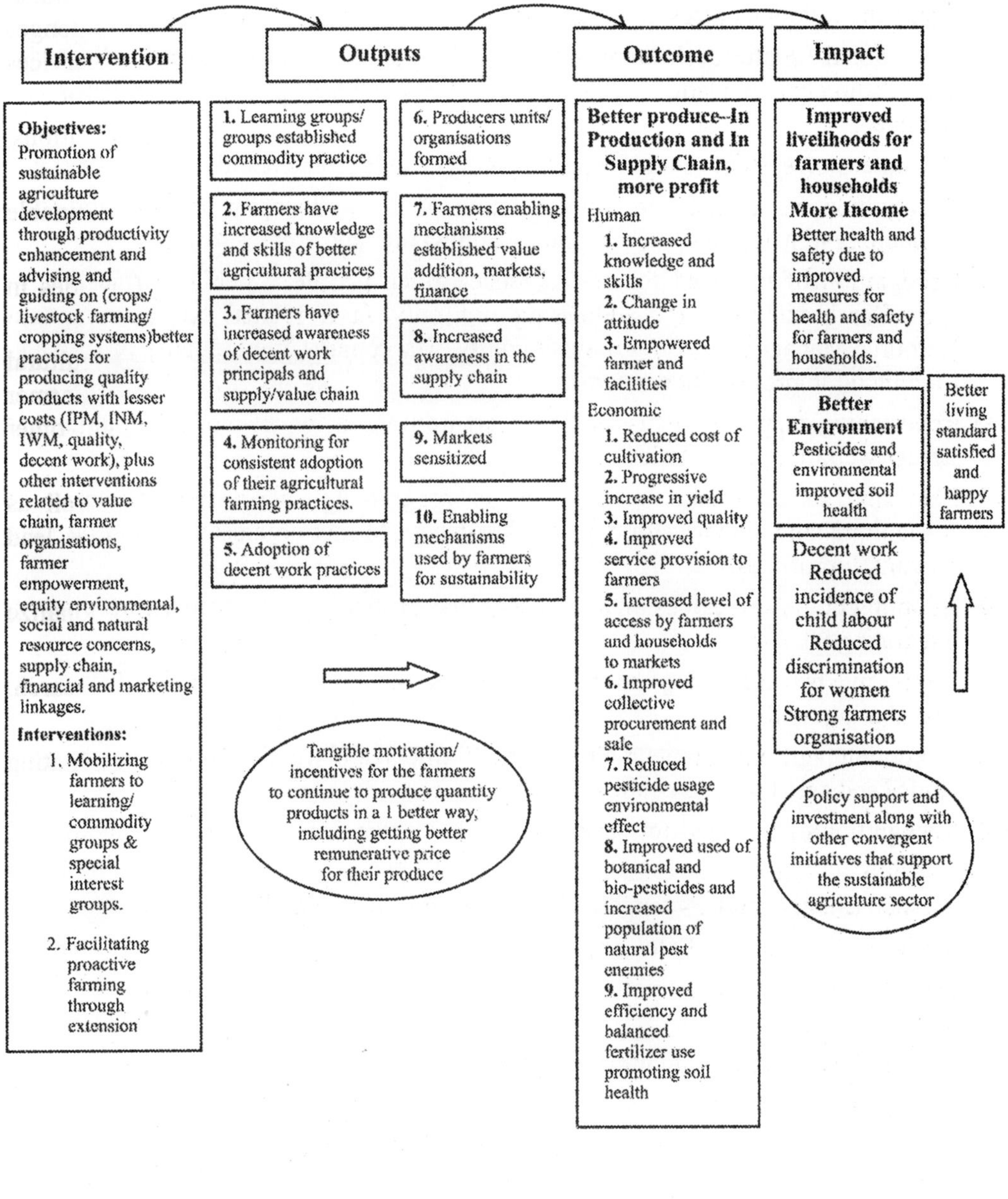

Suggestive Futuristic Framework

A framework of the building blocks that could form the structure of future strategy could perhaps take the following steps, among others:

- ❖ At National / State level, there is need to develop clear-cut extension policy with all stakeholders. A National Policy of "Extension for All" could be enacted and achieved.
- ❖ The extension system should be participatory, result oriented and demand – driven.
- ❖ Recognition of the need for re-orientation of the philosophy of extension-farm technology transfer mode to technology application and empowerment.
- ❖ Recognition of the need for convergence driven private – public partnership in agricultural extension for addressing not only farm related but also Natural Resource Management, Environment equity, Marketing, Credit, quality of Life and Livelihood improvement concerns.
- ❖ Redesigning support systems moving apart from input centric modes to output and farmer welfare centric modes.
- ❖ Extension to be broad – based in its programmes addressing concerns of Agricultural and allied sectors by utilizing groups farmer organisations and farming systems approaches.
- ❖ Adopting pluralistic extension approaches that explicitly underscore the need for an integrating mechanism.
- ❖ Aggressive privatization of extension systems transiting to a demand – driven and cost sharing mode.
- ❖ Promoting agri-entrepreneurship among farmers and rural youth through agri-clinics and agri-business centres.
- ❖ Recognition of the need for strong research – extension – farmer and market and consumer linkages.
- ❖ An increasingly gender - sensitized inclusive and social uptake extension strategy.
- ❖ Strengthening R&D in Extension.
- ❖ Providing training infrastructure to develop extension professionalism in a cost – effective manner.
- ❖ Focussed participatory monitoring and evaluation to strengthen the extension.
- ❖ Agriculture Technologies Management Agency (ATMA) model to be carefully evaluated with some improved interventions at grassroot level such as Village Level Participatory Approach (VLPA), Farmer-to-Farmer Extension through Farmer Field Schools and Unified Extension delivery.

The challenges facing agricultural extension are fourfold: enhanced production and productivity; equality and uneven development, sustainability and enhanced profitability.

These call for developing alternative, viable and sustainable modalities. The extension services in the public and private sectors work without clear policy direction are characterized by uniformity rather than specificity.

Conclusion

In the present context, India`s agricultural sector is faced with severe challenges. These include arresting decline in productivity, producing quality products with less costs for highly competitive external markets. The Nation is marching towards sustainable agricultural development. India requires agricultural growth of 4 percent GDP within an expected economic growth of 7 or 8 percent. Livelihood improvement and Poverty reduction will be possible only when small and marginal farmers and farmers from rain fed areas participate fully in economic growth. Agricultural extension has to play pivotal role in meeting these challenges. We should be able to transform subsistence farming to modernized agriculture and eradicate poverty. This calls for restructuring, retooling and revitalizing the teaching, training, research in agricultural extension as well as retraining the extension service providers. There is greater chance now for extension to bridge the gap between technical know-how and farmers do-how. There is need for instituting rewards and awards for the extension service providers moving on the path of sustainable agricultural production and those who reach the unreached. There is lot of duplication of efforts with multiplicity of agents doing extension work without convergence. Strengthened agricultural extension covering the ailed sectors can play a vital role in ushering ever green revolution and improving the livelihoods. Extension programme can also play an important role in knowledge management and creating public awareness on climate change, conservation agriculture, biotechnology, eco-technology, decent work and other upcoming areas.

The motivated new extension workers and the result-oriented management are bound to meet the futuristic challenges.

Let this glorious profession help the farmers to bring domestic gross happiness. Let us pledge to pursue our policy markets to enact a National policy of ***"Extension for all"*** to be achieved as one of the millennium goals.

References

1. Report of National Knowledge Commission, Government of India, 2009.
2. Working group report on Agricultural Extension, 12th Five Year Plan, Planning Commission New Delhi, 2012.

02 Agricultural Extension and the Sustainable Development Goals : Welcoming the Commitments, Confronting the Complexity

- Daniel J. Gustafson

In September 2015 the member states of the United Nations agreed on a set of 17 goals with 169 targets that set the agenda for sustainable development interventions over the next fifteen years. These Sustainable Development Goals (SDGs) came out of a three-year intensive consultation process involving governments, businesses and civil society. It is an ambitious agenda to be sure but, whatever its shortcomings, the 2030 Agenda represents the most comprehensive, negotiated statement of current global challenges, with concrete targets that countries, individually and collectively, agree to work toward.

As the goals and targets build on experience and a degree of consensus on what needs to be done, it would be expected to find linkages with many areas of practice across the three dimensions of sustainable development, the economic, social and environmental. As in other areas, this intersection is strong for the field of agricultural extension. It is interesting, and hopefully encouraging, to examine how the goals and targets overlap with what the extension profession has been advocating and practicing for a long time. It is also heartening to see the recognition in Agenda 2030 of the complexity and inter-related nature of the sustainable development goals, a mirror in many ways of similar discussions within the extension field.

Unlike the Millennium Development Goals (MDGs), which were drawn up by the United Nations Secretariat and only gradually came to be accepted by countries, the SDGs came out of a long consultation process and were agreed at the UN General Assembly. This difference is more important than simply a change in the process as, unlike the MDGs that only addressed developing countries, the SDGs apply to all countries, with the idea that everyone has a role in meeting these targets. Governments have had a say in their formulation and have committed to achieving the shared goals. This, of course, does not mean that reaching the targets will now be easy or straightforward, but the goals and their targets are powerful reference points of commitment that can and should be used to advocate for adequate attention and resources across programmes and institutions, including extension, that are necessary to achieve them. Making the case for extension as a key mechanism of support for a number of the SDGs should help increase understanding of extension's role by policy makers and funders.

The 2030 Agenda for Sustainable Development: Goals and Targets and Extension Relevance

The United Nations resolution document is titled "Transforming Our World: The 2030 Agenda for Sustainable Development"(A/RES/70/1)[1]. The preamble is lofty in its ambition:

> *This Agenda is a plan of action for people, planet and prosperity. It also seeks to strengthen universal peace in larger freedom. We recognize that eradicating poverty in all its forms and dimensions, including extreme poverty, is the greatest global challenge and an indispensable requirement for sustainable development.*
>
> *All countries and all stakeholders, acting in collaborative partnership, will implement this plan. We are resolved to free the human race from the tyranny of poverty and want and to heal and secure our planet. We are determined to take the bold and transformative steps which are urgently needed to shift the world on to a sustainable and resilient path. As we embark on this collective journey, we pledge that no one will be left behind.*

The pledge *that no one will be left behind* is followed in the *Preamble* by the statement that the agenda is *focused in particular on the needs of the poorest and most vulnerable, and with the participation of all countries, all stakeholders and all people,* followed in paragraph four of the *Declaration* by the statement *And we will endeavour to reach the furthest behind first.*

These statements resonate easily with those in extension who see rural communities and smallholder farmers in particular as among those who have been left behind and who deserve attention, from government and civil society and from businesses that have a lot to gain from reaching this large population. Many extension practitioners in fact define their mission as working with the most vulnerable rural populations. On the other hand, extension services still reach only a small percentage of smallholder farmers. Even more cause for reflection is the fact that most extension systems have generally performed poorly in reaching the *furthest behind*, as the Declaration puts it. Although often relatively poor by global or national standards, many of the most active individuals in producer or rural women's groups are self-selecting individuals with more resources, however meagre, and more social skills than those who remain outside of these activities. Extension services generally focus on those owning land, even if of marginal size. If the extension profession, in all of its configurations, embraces the SDGs and advocates for

[1]https://sustainabledevelopment.un.org/post2015/transformingourworld

more attention because of its role in achieving Agenda 2030, the profession must also recognize the need to *leave no one behind* and in particular those furthest behind.

Before turning to the relevant content of the Sustainable Development Goals and Targets, it is important to call attention to one of the most striking areas of emphasis and overlap with the extension profession. This is the emphasis on *empowerment* and in particular focus on women's empowerment. These themes are highlighted throughout the Agenda, with the following as examples:

- *Realizing gender equality and the empowerment of women and girls will make a crucial contribution to progress across all the Goals and targets. (Preamble).*
- *Realizing gender equality and the empowerment of women and girls will make a crucial contribution to progress across all the Goals and targets. (para 20).*
- *People who are vulnerable must be empowered. (para 23).*
- *We are also determined to end hunger and to achieve food security as a matter of priority and to end all forms of malnutrition....We will devote resources to developing rural areas and sustainable agriculture and fisheries, supporting smallholder farmers, especially women farmers, herders and fishers in developing countries, particularly least developed countries. (para 24).*

Empowerment of individuals and groups is at the heart of extension work, by increasing scientific literacy and technical and managerial competencies to make more informed decisions, with greater confidence to take appropriate action. Furthermore, after many years where women's roles in agriculture went largely unnoticed, the importance prioritising attention on women is now recognized as essential by the profession and is a feature of most extension policy statements, backed up by solid research. Much remains to be done.

In general terms, even a cursory review of the 17 Sustainable Development Goals (Box 1 below) reveals the strong affinity between them and the larger aims of agricultural extension.

Box 1 |	**The Seventeen Sustainable Development Goals**
Goal 1.	End poverty in all its forms everywhere.
Goal 2.	End hunger, achieve food security and improved nutrition and promote sustainable agriculture.
Goal 3.	Ensure healthy lives and promote well-being for all at all ages.
Goal4.	Ensure in clusive and equitable quality education and promote life long learning opportunities for all.
Goal 5.	Achieve gender equality and empower all women and girls.
Goal 6.	Ensure availability and sustainable management of water and sanitation for all.
Goal 7.	Ensure access to affordable, reliable, sustainable and modern energy for all.

Contd...

Goal 8.	Promote sustained, inclusive and sustainable economic growth, full and productive employment and decent work for all.
Goal 9.	Build resilient infrastructure, promote inclusive and sustainable industrialization and foster innovation.
Goal 10.	Reduce inequality within and among countries.
Goal 11.	Make cities and human settlements inclusive, safe, resilient and sustainable.
Goal 12.	Ensure sustainable consumption and production patterns.
Goal 13.	Take urgent action to combat climate change and its impacts.
Goal 14.	Conserve and sustainably use the oceans, seas and marine resources for sustainable development.
Goal 15.	Protect, restore and promote sustainable use of terrestrial ecosystems, sustainably manage forests, combat desertification, and halt and reverse land degradation and halt biodiversity loss.
Goal 16.	Promote peaceful and inclusive societies for sustainable development, provide access to justice for all and build effective, accountable and inclusive institutions at all levels.
Goal 17.	Strengthen the means of implementation and revitalize the Global Partnership for Sustainable Development.

Although different aspects of extension work highlight one or more of the goals, there are eight that appear to have the clearest relevance. They are: Goal 1 on eradicating poverty; Goal 2 on ending hunger and achieving food security improved nutrition and promotion of sustainable agriculture; Goal 4 on lifelong learning opportunities; Goal 5 on women's empowerment; Goal 6 on sustainable management of water; Goal 12 on sustainable consumption and production; Goal 13 on combating climate change; and Goal 15 on sustainable use of land and forests and combating desertification, soil degradation and biodiversity loss.

A case could be made in fact for the relevance of extension in almost all of the 17 goals, especially when examining their specific targets. Below are some salient examples of the relationship between the goals and extension and where it might be relevant in achieving the corresponding SDG targets.

Goal 1. End poverty in all its forms everywhere. The relevance of extension: Promotion of rural livelihoods and income growth, with an emphasis on access to technology, control over land and other resources, while reducing vulnerability to shocks and disasters are all staples of extension work. Applicable Goal 1 targets:

1.1 By 2030, eradicate extreme poverty for all people everywhere, currently measured as people living on less than $1.25 a day.

1.2 By 2030, reduce at least by half the proportion of men, women and children of all ages living in poverty in all its dimensions according to national definitions.

1.3 By 2030, ensure that all men and women, in particular the poor and the vulnerable, have equal rights to economic resources, as well as access to basic services, ownership and control over land and other forms of property, inheritance, natural resources, appropriate new technology and financial services, including microfinance.

1.4 By 2030, build the resilience of the poor and those in vulnerable situations and reduce their exposure and vulnerability to climate-related extreme events and other economic, social and environmental shocks and disasters.

Goal 2. End hunger, achieve food security and improved nutrition and promote sustainable agriculture. The relevance of extension: This integrated goal linking food security, nutrition and sustainable agriculture is the one most easily associated with extension by policy makers and the public. Applicable Goal 2 targets:

2.1 By 2030, end hunger and ensure access by all people, in particular the poor and people in vulnerable situations, including infants, to safe, nutritious and sufficient food all year round.

2.2 By 2030, double the agricultural productivity and incomes of small-scale food producers, in particular women, indigenous peoples, family farmers, pastoralists and fishers, including through secure and equal access to land, other productive resources and inputs, knowledge, financial services, markets and opportunities for value addition and non-farm employment.

2.3 By 2030, ensure sustainable food production systems and implement resilient agricultural practices that increase productivity and production, that help maintain ecosystems, that strength encapacity for adaptation to climate change, extreme weather, drought, flooding and other disasters and that progressively improve land and soil quality diversified seed and plant banks at the national, regional and international levels, and promote access to and fair and equitable sharing of benefits arising from the utilization of genetic resources and associated traditional knowledge, as internationally agreed.

2.a Increase investment, including through enhanced international cooperation, inrural infrastructure, agricultural research and extension services, technology development and plant and livestock gene banks inorder to enhance agricultural productive capacity in developing countries, in particular least developed countries.

Goal 3. Ensure healthy lives and promote well-being for all at all ages. Extension's relevance: The work on integrated pest management and other practices to reduce the inappropriate use of chemicals fits squarely in this goal and it target 3.1.

3.1 By 2030, substantially reduce the number of death sand illnesses from hazardous chemicals and air, water and soil pollution and contamination.

Goal 4. Ensure inclusive and equitable quality education and promote lifelong learning opportunities for all. Extension's relevance: Lifelong learning, improving scientific literacy by farmers and helping learner acquire skills and knowledge form the basis of extension practice. Applicable Goal 4 targets:

4.1 By 2030, ensure that all youth and a substantial proportion of adults, both men and women, achieve literacy and numeracy.

4.2 By 2030, ensure that all learners acquire the knowledge and skills needed to promote sustainable development, including, among others, through education for sustainable development and sustainable lifestyles, human rights, gender equality, promotion of a culture of peaceandnon-violence, global citizenship and appreciation of cultural diversity and of culture's contribution to sustainable development.

Goal 5. Achieve gender equality and empower all women and girls. As mentioned above, while a great deal remain to be done, there has been a significant shift in the past two decades within the extension profession on the centrality of women's empowerment, including in leadership roles. Applicable Goal 5 targets:

5.1 Ensure women's full and effective participation and equal opportunities for leadership at all levels of decision-making in political, economic and public life.

5.a Undertake reforms to give women equal rights to economic resources, as well as access to ownership and control over land and other forms of property, financial services, inheritance and natural resources, in accordance with national laws.

5.b Enhance the use of enabling technology, in particular information and communications technology, to promote the empowerment of women.

Goal 6. Ensure availability and sustainable management of water and sanitation for all. The role of extension in promoting sustainable groundwater management is an important example of its role in achieving this goal. The needed skills and knowledge that extension can facilitate to address this goal will be increasingly important. Applicable Goal 6 targets:

6.1 By 2030, substantially increase water-use efficiency across all sectors and ensure sustainable withdrawals and supply of freshwater to address water scarcity and substantially reduce the number of people suffering from water scarcity.

6.2 By 2020, protect and restore water-related ecosystems, including mountains, forests, wetlands, rivers, aquifers and lakes.

Goal 8. Promote sustained, inclusive and sustainable economic growth, full and productive employment and decent work for all. Extension's relevance: Although many would see this goal as focused on formal employment, as target 8.1 makes clear, it also applies to entrepreneurship, creativity and innovation that apply equally well in on-farm work.

8.1 Promote development-oriented policies that support productive activities, decent job creation, entrepreneurship, creativity and innovation, and encourage the formalization and growth of micro-, small- and medium-sized enterprises, including through access to financial services.

Goal 10. Reduce inequality within and among countries. Extension's role: For many countries, the bottom 40 per cent of the population is rural with agricultural or other natural resource-based livelihoods. The only way this target can be met is though attention to the needs of these smallholder farmers, those with forest-based livelihoods and landless agricultural labourers. Applicable Goal 10 targets:

10.1 By 2030, progressively achieve and sustain income growth of the bottom 40 per cent of the population at a rate higher than the national average.

Goal 12. Ensure sustainable consumption and production patterns. Extension's relevance: Promoting the sustainable management of natural resources, reducing post-harvest losses and the management of agricultural chemicals are all well known aspects of the role of extension. Applicable Goal 12 targets:

12.1 By 2030, achieve the sustainable management and efficient use of natural resources.

12.2 By 2030, halve per capita global food waste at the retail and consumer levels and reduce food losses along production and supply chains, including post-harvest losses.

12.3 By 2020, achieve the environmentally sound management of chemicals and all wastes throughout their lifecycle, in accordance with agreed international frameworks, and significantly reduce their release to air, water and soil in order to minimize their adverse impacts on human health and the environment.

Goal 13. Take urgent action to combat climate change and its impacts. As highlighted by many at the COP 21 Climate Change deliberations in Paris, agriculture has an enormous role to play in both mitigation and adaptation. Applicable Goal 13 targets:

13.1 Strength enresilience and adaptive capacity to climate-related hazards and natural disasters in all countries.

13.2 Improve education, awareness-raising and human and institutional capacity on climate change mitigation, adaptation, impact reduction and early warning.

13.3 Promote mechanisms for raising capacity for effective climate change-related planning and management in least developed countries and small island developing States, including focusing on women, youth and local and marginalized communities.

Goal 14. Conserve and sustainably use the oceans, seas and marine resources for sustainable development.

Goal 15. Protect, restore and promote sustainable use of terrestrial ecosystems, sustainably manage forests, combat desertification, and halt and reverse land degradation and halt biodiversity loss. Extension has a key and obvious role to play in this area as well, across different ecosystems. Applicable targets:

15.1 By 2020, ensure the conservation, restoration and sustainable use of terrestrial and inland freshwater ecosystems and their services, in particular forests, wetlands, mountains and drylands, in line with obligations under international agreements.

15.2 By 2020, promote the implementation of sustainable management of all types of forests, halt deforestation, restore degraded forests and substantially increase afforestation and reforestation globally.

15.3 By 2030, combat desertification, restore degraded land and soil, including land affected by desertification, drought and floods, and strive to achieve a land degradation-neutral world.

15.4 By 2030, ensure the conservation of mountain ecosystems, including their biodiversity, inorder to enhance their capacity to provide benefits that are essential for sustainable development.

15.5 Take urgent and significant action to reduce the degradation of natural habitats, halt the loss of biodiversity and, by 2020, protect and prevent the extinction of threatened species.

Goal 16. Promote peaceful and inclusive societies for sustainable development, provide access to justice for all and build effective, accountable and inclusive institutions at all levels. Extension's role: Although the Goal applies to other levels as well, the role of the extension in promoting community-based organizations, farmers associations and women's group networks is particularly strong, especially among those working in civil society organizations. Applicable targets:

16.1 Develop effective, accountable and transparent institutions at all levels.

16.2 Ensure responsive, inclusive, participatory and representative decision-making at all levels.

Goal 17. Strengthen the means of implementation and revitalize the Global Partnership for Sustainable Development. Although also focused on all levels, including the global, this goal and its target 17highlights the recognition that extension and advisory services in government, the private sector and civil society need to work in partnership, within the agricultural innovation system.

17.1 Encourage and promote effective public, public-private and civil society partnerships, building on the experience and resourcing strategies of partnerships.

Confronting the Complexity of the Interlocking Challenges

Another notable feature of the Sustainable Development Goals is the recognition of their complexity and inter-relatedness. In a memorable blog posting by John McArthur[2] of the Brookings Institution in Washington DC he recounts the story of explaining the adoption of the SDGs to his mother. He, along with many others, had been concerned that the number of Goals at 17 was far too high and too difficult to communicate. When he mentioned this to his mother, however, her reaction was entirely and unexpectedly positive. "Seventeen is a great number…It sounds like they didn't fake it. The world is complicated. If they had come back with some Letterman-style top 10 then I probably wouldn't have believed them."

MacArthur writes how this story has resonated with others, who told him "It's like your mom said—the world is complicated." As he relates, "I heard this remark enough times that I started sharing her vignette more deliberately in other speeches and conversations....It's as if people feel energized by a simple statement that articulates the complexity they already feel." He goes on, "I now have a fresh take on the 17 goals. They are a reflection of a broader truth. Our collective aim should not be to rally around narrow simplicity, but instead to respect people's intellects and their desire to overcome complexity through cooperation."

This sentiment corresponds very well with discussions within the extension profession over the past decade. While extension has key roles to play, other factors including research, credit, access to land, a conducive policy and legal environment and infrastructure, among others, need to be present for extension to be truly effective. Within the extension sphere on its own there has been a growing and sometimes disheartening acknowledgment of the complexity of the challenges and the inability of extension services, for a number of reasons, to deliver on them.

The simplistic view has long been outdated of seeing extension as the technology transfer arm picking up ready-made products and ideas from agricultural research and passing them on to farmers. This model has worked in the past for some technologies and for some farmers but it has not worked well for resource-poor farmers in more fragile environments and it is clearly not sufficient to respond to the emerging challenges in agricultural and rural development and sustainable natural resource management. These include climate change and a backdrop of declining water availability and soil degradation, rapidly changing mixes of on- and off-farm family income and labour availability, fast growing demand for horticulture crops, meat and dairy and stagnant growth in traditional cereal crops, more integrated markets and greater price volatility, among many others.

---

[2] "Why 17 is a beautiful number" 28 September 2015. Brookings Institution. http://www.brookings.edu/blogs/future-development/posts/2015/09/28-sustainable-development-goals-mcarthur

It is now widely accepted that agricultural innovation needs to be seen a system with multiple players working across a spectrum of activities involving government, civil society and the private sector. This "agricultural innovation system" comprises a network of organizations, enterprises and individuals who require capacity and an enabling environment that supports farmers in trying out, adapting and benefiting from innovation. Many supportive actions by a number of actors have to come together to address the challenges facing small producers, and often institutions and individuals lack relevant capacities and the social and economic environment may not be very supportive. The complexity of this chain of support, its lack of integration, and weak capacity to support small family farmers in a more creative and enabling manner is indeed daunting.

The expansion of new extension actors from NGOs and the private sector creates additional complexity, along with the decline in funding and confidence in government extension services. Where these public sector services have recovered from lack of attention, or worse, they have often undergone considerable institutional upheaval as part of devolution processes or other changes. This new extension pluralism is here to stay and, given the complexity of the overall system, it can be difficult to envision moving forward in a coherent and coordinated manner. Nevertheless, this will be fundamental in addressing the complex challenges facing food and nutrition security, rural poverty eradication and sustainable natural resource management, all highlighted in the SDGs.

The formulation of the SDGs may offer some encouragement on this as well. Agricultural innovation is genuinely complex and it is helpful to recognize it as such. There is no point in attempting to oversimplify but rather recognize that many inter-related capacities need to come together, performed by different actors in different roles. The change processes that extension supports in production, marketing, organization and in the adoption and adaptation of new technologies are inherently multifaceted, with solutions that defy simplification.

Like the 2030 Agenda, the agriculture innovation system, and extension's role within it, face a set of challenges where each component is sufficiently unique to merit is own description but where are all connected and all are necessary to contribute to the larger goals. Making sense of these interlocking challenges has been a focus of attention of research and extension analysts for some time. Three documents, among many others, offer illustrations of the need to be comprehensive in describing the system and its complexity.

The first is the *Road Map for Transforming Agricultural Research for Development (AR4D) Systems for Global Impact*[3] that came out of the 2010 Global Conference on Agricultural Research for Development (GCARD).The Conference was organized by the Global Forum on Agricultural Research (GFAR), whose Secretariat is hosted at the headquarters of the Food and Agriculture Organization of the United Nations (FAO) in

[3]http://www.fao.org/docs/eims/upload/294891/GCARD%20Road%20Map.pdf

Rome, in association with the reform process going on at that time of the Consultative Group on International Agricultural Research (CGIAR).

The Road Map recognizes that "The global fragmentation and under-resourcing of public innovation, education and advisory processes and weak linkages with wider development processes and with farmers, NGOs and the private sector, are major bottlenecks constraining the value and impact of agricultural innovation on the lives and livelihoods of the poor." It lays out a plan to address the challenges, building on the consensus that "business as usual" is not an option and states that these aims can only be achieved only if: (i) All stakeholders work more effectively together to address needs identified as most important for the poor and see themselves as true partners in AR4D...; (ii) The capacities and investments required are put in place...; (iii) The millions of resource-poor small farmers in diverse environments, along with all other actors in value chains and food systems, including consumers, form part of innovation processes...; and (iv) AR4D and related knowledge-sharing actions with key outcome-focused themes are embedded in the wider development agenda, with the required enabling environment to transform innovation into development outcomes"[4].

The Road Map then outlines six strategic elements, each with a list of relevant actors, roles, desired outcomes and milestones, all of which need to come together to address the challenges at the national, regional and global levels.

The second illustration is contained in FAO's *Corporate Strategy on Capacity Development*[5], which presents a framework for visualizing and analyzing the component parts of agriculture-related capacity development, combining four categories of functional capacities (policy and normative, knowledge, partnering and implementation) that need to be present in different ways in all three levels of individuals, organisations, and the enabling environment. For example, the individual level relates to knowledge, skills (technical and managerial), and attitudes that can be addressed through facilitation, training, and competency development. The organisational level relates to public, private and civil society organisations in terms of: (i) strategic management functions, structures, and relationships; (ii) operational capacity; (iii) human and financial resources; (iv) knowledge and information resources; and v) infrastructure. The enabling environment level relates to political commitment and vision, policy and legal frameworks, public sector budget allocations, governance and power structures, and incentives and social norms. Needless to say, this matrix of capacities and levels leads to a large number of cells identifying a host of needed capacities. The document also points out seventeen additional critical factors for success that apply to different stages of the capacity development work.

---

[4]Ibid. pp 6-7.

[5]http://www.fao.org/fileadmin/user_upload/newsroom/docs/Summary_Strategy_PR_E.pdf

The final illustration puts these elements together focusing on extension alone, the 2012 paper by Rasheed Sulaiman and Kristin Davis entitled *The New Extensionist: Roles, Strategies, and Capacities to Strengthen Extension Advisory Services.*[6] This new vision for extension and advisory services within the agricultural innovation system similarly encompasses individual roles and capacities as well as those at organisational and system levels. This is necessary, as they put it, to:

link more effectively and responsively to domestic and international markets where globalization is increasingly competitive; reduce the vulnerability and enhance the voice and empowerment of the rural poor, promote environmental conservation; couple technology transfer with other services relating to credit, input and output markets, and enhance the capacity development role that includes training but also strengthening innovation processes, building linkages between farmers and other agencies, and institutional and organisational development to support the bargaining position of farmers.[7]

Their analysis goes on to highlight multiple aspects of support necessary to achieve these outcomes, along with a series of recommendations for capacity development to allow extension to contribute better to agricultural innovation, and "to collectively perform a wide range of roles."[8]

Conclusion

The range and thoroughness of these three frameworks can look intimidating, outlining all that needs to happen, implemented by multiple providers, with increasing capacity at all levels and through partnerships and collaboration. Nevertheless, the complexity is real and contributions from many actors within the extension system and beyond are required. There is no way around this; better to acknowledge the challenges and get on with the work, respecting the roles of each. As MacArthur put it with respect to the SDGs, "Our collective aim should not be to rally around narrow simplicity, but instead to respect people's intellects and their desire to overcome complexity through cooperation." This is excellent advice.

---

[6] http://www.g-fras.org/en/knowledge/gfras-publications/file/126-the-new-extensionist-position-paper?start=20

[7] Ibid. p. 3

[8] Ibid. p. 17

03 Five Pillars of Agricultural Extension Education for Total Development

- Dr. C. Prasad

Preamble

The academic world, broadly speaking, is bifurcated into two divides: (i) Science and (ii) Art – both are important for the society, for the country, nay the World. Then there are two more divides: (a) Basic/Natural Sciences – Physics, Chemistry, Engineering ... and (b) Social Sciences – Sociology, Economics, Extension Education, Anthropology, Politics, Law, Religions and so on. Learning, if sincere and real, is a contagious game; more we learn, more we need. They are a vast arena of academics and one has to decide for a part of these disciplines – Agricultural Extension Education, Agricultural Economics, Political Science etc. for one's career. While the basic knowledge of these academics are generally essential, then one decides for a narrow specific field of study – the so called specialization. The life being short and a lot more to study, learning and doing something worthwhile for oneself, for others, for the society and for the country at large, demands a deeper specialization. The relationships of these branches of studies are equally important; we need to study them, some broadly and some specialized – the latter we decide for our profession for the life long career! Teachers and trainers need to create such a conductive learning environment for the learners/students or the public at large, for changing their values and improving the knowledge. We relate past to the present, and present to the future. Our concerns away from the real and righteousness are bound to create chaos or disasters sooner or later, because of our weak human resources. If we are too bad in understanding the nature or so to say, the God, the Tsunami is the result! So the broad and basic knowledge must precede the specialization(s) and above all the total commitment to one's job and the people we serve.

The World of today itself is a big/large divides: countries/nations, rich countries-poor countries; rich people-poor people; Indian, African, American; Hindu – Muslim; Christians, Churches and Temples and so we go on. And in the absence of deeper, clearer, sincere and committed education and training, we are poorer by the day resulting into much more away from the 'we- feeling' – *Vasudev Kutumbkam!* One could clearly visualize today the challenge to the society at large – Our road map should be clear and pious. We need to follow Swami Vivekananda: *'when our path is right, we are bound to reach our goal!'* We should be afraid of God, who is watching us all the way! This human life is the gift of God – let it be preserved, utilized fully for the needy and respected by society for our next life to come.

Agricultural Extension Education

Agricultural Extension Education – Extension or Agricultural extension, the common terminology, is an applied social science, relatively away from the basic. This has to be understood and acted upon for the good of the working environment; we learn from basic to applied. In our short sightedness and short-cut, we are missing the boat. The pure applied sides do not bring to fore the deeper dimensions of respective basic and core disciplines. In that context, we are the sufferers; we are always going for the short-cut, and that has made us a lighter discipline so to say. We must take an 'U-turn' starting from basics to the applied, building the discipline on a stronger foundation – strong background of Sociology, Rural Sociology, Psychology, Educational Psychology, Agricultural Economics etc. It is always useful to use the total name – 'Agricultural Extension Education' rather than in its part (Extension), for in the long run, they are not correctly understood or else one is required to labour hard to make things at home for the public at large. But commonly people would like to be brief and shorter, much more handy, but are they not likely to be misunderstood in the long-run? We, normally and naturally, should move from broad and basics to the applied – the narrow specialization; must develop a natural habit of reading a lot and enjoying it in practicing in the fields. The taste of the pudding is when you start enjoying your work and making contributions to the society – rural society preferably in our case.

The 'Extension Education' discipline is devoted to changing people and through them the society, the country or the universe – the World! Is not this a very difficult and the challenging task? But more often, we take it lightly to our ultimate peril. These concepts/definitions must be ingrained in the minds of students, the trainers, educators and ultimately the end users – the farmers, farm women and the farm youths. Then from farming, you go to the farm industries, farming as a business, trades & exports – the goal should be A-Z. Give learning a long rope! This is why American teachers commonly insist that the learning must not remain 'skin-deep' rather it should be a 'blood deep' phenomenon– there is no end to learning! So we need very qualified and committed teachers; human resources development leads to improvement in desirable directions. And mind it, in our Third World basically, we are being degraded – growing self-centred; an U-turn is needed, no it is imperative.

Maunder (1973)defined agricultural extension as 'a service or system which assist farm people, through educational procedures, in improving farming methods and technologies, increasing production efficiency and income, bettering their levels of living, and lifting the social and educational standards of rural life. Some people tend to equate agricultural extension with the term *'technology transfer'*. This is incorrect, because technology transfer includes the additional function of input supply and agri-services'. But do we agree to this definition of Transfer of Technology (TOT)? We hope not; agricultural education and the transfer of technology are synonymous words. We need to go beyond this definition: USA promoted three integrated pillars of development – teaching, research and extension; this has to be improved to five pillars – teaching, research, extension, training and management. Technologies cannot be utilized if it does not cross the extension bridge reaching the farmers. In other words, the investments in technology development could be wasted – money spent on research is wasted and farmers ultimately suffer. Fig.3.1 explains in simple terms the importance of extension education; making research findings productive, useful and rewarding.

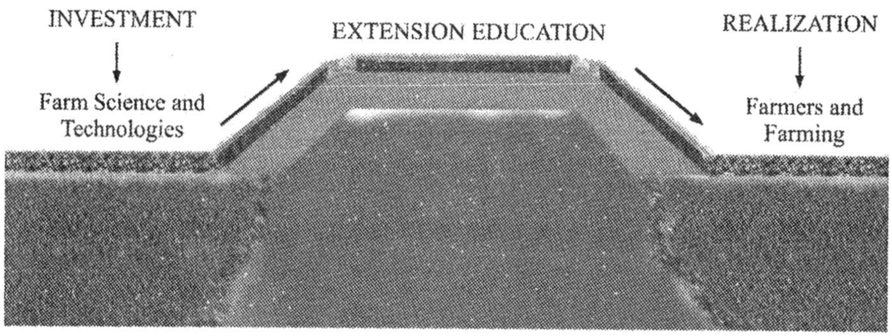

Fig. 3.1 Extension Education – The Bridge between R&D in Agriculture

The bridge of agricultural R&D is self-explanatory; the weak extension/transfer of technology would mean the loss of investments on research, education etc. – it is so serious! But how much the Governments are sold to this would matter?

Evolution of Agricultural Extension

It has been a long time to reach where we are today in creating the discipline of Agricultural Extension and appropriately *'Agricultural Extension Education'*. By 2015, we are in a way, a364 years old discipline, can you imagine? We must look at the history of Agricultural Extension Education, where professionals of the advanced countries have given their best from time to time leading to this stage – a lot of collective efforts and experiences by: Rabelais (1483), Samuel Hartlie(1651), Jean Jaques Rousseau (1712), Benjamin Franklin (1744), Heinrich Pestalozzi (1746), Philip E. Von Fellenberg (1771), Arthur Young (1784), John Lowell (1818), Seaman A. Knapp (1833), N.S. Townshed (1845), Earl of Clarendon (1847), Edward Hitch Cock (1853), Farquahar (1873), Smith-Lever Cooperative Extension Act (1914), Smith Hughes Act (1917), Asia & Oceania (1960), Africa (1960) and Latin America & Caribbean (1990).

During the past several long years, the growth of Agricultural Extension took several routes and sincere efforts: Rabelais (1483-1553) promoted 'organized pupils study nature' as well as the books and used their knowledge in their daily occupations; Milton (1600-70) published a book in 1651 in England; titled as, *'An Essay for Advancement of Husbandry-Learning'*. Milton followed the plan of 'education' involving a broad study of classical literature including 'Agriculture'; the academies of England invariably included 'practical learning in their curriculum studies' having practical learning; Rousseau (1912-70) dealt with promoting 'manuals and industrial activities' in education; Heinrich Pestalozzi (1746-1826), the Swiss educational reformer promoted 'school for poor children' who were commonly engaged in raising farm products, spinning, weaving of cotton; and P.E. Von Fellenberg (1771-1844) promoted at Hofwyl, Switzerland, 'manual training schools' in agriculture and allied areas which considerably influenced the USA.

Also *publications* and *societies* were helpful in promoting agricultural development efforts during seventeenth and eighteenth centuries in the USA, UK, Germany and France as early as 1761 and in Russia in 1765. In 1843, in North America, itinerate teachers were used to improve agriculture – the practical and scientific farmer were also employed to give public lectures on all relevant sciences for agriculture – 'Farmers' clubs' were also promoted. In 1848, the State Agricultural Chemist was appointed in Maryland, USA – he was required to lecture all over the State among the farmers; in 1853, President Edward Hitchcock recommended to open 'Farmers' Institutes'; and in 1914 with the Smith Lever Act formally the agricultural extension was established in the USA.

In 1866, the term *extension* was used in England and it was integrated with the university as the *University Extension* with Cambridge University and then to Oxford University. The term *extension education* was first used in 1873 by *Cambridge University* as an *innovation*, taking university education to the farmers– the people who arrange our food and nutrition. Do not you salute such people?

'The first modern, Agricultural Advisory and Instructional Service was established in Ireland during the great potato famine during 1847-1851'.

Early Extension Work In India

Prior to independence (1947), rural development work in India was a casual affair; sporadic attempts were made here and there both by voluntary and government agencies to help and develop the rural poor – the farming communities basically. For instance, Gandhian Experiment in Rural Reconstruction (1920), Sriniketan Project (1921), Marthandom Scheme (1921), Gurgaon Project (1927), Rural Reconstruction Movement (1932), and Indian Village Upliftment Scheme (1945), were some of the early pioneering ventures in this direction had taken place.

Soon after independence, the attention of the Government was drawn to the development of villages – rural areas. Some of the initial rural development attempts in later 1940s were: Grow More Food Campaign (1942), Etawa Pilot Project (1948),

Nilokheri Experiment (1947), and Sarvodaya Programme (1948). These efforts, though very meager to match the needs, provided valuable experiences which, in a great measure, influenced the present Community Development (C.D.) and Extension Education Programmes. Some of the lessons worthy of emulation from these earlier efforts were:

- The multi-purpose Village Level Workers (VLWs) were needed for rural upliftment in order to cope up with the multiple tasks and problems with which villagers were confronted. The VLWs must reside in their respective village circles so that they could maintain frequent contacts with the clientele and provide timely guidance and regular information support.
- An integrated approach to resolving complex and inter-woven rural problems was essential for counteracting barriers to change and for promoting balanced growth.
- A two-way communication network between the villagers and the rural development workers and authorities was a must; bottom-up communication was more important than the usual top-down communication.
- The farmers would mean farmers, farm women and farm youth like in the USA – these three groups have been the clientele of the extension professionals.
- Effective functioning of village institutions – Panchayats, Cooperatives, youth clubs, NGOs was essential.

There were four major Projects of the Indian Agricultural Research Institute (ICAR) in 1960s viz. (i) National Demonstration Project (1965), (ii) Operational Research Projects (1972), (iii) Krishi Vigyan Kendras (KVKs) and Trainers' Training Centres (TTCs) in 1974 and (iv) the Lab-to-Land Programme (1979). Later the projects launched by the Agricultural Ministry were: (i) Intensive Agricultural District Programme (1961), (ii) Intensive Agricultural Areas Programme (1964), (iii)Training and Visit (T&V) System in 1974-75, (iv) National Agricultural Extension Project (NAEP) in 1979 and (v) National Agricultural Technology Project (NATP) in 2002.

In fact, the National Demonstration project (1965) initially started by the Union Ministry of Agriculture, but when the Agricultural Secretary was Mr. C. Subramaniam (IAS), he opined the basic philosophy of agricultural extension: 'those who generate the technologies – the farm scientists, they were in the best position to demonstrate them to the farmers – the farming communities. Thus this project was transferred to the ICAR for promoting it in the larger interest of the farmers'. This was no doubt, a great and deeper thought. Since, this was practically possible, the subject-matter specialists were used in conjunction with the agricultural extension educators.

This philosophy is very deep, which the developed as well as the developing nations should understand and practice. And unfortunately, we have not done this to our sufferings. This proposition makes the extension education, may that be in agriculture or industry, a subject of the scientists – not any professionals, not really the other classes devoted to agricultural development strategy. Then they may remain in the extension system for whole of their lives, and their promotion in a fixed schedule – after five years or so. This has to be realized by the governments and the non-governments institutions and agencies devoted to development – more so the development of agricultural and rural areas.

The use of extension educators and the subject-matter specialists in conjunction have done extremely well.

Government Policy Directions

Planned development has been a major concern of the Government since Independence (1947). The First Five Year Plan was launched in 1950-51, and now the Twelfth Plan is about to be over by March, 2017. In these plans, agriculture and rural development, by and large, have been given priorities.

The Article 39, Clauses (b) and (c) of the Indian Constitution, has emphasized on balanced and progressive socio-economic development with social justice. Mahatma Gandhi gave the concept of *Antyodaya*; in his view, the utility of all development programmes must be measured against one yardstick viz. the real benefits to be delivered to the poorest persons. The Jayaprakash Narayan Committee (1961) recommended for special attention for weaker sections of the rural community which in 1970 was promulgated as '*Growth with Social Justice*'.

The Five Year Plans have been insisting on growth with Social justice. While addressing the first meeting of the Planning Commission in 1951, Jawaharlal Nehru pronounced, "*The objective of the plan, gentlemen, is growth with justice. It is your job to combine them I do not know how?*".

One of the objectives of the Fifth Five Year Plan (1974-78) was to providing minimum consumption for the bottom30 percent of the population. The Government resolution constituting the National Commission on Agriculture (NCA),between 1970-75, underlined the need to ensure, "*development of agriculture for the welfare of the vast multitude of population living in the rural areas*". The Commission was directed to indicate measures which could secure effective participation of the bulk of the Indian peasantry in accelerating production and ensure that the benefits of the new technologies were widely shared by all categories of farmers.

The Scientific Policy Resolution of the Government of India (1958) stressed on promotion of application of science for development. The Science and Technology Policy Statement of the Government of India (1982) further encouraged absorption and adaptation of improved technologies by the farmers. The Ministry of Rural Development, Government of India, has encouraged rural development efforts by introducing section 35 cc/35 ccA of the Income Tax under which Companies, Co-operative Societies, Associations, Institutions are entitled to a deduction in the computation of taxable profit of the expenditure incurred by them on any approved programme of rural development.

The 20 points Development programme of the Government had given added attention to agricultural and rural development and alleviation of rural poverty.

India got independence in 1947 and after a while the 'Community Development Programme (CDP) was launched in 1952, but right in next year in 1953, the 'Agricultural Extension Service (AES)' started, for it was realized that development is a long range process and a continuous game, and not a relatively short approach like the CDP that was thought of – lasting for three years area-wise. The philosophy pronounced by Hon'ble S.K. Dey duly supported by Pandit Jawaharlal Nehru, Hon'ble Prime Minister of India then, was that our development approach should be based on the *Socialistic pattern of Society*" and not like the *materialistic approach* of the Western World. But we have followed, unfortunately, the *materialistic one* even at the cost of our society at large. The USA is a strong Country both *socially* as well as *materially*, for their balanced approach on the overall education and extension (social sciences) and research – the natural/basics sciences. Whereas, they integrated their values, education, research and extension; we adopted from the West the lucrative sides – the money as well as the political powers which has dominated in our Country, creating a big gulf between rich and poor. It is rightly said that money begets more money leading to money madness, and power corrupts people and absolute power corrupts them absolutely. This is by and large the order of the day not only in India but in almost all the third world – the developing nations.

Broadly speaking, we have two extension systems in the country: (i) the first-line/front-line system of the ICAR – a smaller in size but with a critical and catalytic efforts and influencing both the farmers as well as the extension staff of the Ministry of Agriculture, Government of India; and (ii) the Agricultural Technology Management Agency (ATMA) created in 1998, the approach of the Department of Agriculture & Cooperation is the major extension body in agriculture, but are far from the crucial roles they are expected to play in the Country. Basically, the Ministry of Agriculture is performing the role of agricultural extension services (AES, 1953), rather than the extension education functions – the informal education of the farmers or for that matter, any other clientele system(s). The extension education function, in fact, should have been a part of the ICAR system focused through State Agricultural Universities (SAUs) or the ICAR Research Institutes like the USA. Or ideally, at this point of the Agricultural Extension Education system, it could be an Independent Body (Autonomous Agricultural Extension Council), for India is a big agricultural country with huge rural population. The Agricultural Extension Education Council as suggested above may be suitably and closely linked with the ICAR – the research system on one side as well as the Ministry of Agriculture – the agricultural extension services system on the other – the triple organic correlates!

Fig. 3.2 The Socio-Economic Imbalance – Skewed Development

Socially we have gone weaker, but no doubt economically we have made a mark. But because this imbalance, there is skewed development – more economic than social (Fig. 3.2); and thus, the poverty and hunger for at least 30 per cent of our population continues unabated. Look at the Scheduled Castes (SCs), Schedule Tribes (STs) and Minorities. In fact, socially, the society is getting disintegrated – no effective working of Panchayats or Cooperatives or Farmer/Women Clubs; and thus we have become caste and creed ridden society. One may see in our Newspapers as well as on T.V. the conditions of our society – full of scams and such other social maladies and disorders. Since the social sciences are the foundation for *building people – the human resource,* who then builds the nation. But this base has been weakened and, therefore, the proper balance of social education and the basic science education should be developed like the USA. This is why, we need to promote the concept of *the Social Sciences First;* it is going to take a long time for an 'U' turn unfortunately.

During September's UN Summit on Climate Change (2005), NGOs issued an appeal emphasizing the *moral and ethical dimensions of global warming*. Tahrin Naylor, a Bahá'í representative to the United Nations, said the purpose of the document is to call attention to the fact that Climate Change is more than a political, economic, and scientific problem. "There is a moral and ethical dimension to climate change that must be addressed", said Ms. Naylor. "For example, we know that wealthy nations have contributed more to climate problems than the poor nations, and so there is an element of justice that must be considered in any long-term solution".

Five Pillars of Development

By and large, we talk of agricultural R&D – Research and Development, whereas we need to consider, the five important pillars of development referred to in the very title of this chapter: (i) Education/Teaching, (ii) Research, (iii) Extension – the triple functions as they are known usually, (iv) Training, and (v) Management (Fig. 3.3). Unfortunately, among all these pillars, we are quite successful on *research*, but on all other four pillars, basically the social sciences we are very relatively weak. In our case, they are the *applied social sciences;* we are facing them in/for application, not only for an academic exercise. And *management* for us, it seems, is not at all a subject unfortunately, whereas it is omnipresent everywhere and that also as an oldest discipline – extremely important. Do we give enough meaning to efficient and effective management in our day to day working at all levels? Management exists as a common sense in fact, when it should be based on proper training on management at least on basic dimensions.

Fig. 3.3 The Five Important Pillars of Development

Look at the five pillars of agricultural development; except the agricultural research, all remaining are applied social components. See the agricultural education – 73 Agricultural Universities and Scientists & teachers in lakhs, but there is not a single 'Journal on Agricultural Education' in the country, whereas we claim to influence the developing nations. The ICAR, in the past, tried twice to publish a Journal on Agricultural Education, but they failed surprisingly. Teachers/Professors/Scientists only teach, but they cannot write on this vital area, whereas there are numerous Journals in Agricultural Research – discipline wise. What this means to us? You know it better – we do not care much for 'education and training' which are the foundation for human resources. This is why, today we have distorted more personalities, self-centered, selfish! Fig. 3.4 depicts the relationship between education and training: 'education' is more knowledge oriented, where 'training' is skill-oriented, the job-led.

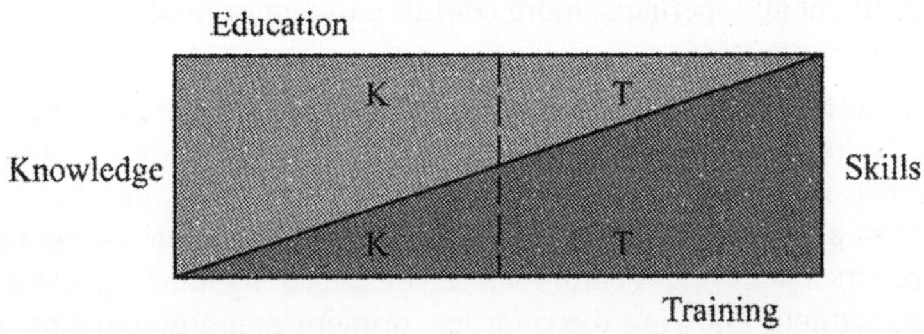

Fig. 3.4 Functional Relationship between education and Training

Management Omnipresent

Talking on social sciences in agriculture, let us look at a few values and theories in action in social sciences for Agricultural R&D. Let us look at Fig. 5– the 14 components of management, which we do use, though, sparingly, and perhaps without knowing their deeper values, scope and importance.

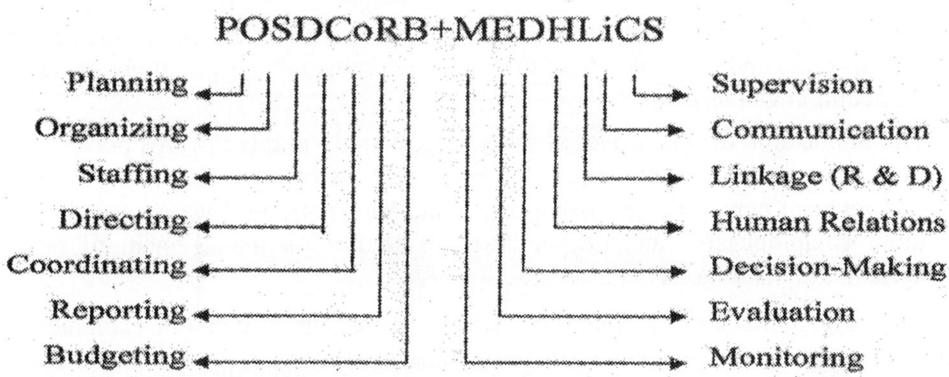

Fig. 3.5 Fourteen Processes of Management (Luther Gullick, 1937) (C. Prasad, 1993)

Prasad made a study on the 'Dynamics of research – Extension Linkages for strong and sustained Farm Transfer of technology System – evolving a Management Model' supported by the NATP Project of the ICAR – World Bank funded (2002-05). The 14 components of management, were studied in relation to the progressiveness of the select States/Institutions. Let's see what this study points out. As a part of this study, the "effectiveness" and "efficiency" of 18 States, SAUs and Research Institutes were studied on these14 components (Fig. 3.5). It was very surprising that the respondents, mostly scientists as managers, could not understand all those 14 components and the results thus were not as accurate rather distorted. The overall outcomes of the linkage study were:

1. In case of Punjab, there were more positive responses on independent variables relating to 'effectiveness' and 'efficiency'. In other states, the responses appeared very indifferent and, perhaps, more confused owing to relatively less understanding of the manage processes.
2. The respondents did not seem to understand clearly, perhaps, management as a serious discipline or business, what to call of the 14processes or components? This was expressed by the positive and negative responses on the 14 independent variables in all cases. This was also clear at the face value of the management of the Departments or the Institutions as observed by our eye-estimates. We are managers without knowing the rigorous, principles and practice of management it appears. Frankly, where our scientists been academically trained in agricultural management, they have done well. We have management institutes today the same reasons.
3. In several cases, the Standard of Errors were being so high that, one takes it that there were a lot of confusion and misunderstanding about those management processes. We need to study management – an omnipresent discipline.
4. There were limited components of management effectiveness and efficiency which were significant at 5 per cent level; some were at 10 percent level and most were beyond that.
5. The ICAR Institutes appeared slightly better in perception of management effectiveness and efficiency than the State Departments of Agriculture or the State Agricultural Universities.
6. The implication of this aspect of management study points towards poor concerns in this respect: (i) more education and training of agricultural scientists and development personnel on 'management and administration' were needed, and (ii) more comprehensive and deeper research studies on management of different agricultural institutions/departments.

Through Fig. 3.6, one of the important theories of management can be explained – *the relationships between the incumbents and the Managers/Masters.* This explains how one has the best of the work and satisfying activities, when the incumbents get their appropriate dues and recognitions and the Masters get the best results. The values of the Masters and the Incumbents must reciprocate each other for the highest productivity, but usually they don't or are they?

Fig. 3.6 Fusion Theory of Management

In management, there are certain basic principles, let us examine one for example: it is said that there are two very important words/principles in administration/management relating to *(i) authority* and *(ii) responsibility*. Both the acts must be clearly assigned as well as well-defined so that the incumbents, whether higher and lower in the hierarchy, may understand their work/activities and act accordingly. This, many times do not happen, which hampers the work and the progress. For a progressive results, this condition must exist. The other basic principle of management is that the *responsibilities* assigned to an individual/incumbent must commensurate with the related *authorities*. This often does not happen in our Institutions/Organizations; the authorities are vested usually in the Senior Managers and the incumbents have the responsibilities to work on. And, therefore, this clash of interests and assignments often occur and hence the work suffers immensely.

Assessment of Progress of a Nation – GDP VS GNH

The denominator of success of the Government has commonly been the parameter of annual GDP (Gross Domestic Product) – it is the measurement by the production in terms of money, assets – the materialistic measure in fact.

How much such an effort touches the hunger and poverty? Village Panchayats and Cooperatives were launched as a societal organs to develop the people from below on the 'Antyodaya' concept of Mahatma Gandhi. But unfortunately, they have become a part of the politics in elections of the Panchayats on the same parameters; usually, not the local leaders who are welcomed by the communities for their sincerity and leadership are elected!

One would be happy to note that the two basic principles of *Bhagwat Geeta* and *Mahatma Gandhi* are being put into practice in a *tiny Country* like *Bhutan* –our neighbour. They measure their progress not by *GDP*, but by *GNH*– the *Gross National Happiness!* And when this happens, there is a fair distribution of wealth in a manner that no one suffers at the cost of others! This is what we call as the socialistic approach to development vis-à-vis the materialistic one that we pursue. In *Bhagwat Geeta*, it is said that *one is entitled toonly that amount of wealth, which is enough for ones upkeep; the rest has to be shared and if one does not do so, the same can be taken/snatched away, and that cannot be considered as a sin*". What did Mahatma Gandhi said in this context: *the approach to development should be Antyodaya (development of those who are at the end – the lowest or poorest)*, but we do the reverse. Bahá'u'lláh (Iran)goes even beyond this approach when he states: *'The Earth is But One Country, and Mankinds its Citizens'*.

Sustainable A-Z Extension Model

Historically, extension professionals have proposed and practiced varied extension models but with only partial success; some extension models have been successful on the ground sustainably: (i) FELDA Model of Malaysia; (ii) Agro-industrial complex of Bulgaria; (iii) Cooperative Extension Work of the USA; (iv) IADP – Intensive Agricultural District Project of India leading to Green Revolution; (v) AMUL – India; (vi) E-Chaupal of the ITC, India; (vii) SHG – Self-Help Group of Bangladesh; (viii) Precision-farming of Tamil Nadu, India; (ix) National Demonstrations Project (NDP) of ICAR – India, the Krishi Vigyan Kendras (KVKs)of ICAR, India and Farm Schools recommended by the National Commission on Farmers by Dr. M.S. Swaminathan, India. This list is not exhaustive but illustrative only (2006).

How and why these extension models have been successful as well as sustainable; we need to examine the parameters for the future of the extension education discipline? This better be explained on the Five Pillars of Development through the A-Z model (ERETAM) – Education, Research, Extension, Training & Management. This model insists on good *education* for building the quality and committed *human resources,* both the farmers on the ground and extension professionals on the top; *research* – action research on the on-going extension field programmes and applied or basic research on the performance of the model(s); *field extension education,* the most difficult but the very satisfying and rewarding part; *job-led training* the generic of education but less theoretical and academic, and more skill-oriented – job led; and the *management* is the central pillar working everywhere and every time in and through the four pillars. A few dimensions of management have been highlighted earlier in the text.

Wherever in the globe, these basics of extension education or for that matter, the types of development are taken into account day-in and day-out, the success had backed sustainability and thus the most satisfying and rewarding.

This is why the author repeats – the Social Science First!

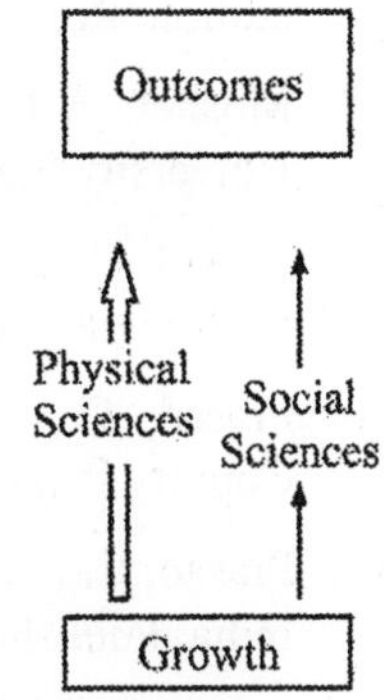

The day will come

when nations will be judged

not by their military or economic strength

nor by the splendour of their capital cities and public buildings,

BUT

by the well-being of their peoples;

by their levels of health, nutrition and education;

Growth

by their opportunities to earn a fair reward for their labours;

by their ability to participate in the decisions that affect their lives; by the respect that is shown for their civil and political liberties;

by the provision that is made for those who are vulnerable and disadvantaged; and by the protection that is afforded to the growing minds and bodies of their children.

– UNESCO, Paris

We must accept finite disappointment,

but never lose infinite hope !

– Martin Luther King Jr.

References

1. Axinn, G.H. & Thorat, S. (1972). Modernising World Agriculture: A Comparative Study of Agricultural Extension Education Systems. New York: Praeger.
2. Ban, A.W. Van Den (1979). Inleiding TOT De Voorlichtingskunde [Introduction to Extension Education]. Meppel, the Netherland: Boom.
3. Benor, D. & Baxter, M. (1984). Training and Visit Extension, Washington, D.C.: The World Bank.
4. Burton E. Swanson (Editor) (1984), Agricultural Extension – A Reference Manual, Second Edition, FAO, Rome.
5. Gowda, K. Narayan & Narayan Swamy (2010). Innovative Extension Model for Sustainable Development of Rural People, NAARM-IFPRI Workshop, NAARM, Hyderabad.
6. Leagans, J.P. (1971). Extension Education and Modernization. In J.P. Leagans and C.P. Loomis (Eds.) Behavioural change in Agriculture (pp 101-147). Ithaca, N.YU.: Cornell University Press.

7. Maunder, A.H. (Ed.) (1973). Agricultural Extension: A Reference Manual. Abridged edition. Rome: food and Agriculture Organization of the United Nations.
8. Mosher, A.T. (1978). An Introduction to Agricultural Extension. Singapore: Singapore University Press.
9. Prasad, C. & Babu, Suresh (2008). Social Science Perspectives in Agriculture – A Thrust for IntegrationNew Delhi: VARDAN, IFPRI, ISEE.
10. Prasad, C. (1978). A Study of Agricultural/Rural Extension Experiences in India, A Country Paper for FAO of the U.N.O., Presented in an International Meet, Thailand.
11. Prasad, C. (1981). Elements of the Structure and Terminology of Agricultural Education in India, Published by UNESCO, 7 Place of de Fontenoy, 75700 Paris.
12. Prasad, C. (2005). Dynamics of Research-Extension Linkages for Strong and Sustained Farm Transfer of Technology System – Evolving a Management Model, NATP, ICAR, Final Report.
13. Prasad, C. (2010). Redesigning Agricultural Extension in India – Challenges and Opportunities, NAARM-IFPRI Workshop on Redesigning Agricultural Extension, NAARM, Hyderabad.
14. Prasad, C. (2015). Basics of Extension Education for Agricultural Research and Development, Paper presented in National Symposium on Dynamics of Crop Protection – Challenges in Agri-Horticulture Ecosystems Facing Climate Change, MPUAT, Udaipur & Society of Plant Protection Society.
15. Rogers, B. (1980). The Domestication of Women: Discrimination in Developing Societies. London: Tavistock.
16. Röling, N. (1979). Basic Extension Strategies for Small Farmer Development. Approach (Netherlands) 5, 3-11.
17. Swaminathan. M.S. (2006). Serving Farmers and Saving Farming. National Commission on Farmers, Ministry of Agriculture, Government of India, New Delhi.
18. Swanson, B.E. & Rassi, J. (1981). International Directory of National Extension Systems. Urbans, III.: Bureau of Educational Research, College of Education, University of Illinois at Urbana/Champaign.
19. Vijayaraghvan, K. & Padaria, R.N., (2010). Innovative Extension Approaches of IARI: Experiences and Lesson Learnt, NAARM-IFPRI Workshop, NAARM, Hyderabad.

04 Agricultural Extension at the New Age of Change for Accelerating inclusive Growth

- *R.K. Samantha*

Prologue

Food production, hunger, poverty, economic growth, human resources development in agriculture and natural resource upgradation will continue as the major challenges throughout the world especially in the few developed and developing nations. These challenges could only be met by policy changes in agriculture, strategic reorientation of agricultural extension and ofcourse in political will of the concerned national governments (Samanta and Sontakki, 2005).

Agricultural extension, as widely recognised, plays a significant and important catalytic role in agricultural and rural development process in the nations predominantly dependent on agrarian economy like ours. Agriculture, perhaps the most oldest source of occupation continues to be the major livelihood option for more than 60 percent population in developing nation. Therefore, agriculture extension should play the most pivotal role in enabling these national to improve their agricultural production and productivity leading to country's improved economic boom encompassing inclusive growth in all sectors and sections of the society. The relevance and effectiveness of future of technology transfer systems will be determined by how quickly the extension systems adapt to the knowledge economy and associated information and communication technology (ICT) tools. It will also be determined by how quickly and effectively they provide feedback to the research systems.

As the Green Revolution has been the cornerstone of India's spectacular agricultural achievement, transforming the country from food deficiency to self sufficiency and the model of India's green revolution has well been adopted to accomplish the national agricultural objectives of many other agro-asian nations. However, the experiences and lessons learned with the green revolution is that the different elements should be well integrated is a system to irrigate dryland will not work well and multiple cropping in irrigated in rainfed land will not fetch desired results, if farmers are not taught to make optional use of the irrigation water and managing use of national resources judiciously. Reoriented green revolution strategy combining with New Age agricultural extension will increase the rate of return on investments in irrigation, as well as investments in other development policies focussing inclusive growth.

Further, Agricultural Extension must provide a well organised channel through which operational problems can be identified for research and also for the modifications of national agricultural policies to the benefit of the farming and rural communities. Moreover, the New Age Extension system must provide a framework through which farmers and rural peasants could be organised into a more functional and action oriented groups in order to gain access to essential production resources like inputs, credit, post harvest technologies, marketing and information services on government sponsored development programmes and Farmers Empowerment programmes of well organised non government organisations (NGOs).

Agriculture has Changed so has to Agricultural Extension

It is true that the green revolution has resulted in a large increase in the yields of cereals and therefore in a major decrease in the number of people suffering from hunger. This does not mean, ofcourse, that hunger has been abolished. In large part of south Asia still more than 40 per cent of pre-school children are underweight (Sanchez and Swaminathan, 2005), but this is more because there family is too poor to buy enough food for children than the non availability of food in the market.

The increase in yields has resulted in all increase in agricultural production and therefore in a decrease in prices. For an increase in agricultural productivity many people think mainly of an increase in yields, but there are more possibilities to increase labour productivity. According to estimates of the World Bank the added value per worker in agriculture is in many Asian countries less than 1 percent of this value in the countries with the highest labour productivity: France and Netherlands (World Bank, 2005).

In recent decades there has been a large increase in the international trade in agricultural products, because:

- the transport costs have decreased,
- the demand for high value products, like fruits and agriculture products, has increased,
- multinational companies and super market chains play an increasing role in this trade,
- the development of ICT makes it much easier to discover where one can buy good quality products for the lowest price, and
- trade barriers have decreased as a result of the WTO and they may decrease even more rapidly in the next decade as the political power of farm growers in rich

countires decreases with the large decrease in the number of farmers there (Van den Ban and Samanta, 2006).

A result is that Asian farmers now have to compete not only with farmers in neighbouring villages, but with farmers all over the world to be successful in this endeavour they have to be well informed about consumer demand in other countries, especially the quality they demand and they have to be served by an efficient and effective marketing system.

To ensure this to happen extension has to change and good thing is that the change has already begun at its own because of demand of farmers & consumers all over the world. Now only experts and policymakers have to give proper shape and direction so has to make and help extension to march ahead with right objectives and notions.

Major changes in the agricultural extension in the recent past, we observe particularly in Asia, partly as a result of the change in society mentioned above, are:

- From extension provided by a government extension service to a pluralistic extension system, which includes extension by NGOs, farmers' organisations, consultants and business houses/ companies providing inputs and marketing agricultural products and is financed from other sources than the tax payers money. Often agricultural development problems cannot be solved by one of these institutions, but only by solid collaboration, coordination and partnership of several of them.
- From extension which tries to increase crop yields and production per animal to extension which also helps farmers to produce products having huge market demands and inspires farmers to increase their power in the system of input supply and marketing. This implies a change from all extension system which transfers production technologies to an extension system which make farmers to decide (education and empowerment) what is in their situation and optimal farming system to realise their goals.
- From extension which transfers technologies developed at Agricultural Research Institutes (Universities/ Research Labs) to farmers to extension which stimulates and motivates farmers to experiment and to learn from their own experiences and the experiences of their fellow farmers.
- From a top down approach in extension to a participation approach by scientist researcher, extension personnel and farmers.
- Towards an extension service which tries to help to alleviate poverty and ensure livelihood security. The extensionists must remember that vast majority of the poor people in Africa, Asia and Latin America are small farmers and farm labourers.

Therefore, more innovative and out of box thinking are to be employed to change agricultural extension in the New Age of change of this 2nd decade of 21st century.

Innovative Extension strategies in the New Age of change

At the advent of technological and knowledge revolution in agriculture and its application in the broader context of WTO and IPR regime, agricultural extension should be of 360° extension in its reorientation, redesign and remodelling of its strategy to help agriculture production process enabling the farmers to make more profit from their farming operations.

The innovative extension strategies, therefore, should utilize the revolutionary changes taken place in ICT and modern and near perfected agricultural production technology sector for its overall and inclusive growth ensuring livelihood security encompassing all sectors and sections of the society. Some of the areas, issues and concerns relating to these could be:

(i) **Farmers' Field Level Issues:** Reducing the yield gap on farmers fields, increasing input use efficiencies and reducing cultivation costs (through packaging knowledge of agricultural production process along with inputs including post harvest technologies and marketing information).

(ii) **Globalisation and Rural Livelihood concerns:** Improving ensuring livelihood security by increasing income and providing employment require enlarging the research-extension-farmers interface to research-extension-farmers-agribusiness interface for marketing and value addition (through facilitating dynamic linkages and feedback between research-extension-farmers and agribusiness).

(iii) **Sustainability in Agriculture Issues:** Building farmers' knowledge base on sustainability of natural resources and ensuring adaptation of technologies like natural resource management (NRM) and Integrated Pest Management (IPM) to the minimum threshold levels of special scale and precision farming.

(iv) **Indigenous Technical Knowledge (ITK) and Intellectual Property Right (IPR) Protection Concern:** Documenting ITK or traditional knowledge and biodiversity to ensure their protection and its appropriate use in farming operations to produce more and better quality crops to help farmers to have more income.

(v) **Bio-safety concerns:** Creating awareness of bio-safety and risks of human and animal health hazards associated with adopting new transgenic crop varieties.

(vi) **Farmers and Rural Women Empowerment Issues:** Empowering farmers and rural women with socio-political and economic independence can make valuable contributions to agricultural development and to the welfare of the farm families and other rural women and ultimately help them to resist onslaught of economic power of multinational companies (Vanden Ban and Samanta, 2006).

(vii) **Mainstreaming women in Agriculture:** Gender concerns required to be mainstreamed in agricultural extension or technology dissemination process through improving access to extension and training, redesigning extension services to reach women farmers and farm women effectively and expanding the sphere of women extension workers.

(viii) **Public-Private Partnership (PPP) issues in agricultural development process:** Involving public and private sectors in agricultural research education and extension and integrating their activities with public sectors to enable extension system to be more strengthened to help farmers in their varying needs of farming operations.

(ix) **Changing strategy in Transfer of Technology (TOT):** In the context of global agricultural scenario the needs of the farmers have been changed. To reach out effectively and holistically the farmers' needs are to be ascertained and understood in the premise of present day open market economy. And hence, extension agencies and extension personnel must emphasise to prepare them to become competent decision makers in what they should produce and where they market their produce in the global competitiveness.

(x) **Innovative Human Resources Development (HRS) issues in Agricultural Extension:** There is no denying fact that extension call contribute to over overall agricultural development through both technology transfer and human resources development particularly among small and marginal farmers of developing nation. As the human resources in agricultural extension includes largely the extension personnel, input dealers and farmers the manpower management and development of it has to be planned and designed for their capacity building for skill and knowledge upgradation to help and facilitate extension and client system for inclusive growth in modern globalized agriculture.

Extension and need of its strategic reorientation and redesign to fit in to changed scenario of 21st century Agriculture

To day's farmers irrespective of their belongingness to the country or region and/or expertise in farming are well informed and know better what they want in their farming endeavours and making profit. So, only free inputs or technological information will not suffice their purpose.

Further, most of the extension scientists, consultants, practioners and agricultural development administrators have widely accepted the view that the extension must provide the support and educate the farmers to become competent decision maker in their endeavour. And therefore, there is an urgent need today to think and decide how extensions' objectives could be reoriented so that its long standing paradigm could be shifted from just providing knowledge and supplying inputs to making farmers a ideal managers of farm operations keeping in view of their resources available with them, market demand and farm operations' support systems (Samanta, 2008).

Today extension programmes must change because of production problems in irrigated areas, location specific problems in rainfed areas, opportunities possibilities and competitions in markets, extension services of Govt. various research organisation and NGOs and availability of production inputs and its supply procedures to farmers' sustainability. It is also to be kept in mind that extension programmes requires decision on goals, target groups, contents, methods, organisations, funds and time framework.

In the present dynamic time nearly everything is changing rapidly including the process of agricultural development and the role extension agencies can play to support this development. In Asia, the demand for agricultural products is rising more rapidly than in recent decades as a result of rising incomes. At the same time it is difficult to maintain the present rate of growth in agricultural production, because many of the present farming systems are not sustainable. It seems that more location specific solutions for farming problems are needed which have to be developed by farmers, extension personnel and researchers working together as a team. This will be most needed in the rainfed areas, where poverty and backwardness are common.

In this situation, it becomes less the task of the extension agencies to transfer technologies developed at the research institutes to the farmers. But more to increase their capability to participate in the development of better farming methods, to learn more from each other and to help the farmers to make a choice out of the buckets of options available to them. This bucket does not only contain more productive and at the same time sustainable technologies, but specially new farming systems which the new age markets made possible. Again, the blend of techniques and process of management will play an significant role in bringing vital change in extension methodology and strategy for transferring the technology to its clients purposefully and meaningfully.

It is desired that extension must be managed in such a manner that it will be marketable and promotion of extension programme will be easier. And perhaps clients will come forward to pay for the extension service, they receive from the extension agents and their organisations. Management of extension at the present day context therefore may be defined as the process by which people (farmers, extension personal, input dealers, retail chain managers) technology, job tasks and other resources are combined and coordinated so as to effectively achieve extension objectives for agricultural development.

Further, examination of agricultural extension in national and international contexts raises various important issues. Like professionals in agricultural development are gradually realising that modern agricultural science and technology have a certain bias which causes a different impact on development in different region and areas. Due to the growth of the population the low price for agricultural produce, structural change, nomadic and sedentary agriculture as well as apparent short term success of introduction of modern technologies no longer seem to satisfy the needs of the people.

The new age extension service, therefore, is to function as part of an interdependent technology development, transfer and utilisation system and must achieve a two way blow of information (i) information on farmers' needs, problems and priorities, (ii) information on solution to farmers' problems to enable to empower them to take correct decisions. If agricultural extension service takes care of these two, perhaps many of the limitations of extension can be tackled easily ensuring inclusive and holistic growth for livelihood security of vast majority of the country's 1.25 billion population.

Pluralisation of extension to combat larger problems of Farmers, Extension Agencies, Extension education, research, teaching and training

Extension has often been analysed and concerned as a technical issue with insufficient attention paid to how the effectiveness of extension resource is influenced by the over all policy environment. Monitoring and evaluation of extension requires reference to the broader policy context. Key aspects of the policy context upon which success relies include the existence of an enabling environment for agriculture sector development and sound agriculture policy. Extension pluralism helps in creating enabling environment in which farmers learn and exposed to various opportunities and prospects to accomplish their objective of farming operations. Hence, Government extension agencies must play significant roles in determining the extension strategy, its resources, reorientation and future direction to benefit farmers. However, it does not happen ever as expected or desired. And therefore, at the new age of change in this second decade of 21st century pluralism in extensionfinds a place in the new vista of agricultural extension services to the farmers and rural peasants in most of the developing nations.

Pluralism is an inherent feature of extension. The single problem of farmers are combated for its possible solution from different angles/view points by different sources, agencies or individuals. Extension pluralism helps in meeting the rapid changes in agriculture scenario in the context of WTO and IPR regime as it takes care of all facts of technology transfer process and also helps to meet farmers' varied needs. The concern for pluralism in extension is justified as it is being increasingly realized that in the first place extension alone cannot solve all the problems of rural and peasant population in the developing nations particularly in Asia. And the conventional state sponsored and public funded extension cannot handle it suitably either. Hence, it is very much significant and important that we recognise pluralism in extension and analyse preconditions and implications of this on the performance of extension in this new age of change.

Further, various pluralism inherent in extension make it difficult to assess the performance of extension. Being a fence sitter, extension by virtue of its role in development and also because of its innate pluralities often fail to produce tangible evidences of its efficiency and effectiveness. This is precisely why extension has always been amenable for debate and criticism. One must realize that extension alone cannot do anything as it depends for its very existence on at least two partners of the innovation system (research system and indigenous knowledge system) and the recipient system (rural and farm families). The other partners of rural and agricultural development gamut like the government market and input systems significantly influence its performance. It is a known history that extension cannot and should not work in isolation. Therefore, any judgement on its performance has to consider it as a part of the whole and naturally and holistic approach then becomes imperative in such an endeavour (Samanta, 2008).

Generally, when there is a successful attempt in improving livelihood of rural and agrarian population, the credit and praise are showered on other more tangible or easily quantifiable factors (such as better market, conductive policy environment, useful innovations emanating from research system and so on) but not on extension. For any failed attempt, on the other hand major blame is likely to the inadequacies of extension intervention or system. Not-withstanding such observations there is still hope for reviving extension in terms of its performance. The sheer fact that extension is not totally done away with albeit the shrinking funding is all ample proof that extension is a must to address the fast changing global agricultural and rural development scenario. Extension will continue to be a major policy instrument to foster agricultural and rural development in its inclusiveness, but with major reforms and amendments in its structure and functions.

Once we realize that pluralism is a way of dealing with future concerns of agricultural and rural development, then it means that there is both space and time for each player of the holistic farmers centred agricultural research – extension system (Fig. 4.1).

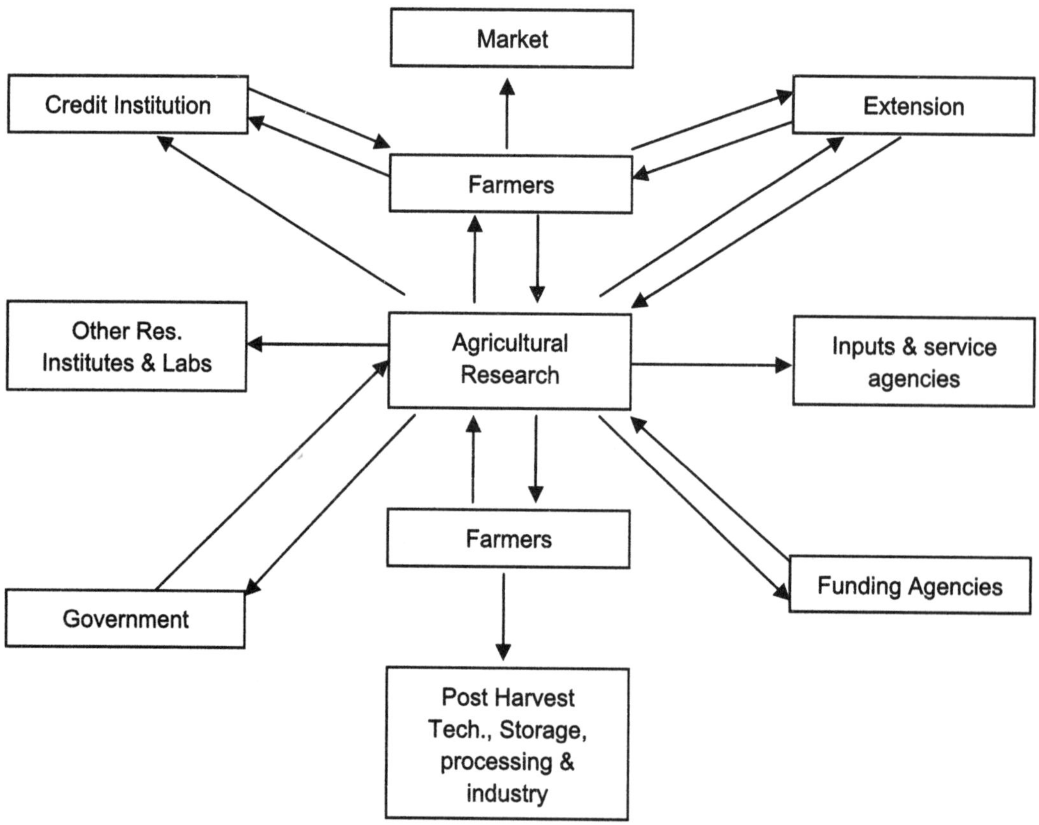

Fig. 4.1 Framework for Farmers Centred Research and Extension

It, however, has to be rationalised how these multiple service providers can serve the vast and varied clientele in agrarian and rural societies without contradicting and confronting each other. It should be clear that different service providers have divergent ideologies, interests and ways of working. While the state sponsored and public funded extension continues to be the major player, their has to be radical changes in the way this player approaches extension. If anything has to be learnt from rich experiences of 65+ years of administering agriculture and rural development in Independent India, the conventional extension structure should confine itself to the very basic objective of extension i.e. human development for empowerment through education, teaching and training process encompassing of course, modern agriculture and technological revolutions.

Epilogue

To make agricultural extension effective, meaningful and purposeful and also compatible with the changing scenario of global agriculture, strategy for agricultural extension and technology transfer system have to be reoriented and redesigned. Keeping in view the producers demand and the market opportunity of the produces, the existing situations, its implications and suggested measures as discussed above should be followed to achieve the desired goals of farmers, extension agencies and consumers alike. Further, in the context of meeting the holistic needs of increasing agricultural production on a sustainable basis, the agricultural extension has a crucial role to play to alleviate rural poverty and ensuring inclusive growth for better livelihood securing for all by engaging all the stakeholders in the process with maximum earnestness. In this regard, following issues are most significant and important which are to be addressed equally by policy planners, agricultural development administrators, scientists, researchers, NGOs, Govt. functionaries, extension personnel and farmers alike in the developed and developing nations of Asia and Asia-pacific regions. The issues are:

- Policy Reforms in Agriculture.
- Organisational restructuring for purposeful delivery mechanism.
- Reforms in Extension management systems and Processes.
- Strengthening and smoothening research, extension, and farmers linkages in TOT systems.
- Capacity building and skill upgradation of extension personnel and farmers.
- Mainstreaming and empowering rural women in agriculture.
- Empowerment of farmers, farm women and agricultural labourers.
- Use of new communication channel and innovative information technology in TOT systems.
- Market and financial sustainability for improved production and productivity.

- Instilling pre-acting role of Govt. and Non-Govt. organisations to facilitate agricultural production process.

Lastly, it is suggested that the systems which are demand driven through active client participation in programme formulation and evaluation are more effective than top down or supply driven systems. Programme priorities should be based on both assessment of client needs and client participation in programme development. Moreover, farmers' group, farmers' associations, commodity groups, farmers' cooperatives, credit societies, women farmers' groups (like SHGs), young farmers clubs so far are effective because of newly devised creating of a demand driven agricultural research – extension system. And that should be encouraged further to scaleup. Hence, extension can play a vital role in organising this pressure and support groups to provide appropriate technology packages and input supplies together with credit etc. to the farmers to achieve their objectives more meaningfully and transforming the TOT systems into a formidable one.

References

1. Samanta, RK and Sontakki, B.S. (2005). Emerging issues in management of agricultural research in the changing global perspective in 'Perspectives of Agricultural Research and Development' (By C. Ramasamy, S. Ramanathan and M. Dakshinamurthy (Ed.), Coimbatore, TNAU.
2. Samanta, RK (2008). Management of agricultural extension in the 21st century in proceedings of "APEC workshop on Innovative Agricultural Technology Transfer and Extension Systems for Enhancing Productivity and Competitiveness in APEC Member Economics" held in Seoul, Republic of Korea on July 1-3, 2008.
3. Sanehez, P., M.S. Swaminathan *et al.* (2005). Halving Hunger: it can be done. UN Millennium Report Task Force on Hunger, London, Earthscan.
4. Van Den Ban, A.W. and Samanta, RK (2006). Changing roles of Agricultural Extension in Asian Nations (Ed.) New Delhi, BR Pub. Corporation.
5. World Bank (2005). Agricultural Growth for the poor: An agenda for development, Washngton DC, World Bank.

05 Extension Reforms in India

Dr. M.N. Reddy

The ATMA model was conceived and pilot tested to promote decentralized decision-making, using an integrated developmental approach based on a participatory planning process, with farmers being central to setting the agenda for research and extension, and making the public extension systems both farmer-driven and farmer-accountable under Extension Reforms. In addition, the ATMA model was created to establish an extension system that was both decentralized and market-driven to increase farm income and rural employment. The Strategic Research and Extension Plan (SREP) at the district level and the State Extension Work Plan (SEWP) at the state level are the master documents that provide the broad guidelines for those responsible for implementing the plans. The plans reflect the needs of the farmers within the available resources. The major thrusts in the ATMA model are 1. to promote the public-private partnership, 2. to ensure the mainstreaming of women and disadvantaged groups, 3. to provide a single window to deliver services, 4. to concentrate all developmental efforts at the district and state levels, and 5. to maintain flexibility in operational procedures. The strategy to realize the intent of ATMA model is further tested and refined which has become the current approach.

India's Agriculture Sector has an impressive long term record of taking the country out of serious food shortages despite rapid population increase. This was achieved through a favourable interplay of infrastructure, technology, extension, and policy support backed by strong political will. The main source of long run growth was technological augmentation of yields per unit of cropped area. This resulted in tripling of food grain yields, and food grain production increased from 51 million tonnes in 1950-51 to 260 million tonnes in 2014-15.With availability of land and water fixed, growth in Agriculture can be achieved only by increasing productivity per unit of these scarce natural resources through effective use of improved technologies. The research system has so far focused mainly on breeding varieties that increase the yield potential of individual crops by enabling more intensive use of inputs.

Although such research did increase potential yields substantially in the past, it put less emphasis on efficient and sustainable use of soil nutrients and water,or on identifying location specific farming systems with proper mix of crops and livestock, especially for rain-fed areas. Besides, the potential yields of new varieties being released seem to have plateaued suggesting that the current system is no longer leading to adequate outcomes. This technology fatigue has to be countered by changing research priorities and extension approaches suitably. At the same time, frontline trials of various research departments provide clear evidence of large gaps between what can be attained at the farmer's field with adoption of available technology as compared to what is obtained with existing practices. Moreover, since yield gaps vary considerably from crop-crop and from region-region, the strategy must enable specific plans for each agro-climatic region. This will also require much stronger links between research, extension, farmers, and markets.

For growth to be at all inclusive, the agricultural strategy must focus on the 85% of the farmers who are small and marginal, increasingly female, and who find it difficult to access inputs, credit, and extension or to market their output. The challenge before extension agencies is how to deliver knowledge to large number of farmers,and especially how to involve and motivate the resource poor farmers with a holding size below 1 hectare to take command of their situation and reduce the innovation adoption period.

Public sector extension is a state level responsibility that has undergone several transformations since independence. By the early 1990's and the completion of the third National Agricultural Extension Project (NAEP-III), there was growing recognition that the T&V extension approach has made an important contribution to agricultural development, but that it needed to be overhauled in meeting the technology needs of farmers during the 21st century. First, it was recognised that extension should begin to broad-based its program, by utilising a farming systems approach. For example, attention should be given to the needs of farmers in rain fed areas, and to diversifying extension programs into livestock, horticulture, and other high value commodities that would increase farm income. Also, issues of financial sustainability, lack of farmer participation in program planning, and the weak links with research have all been mentioned as serious constraints facing the current extension system.

During this period, the agricultural extension system, under the institutional leadership of the Indian Council of Agricultural Research (ICAR), has been strengthened through two parallel National Agricultural Research Projects. Therefore, a National Agricultural Technology project (NATP) was implemented with a purpose to consolidate earlier investments and address specific system constraints, weaknesses and gaps that were not addressed in previous research and extension projects. It was therefore, expected to test Innovations in Technology Dissemination (ITD) that would begin to delineate the future direction of the extension system and , at the same time and bridge serious research-extension – farmer (R-E-F) linkage problems that currently constrain the flow of appropriate technology to farmers.

Based on background studies, carried out by Dayal (1995) in the preparation of this project, key system constraints have been identified that need to be addressed by ITD component of National Agricultural technology Project. The major constraints are listed below:

1. Multiplicity of Technology Transfer Systems.
2. Narrow Focus of the Agricultural Extension System.
3. Lack of Farmer Focus and Feedback.
4. Inadequate Technical Capacity within the Extension system.
5. Lack of Local Capacity to validate and refine technologies.
6. Need for more farmer training.
7. Weak Research-p Extension Linkages.
8. Poor Communications Capacity.
9. Inadequate Operating Resources and Financial Sustainability.
10. Institutional and Operational Reform- A high priority.

Other problems plagued the system. The DOA experienced financial difficulties because of the large increase in the number of new extension workers that were added under the T&V extension system. In addition, because the central government supplied most of the program funds, extension priorities were planned from the top-down, excluding farmer input from the planning process. The dominant focus on food production meant that extension focused on the major cereal crops and was supply-driven. Hallmarks of the "market-driven" system, i.e., increasing farm income and promoting crop diversification, were not a priority. The T&V Extension management system was successful and crop yields increased; but, as a result, crop prices fell and farm incomes fell with them. Other problems were present including 1) the fact that the line department staff became accountable to the central government, not the farmers they were meant to serve; 2) the central government, being involved in supplying inputs to farmers, were prone to view private sector input dealers as part of the problem, not part of the solution; 3) there was a deterioration of the research-extension linkage because extension programs were centrally planned; and 4) on the whole, extension was not mandated to empower farmers by forming farmers groups.

To address the situation, in the late 1990s, the Government of India (GOI) and the World Bank pilot-tested a new, decentralized, market-driven extension model under the National Agricultural Technology Project (NATP). This new approach was designed to help farmers diversify into high-value crops and livestock enterprises, the products of which they could then sell, as a means of increasing farm income, rural employment and poverty alleviation. The Innovations in Technology Dissemination (ITD) component of NATP was implemented in 28 districts in the country. It is currently operational in 652 districts in 29 States & 3UTs.

The key institution in implementing this new approach was the Agricultural Technology Management Agency (ATMA) which was responsible for facilitating and coordinating "farmer-led" extension activities within each district. The key elements of the ATMA model include: 1) taking "farming-systems" approach which required the integration of extension activities across the different line departments; 2) organizing small-scale farmers, including women, into Farmer Interest Groups (FIGs); 3) decentralizing extension decision-making down to the district and block levels including farmer input; and. 4) linking the farmer-interest groups to markets for the purpose of increasing farmer income and creating rural employment.

ATMA focused on a bottom-up planning process in order to make the entire extension system farmer-driven and farmer accountable. This has helped to strengthen research and extension capabilities, restructure public extension services and test new institutional arrangements for technology transfer with the involvement of all the stakeholders of Government and Non-Governmental agencies at the district level. New institutional arrangements were created at different levels to put the project into operation.

The Key Elements of the ATMA Model

Decentralizing Extension

Within the ATMA structure there are two sets of mechanisms that integrate extension activities at the district and block levels. The ATMA Management Committee (AMC) decentralizes and integrates decision-making at the district level while the Block Technology Team (BTT) organizes and integrates extension activities across each block. The key mechanisms for "bottom-up" planning and for stakeholder participation in decision-making are the ATMA Governing Board at the district level and the Farmer Advisory Committees (FACs) at the block level.

As a registered society, ATMA has more flexibility than government line departments. They can receive funds from both government and non-government sources, enter into contracts; maintain revolving accounts, charge for services and recover costs from farmers or other service recipients. In terms of institutional ranking within a district, the ATMA ranked above the line departments; therefore the Project Director (PD) was able to mobilize extension resources across all of the line departments especially extension staff at district level through BTT convener at the block level. In addition, each PD had access to project funds that could be used for a variety of different extension activities as approved by the ATMA Governing Board (AGB), headed by the District collector.

The key to successful project implementation began with project leaders who fully understood and were committed to implementing the ATMA concept and procedures. In addition, there had to be effective leaders and managers who could transform these concepts into useful programs, especially in motivating the BTTs and in activating the farming community.

Strategic Planning

Departing from the traditional top-down practice, the planning process began with the development of a Strategic Research and Extension Plan (SREP) for each pilot district, which was prepared at district level after the systematic assessment of technological gaps, issues, success stories and problems pertaining to various farming systems prevailing in the district.

The district core team in consultation with the district department heads, along with scientists of Zonal Research Stations (ZRS), identified the major agro-ecological systems (AESs) in each district. In addition, the major farming systems under each AESs were identified. Then, representative villages for each of these major farming systems were selected by visiting the villages under each AES. The core team was further divided into interdisciplinary sub teams depending upon the sectoral requirement of the Agro-Eco Situation (AES) within the district. The data was collected on adoption gaps, technological gaps, and institutional gaps by involving different categories of farmers including resource poor and women.

Market-Driven Extension

Strategic planning was the first step towards transformation of 'target-driven extension system' into 'demand-driven extension system'. Participation of farmers on the GB, the AMC and the FAC gave them an opportunity to identify various problems facing the farming community. In addition to giving feedback on action plans prepared by extension official, farmers also brought up different issues of wider relevance. In these ways, farmers played an important role in setting extension priorities within the district. With accountability to solve farmers' problems and built-in operational flexibility, ATMA made suitable interventions.

Farming Systems Approach

Where possible, farmers take up more than one enterprise, based on their resource base, to make their farming economically viable. To operate more profitably, farmers need to integrate multiple enterprises, based on their resources, by diversification and intensification. To make this possible, the strategic planning process promoted in this project focused on identification of popular farming systems followed in the various agro-eco situations. This became the basis for the analysis of gaps in technology adoption, managerial aspects and institutional support systems.

Broad-Based Extension and Integrated Delivery of Services

The integrated delivery of services was a direct result of the integrated/holistic planning process, focused on the existing farming system in an area. Once the plan was in place, individual line departments took up their portion of its implementation individually under the coordination of the AMC and the BTT. Thus, line departments maintained their individual identities but joined together to implement various extension programs that were identified through the bottom-up planning process.

Research-Extension-Farmer-Market Linkages

ATMA has provided a useful administrative framework to effectively integrate research and extension activities at the district level. The project interventions have improved the R-E-F-M linkage and feedback process which began when they cooperated in carrying out the field assessment for and preparing the SREP.

The ATMA Governing Board and Management Committee have provided common platforms for regular and personal interaction among scientists, extension administrators and farmers. On one hand, it has improved the awareness level of farmers, while on the other it has enabled the scientists and extension administrators to more clearly understand the farmers' needs and problems.

Mobilization of Communities

Commodity-oriented Farmer Interest Groups (FIGs) are promoted at the block/village level to make the technology generation/dissemination both farmer-driven and farmer accountable. These Village-level FIGs are organized at the block/district level and represented in the FACs and on the GB. To address the extension needs of these groups, ATMA has reached out to establish close linkages with various players operating at the cutting edge level (viz., public, private, NGOs, Para extension workers, input dealers, etc.).

Team Work

An effort was made throughout the pilot project to use a teamwork approach at all levels to bring together resources and to address farmers' problems in an integrated manner.

Public Private Partnership

As the agricultural private sector became increasingly involved in meeting the many demands of the farming community, a Public-Private Partnership between the private sector and extension provided an opportunity to work together to promote extension efforts. This partnership has emerged as one of the crucial areas in agricultural extension. For example, a large number of ATMAs have taken initiatives to develop partnerships with the private sector in the processing industry, farmer's organizations, cooperatives, corporate bodies etc. in different areas. This partnership facilitated the dissemination of technologies, the supply of quality inputs (seed, fertilizers, micro-nutrients, bio-fertilizers, pesticides and bio-pesticides and other technological tools) and marketing of farmers' produce.

Impact of ICT Interventions in ITD

Agricultural information kiosks were established by different ATMAs in selected pilot districts. Efforts were also made to digitize appropriate content and provide farmers with central information through these kiosks on an on-line basis.

Gender Sensitization

Women's participation in agriculture has been widely recognized by all the development agencies, and women farmers were included at every level of ATMA participation. Women were involved in the decision-making system from the Central level down to the FAC. Two, non-official members representing the interest of women farmers and a NGO were represented at the central level. ATMA also was careful to honour the provision of nominating 30 per cent, non-official, women representatives on the GB.

Up-Scaling the ATMA Model–Current Approach

After NATP's implementation period (1998–2005), when the ATMA model was pilot tested, it was decided to expand the system to cover a wider area, based on the fact that the strategy used resulted in an increase in farmers' income. It is essential to identify the important lessons that were learned from the pilot study in order to refine the concept and the ATMA model before it is applied to all the districts covering in the country. The lessons learned, and which were put in place in the present ATMA model, were both of a practical and qualitative nature. A description of these changes is given as below.

Structural Changes

The Governing Board of ATMA has been strengthened by including public representatives as ex-officio members. As a result, ATMA activities are more widely accepted, indicating that the needs of the farmers are understood and consequently greater accountability of all the stakeholders is given. In addition, the Governing Board has gained greater flexibility as various schemes being run by the State Development Machinery have concentrated at the district level and funds given to implement developmental activities are channelled to a single source.

The service providers for irrigation, the power supply, credit and marketing have been made important stakeholders in the decision making process, with significant results. Inclusion as stakeholders has facilitated the delivery of services to the farmers, for example, when to release water, at what frequency, devising cropping pattern for the given area, gathering market intelligence favouring farmers and streamlining the credit flow. These services have improved both in terms of quality and speed, giving greater leverage to the farmers and enabling them to use their resources at the optimum and to sell their product for the highest price, thus increasing income.

Functional Changes

In the initial phase, the group approach (Commodity Groups, Women Groups, Self Help Groups, and Farmers Interest Groups) developed slowly. After 4–5 years, however, the groups have grown and have now emerged stronger in terms of their negotiating skills. They have developed strong linkages with research institutions, input agencies, credit institutions, insurance groups, private sector agribusiness companies, cooperatives and marketing agencies. These groups, which were initially found only at the village level, have now organized at the block, district and state levels. These organizations have emerged as strong farmers' bodies capable of putting pressure on the system with the ability to articulate their needs and demands for various products and services.

The developmental efforts which were initially made by the individual departments are now being integrated, and all of the departments have been brought together at the block, district and state level. As a result, duplication of effort is avoided and the State Extension Work Plan (SEWP) is being used. The SEWP documents are available and recognized as the blue print for the agricultural development of the State.

The Strategic Research and Extension Plan (SREP) prepared for a district has been expanded to include information on: marketing opportunities, processing facilities, involvement of agribusiness companies, storage facilities and information on how to create other linkages to add value along the entire supply chain.

Policy Changes

No budgetary allocations were made for either the public-private partnership or the streamlining of women into agriculture in the first pilot testing of the ATMA model. Based on the pilot study experience, it was recognized that without the full participation of women, complete agricultural development cannot take place as women make up a substantive portion of agricultural work force. Accordingly, 30 per cent of the budgetary allocation was made in the Extension Reforms Agenda for the inclusion of women. The introduction of such provisions in the agricultural development programs would certainly boost the participation of women in agriculture, ultimately becoming part of the SREPs and SWEPs.

In the pilot-testing phase, the public-private partnership, though recognized as important, did not become a reality because of various legal, financial and procedural hindrances. During the last year, however, the stakeholders have realized that agriculture development cannot proceed properly unless this partnership is in place. The reasons a public-private partnership must be in place are easy to understand. All needs, i.e., the knowledge base, skill base, human resources and financial assistance, cannot be provided by the public extension system alone. These needs should also be met by the Private Sector, NGOs, Cooperatives, Agri-prenuers, Farmers' Organizations and Federations. Toward this end, a 10 per cent budgetary allocation was made for public-private partnership in the Extension Reforms Agenda. While preparing the SREPs, the private sector is encouraged to contribute and partner in the entire value chain to enhance the income of the farmers. There is a built-in mechanism created in the SREPs for providing market intelligence, setting up of market network and physical infrastructure, setting up of kiosks, networking with process industries, storage facilities, development of communication networks, linkages with credit institutions, risk coverage etc. for farmers.

Human Resource Interventions

If the ATMA model is to be up-scaled to all the development districts in the country, it is imperative to create a large pool of trained manpower in the country. To realize this objective, the concept of National Facilitators was created by MANAGE. These national facilitators are identified from the State Agricultural Universities, Extension Education Institutes, State Agricultural and Developmental Departments, NGOs, Krishi Vigyan Kendras, ICAR Institutions and other Centers of Excellence engaged in agricultural development. The group of facilitators is made up of specialists in different technical subjects who also have adequate training and exposure in Management and HR issues. Also, they occupy senior positions in their respective departments which help them to have an adequate influence on decision making machinery at district and state level. Once selected, the national-level facilitators were trained in Extension Reforms Agenda, preparation of SREP and SEWP, Institutional Building and Network, the Participatory Planning Process, Project Management Techniques and such other issuesin the context of the State necessary for implementation of ATMA model.

The Scheme 'Support to State Extension Programmes for Extension Reforms' aims at making extension system farmer driven and farmer accountable by disseminating technology to farmers through new institutional arrangements viz. Agricultural Technology Management Agency (ATMA) at district level to operationalize the extension reforms on a participatory mode. This Scheme shall focus on the following key extension reforms:

1. Encouraging multi-agency extension strategies involving Public/ Private Extension Service Providers.
2. Ensuring an integrated, broad-based extension delivery mechanism consistent with farming system approach with a focus on bottom up planning process.
3. Adopting group approach to extension in line with the identified needs and requirements of the farmers in the form of CIGs & FIGs and consolidate them as Farmers Producer Organisations.
4. Facilitating convergence of farmer centric programmes in planning, execution and implementation.
5. Addressing gender concerns by mobilizing farm women into groups and providing training to them.

The 12th Plan Approach Paper identifies several challenges faced by the agricultural extension and also gives suggestions to deal with the same. Some of these include integrating KrishiVignan Kendra's (KVKs) problem solving skills and the feed-back they provide to State Agriculture Universities (SAUs) and National Agriculture Research System (NARS) with ATMA and strengthen district level planning; using technology to reach out to the farmers, raising capability of rural poor to conserve and manage their livestock and fisheries resources and derive sustainable incomes; link small farmers to markets; promote decentralized participatory research as well as knowledge intensive alternatives in rain-fed regions.

The above key reforms shall be met through strengthened institutional arrangements, dedicated manpower, Innovative Technology Dissemination component and revamped strategy. ATMA will be implemented through the institutional mechanism as detailed below:

State Level

(i) The State Level Sanctioning Committee (SLSC) set up under Rashtriya Krishi Vikas Yojana (RKVY) is the apex body to approve State Extension Work Plan (SEWP) which will form a part of the State Agriculture Plan (SAP). In those cases where SLSC meeting cannot be held in time or there is any other administrative reason, SEWP can be approved by the Executive Committee of SMAE the recommendation of Inter Departmental Working Group (IDWG).

(ii) The SLSC will be supported by the Inter Departmental Working Group (IDWG) which is responsible for day-to-day coordination and management of the Scheme activities within the State.

(iii) The State Nodal Cell (SNC), consisting of State Nodal officer, State Coordinator, State Gender Coordinator and supporting staff will ensure timely receipt of District Agriculture Action Plans (DAAPs), formulation of State Extension Work Plan (SEWP) duly incorporating farmers' feedback obtained through State Farmer Advisory Committee (SFAC) and its approval by the SLSC. The SNC will then convey the approval and monitor implementation of these work plans by State Agricultural Management & Extension Training Institutes (SAMETIs) and ATMAs. The SAMETIs will draw-up and execute an Annual Training Calendar for capacity building of the Extension Functionaries in the State. While doing so, the SAMETI will check duplication and overlapping of training content, training schedule, and trainees themselves.

District Level

ATMA is an autonomous institution set up at district level to ensure delivery of extension services to farmers. ATMA Governing Board is the apex body of ATMA which provides overall policy direction. ATMA Management Committee is the executive body looking after implementation of the scheme. District Farmers Advisory Committee is a body to provide farmers' feedback for district level planning and implementation. With dedicated staff provided for the ATMA, it will continue to be the district level nodal agency responsible for overall management of agriculture extension system within the district, including preparation of Strategic Research and Extension Plan (SREP). The process flow for formulating Action Plans has been described below:

The following Committees shall be set up at various levels:

State Level

Inter- Departmental Working Group (IDWG), (ii) SAMETI Executive Committee (iii) State Farmers Advisory Committee (SFAC).

District Level

(i) ATMA Governing Board (ii) ATMA Management Committee (iii) District Farmers Advisory Committee (DFAC).

Block Level

(i) Block Technology Team (BTT) (ii) Block Farmers Advisory Committee (BFAC).

The organizational structure of ATMA at various levels diagram:

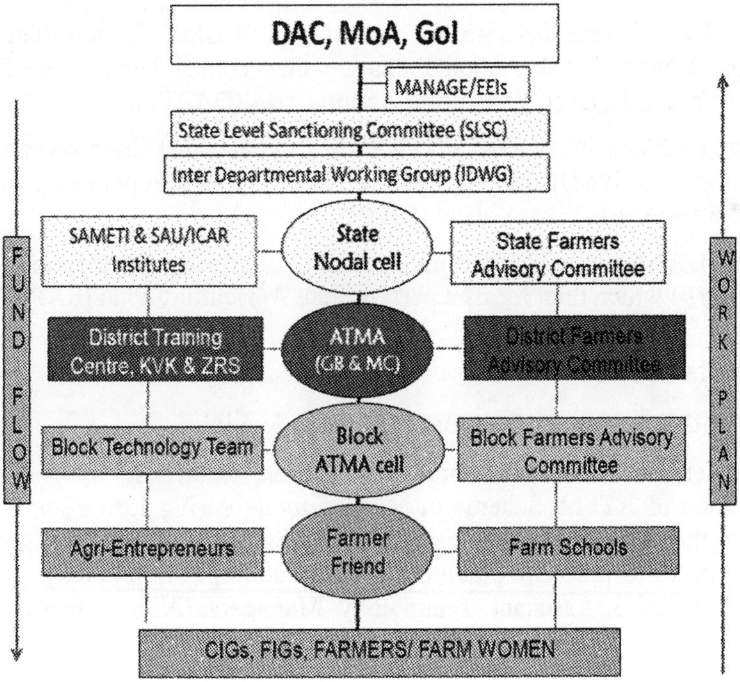

Process Flow for Action Plans

SREP is a comprehensive document identifying research/ extension priorities for district, keeping in mind agro-ecological conditions and existing gaps in technology generation & dissemination in all agriculture and allied sector areas/ activities including Horticulture, Animal Husbandry, Dairying & Fisheries. All extension related activities in the state and district will form part of SREP. SREPs will be prepared in coordination with the line departments, NGOs, KrishiVigyanKendras (KVKs), Panchayati Raj Institutions (PRIs), Private Sector, farmers and other stake-holders at the district level.

Training & field extension related components in other programmes of Department of Agriculture & Cooperation (DAC) and State Governments will also be implemented through ATMA.

Funds earmarked for activities under different Sub-Missions of National Mission on Agricultural Extension and Technology(NMAET), Missions & Schemes / Programmes will be utilized through ATMA. Such convergence arrived through SREP / SEWP will avoid duplication and ensure wider coverage in terms of outreach to farmers and gamut of activities.

Various Action Plans shall emanate from SREP as follows:

(i) These SREPs are Five year vision documents which shall be revisited to accommodate newly identified gaps and emerging areas of importance.

(ii) SREPs will form the basis for formulation of Block Action Plans (BAPs) on an annual basis. Block Action Plans are then consolidated at the District level to prepare the District Agriculture Action Plans (DAAPs).

(iii) District Plans are worked out in such a manner that these serve as subset of the Comprehensive District Agriculture Plans (CDAP) prepared for the District under RashtriyaKrishiVikasYojana (RKVY).

(iv) The DAAPs will be consolidated in the form of State Extension Work Plan (SEWP) which then forms a part of State Agriculture Plan (SAP).

Components of the Scheme

Specialist and Functionary Support at various levels

The component of manpower support has been further strengthened to accelerate implementation of ATMA Scheme in Mission mode during 12th Plan. It is proposed to re-designate the Subject Matter Specialists (SMSs) as Assistant Technology Managers (ATMs). In order to rationalize number of cultivators per functionary at block level, an average Figure of 3 Assistant Technology Managers (ATMs) per block have been envisaged.

Cafeteria of Activities

Optional items form the ATMA Cafeteria from where the States can choose activities as per its priorities. The Cafeteria also contains mandatory components which include manpower, training of extension personnel, organization of Farmers' Advisory Committees, Farmer Friends, joint visits with scientists, low cost publications etc.

Innovative Technology Dissemination Activities

Use of interactive & innovative methods of information dissemination like Pico Projectors, low cost films, hand held devices, mobile based services etc. and other innovative extension approaches (e.g. Kala Jatha) are included as mandatory activities.

Other Operational Matters

Common Issues related to ATMA Cafeteria 2014

(i) Minimum 30% of resources meant for programmes and activities are required to be allocated to women farmers and women extension functionaries. Specific documentation of expenditure and performance for women may be kept.

(ii) No expenditure shall be incurred from extension work plan allocation on in-eligible items. In the event of any such expenditure, the in-eligible expenses shall be deducted from the State's allocation, next year.

(iii) Operational Expenses under State Level, District & block level Activities and Innovative Activities may also include library, internet, telephone and other contingencies.

(iv) Any sub-detailing not specified under ATMA Cafeteria such as for Demonstrations, Trainings and Exposure Visits, would be as approved under an appropriate scheme of the Central/ State Govt. Otherwise, prior approval would have to be obtained from IDWG.

Unless otherwise specified under some compelling circumstances, or in case of States which are not eligible for RKVY funding, the SLSC set up in the State for considering proposals for RKVY funding will also consider and approve SEWP and carry out periodic monitoring. In exceptional cases, the existing procedure for approval of SEWP through IDWG at the State level and Executive Committee of SAME will continue to be followed. The IDWG under the chairmanship of APC/PS (Agriculture) may continue to perform monitoring of the Scheme to ensure that the Extension Reforms are executed as per these Guidelines. The IDWG shall meet at least once in every quarter. Programmatic funds should be spent on rain-fed areas at least in proportion to the extent of rain-fed areas in the district.

Hierarchical structure of officials under SMAE (including ATMA and NeGP-A) is as follows:

State Level

(i) State Nodal Cell: the State Nodal Cell (SNC) will consist of State Nodal Officer, State Coordinator, Gender Coordinator and supporting staff. In order to carry out State level activities, as specified in ATMA Cafeteria, and to ensure convergence with various departments at State level and to assist the State Nodal Officer (i.e. Director / Commissioner of Agriculture) in overall management of agricultural extension system within the State, one State Coordinator has been approved for each State/ Union Territory.

The State Nodal Officer (SNO) shall be designated by the State Government, which will also provide requisite ministerial support. The State Coordinator is to be engaged on contract basis under this Scheme. The functions of State Coordinator and Gender Coordinator are given in Annexures. The State Coordinator and Gender Coordinators will function under the overall supervision of State Nodal Officer. The Gender Coordinator will nevertheless keep on sharing all gender related findings and strategies with the State Coordinator as well.

SAMETI

In order to ensure regular training and skill upgradation of State and District / Block level extension functionaries and for reaching out to the grass root level extension functionaries and farmers through field visits, the following manpower is provided for SAMETI in each State. Director, SAMETI shall work under the overall guidance of the State Nodal Officer identified under ATMA scheme. However, in cases where the State Nodal Officer is not an officer of equivalent or higher rank than Director, SAMETI, may work under the overall guidance of the officer under whom State Nodal Officer is placed. The Faculty Members (Deputy Directors) of SAMETI shall report to Director SAMETI. The duties of Director SAMETI/ Faculty are given in Annexures.

District Level

Each ATMA Unit consisting of the core staff of five persons, under the overall supervision of PD, ATMA, will be responsible for management of agricultural extension services within the District including holding of regular meetings of ATMA Management Committee (MC) and ATMA Governing Board (GB)i)Project Director-1; (ii) Deputy Project Directors – 2; (iii) Accountantcum-Clerk – 1; and (iv) Computer Programmer/ Operator–1 The Project Director ATMA shall report to the Chairman, ATMA GB and also function as Chairman of ATMA Management Committee. The two Deputy PDs would work under the administrative control of PD, ATMA. The duties & responsibilities of PD & Deputy PD are given at Annexures.

Block Level

(i) One Block Technology Manager (BTM) is provided in each Block to co-ordinate the ATMA related activities of the BTT and BFAC. BTM will work under the overall supervision of the BTT Convener for all ATMA related activities.

(ii) On an average three Assistant Technology Managers (ATMs) are to be placed in each Block (2 to 4 depending on size of the Block) exclusively for delivery of extension services in agriculture and allied sectors as per priority areas for various Blocks. The areas of expertise (i.e. Agriculture, Horticulture, Animal Husbandry, Dairying, Fisheries etc.) of these ATMs will be decided based on priorities for various Blocks. These ATMs shall be provided requisite connectivity and mobility to perform the requisite functions. The duties of BTMs and ATMs are given in Annexures.

(iii) Block level and District level manpower available under various schemes will be pooled for regular coverage of groups of Villages/Panchayats for extension related expertise. However, domain expertise of various extension personnel in a Block / District will be suitably used to ensure all-round outreach to farmers.

(iv) Panchayat-wise roster giving the name and mobile number of the extension worker's availability on the pre-decided days (giving day of a week/fortnight etc.) will be displayed on a Board at vantage points in various villages. This Display Board will also contain contact details of other agricultural functionaries of the area.

Process Flow for Action Plans

SREP is a comprehensive document identifying research/ extension priorities for district, keeping in mind agro-ecological conditions and existing gaps in technology generation & dissemination in all agriculture and allied sector areas/ activities including Horticulture, Animal Husbandry, Dairying & Fisheries. All extension related activities under other Sub-Missions will naturally form part of SREP. SREPs will be prepared in coordination with the line departments, NGOs, KrishiVigyanKendras (KVKs), Panchayati Raj Institutions (PRIs), Private Sector, farmers and other stake-holders at the district level.

Illustrative list of farmer centric trainings & field extension under other sub-missions to be implemented through ATMA, Seed Village Programme, Farm Schools, Demo Plots, Trainings, Exposure Visits.Similar training & field extension related components in other programmes of DAC and State Governments will also be implemented through ATMA.

Various Action Plans shall emanate from SREP as follows

These SREPs are Five year vision documents which shall be revisited to accommodate newly identified gaps and emerging areas of importance.

(i) SREPs will form the basis for formulation of Block Action Plans (BAPs) on an annual basis. Block Action Plans are then consolidated at the District level to prepare the District Agriculture Action Plans (DAAPs).

(ii) District Plans are worked out in such a manner that these serve as subset of the Comprehensive District Agriculture Plans (CDAP) prepared for the District under RashtriyaKrishiVikasYojana (RKVY).

(iii) The DAAPs will be consolidated in the form of State Extension Work Plan (SEWP) which then forms a part of State Agriculture Plan (SAP).

Support for Innovative Activity at Village Level (Farmer Friend)

(i) It is necessary to identify and groom progressive farmers (including women farmers) with requisite qualification (Senior Secondary/ High School) & experience as Farmer Friend (FF) ,one per two census villages. In case of non-availability of progressive farmers with requisite qualifications, at some places, a progressive farmer, with required oral and written communication skills, may be considered. Only experienced and achiever farmers (say, with an age of more than 40 years or so, as the State Government may deem appropriate) are to be designated as Farmer Friends. Farmer Friend will lead by example and is expected to have up-graded skills and would be available in the village to advice on agriculture and allied activities. The Farmer Friend will be identified by BTM on a resolution of Gram Panchayat (GP), which will, in turn, consult CIGs /FIGs working in the Panchayat area.

(ii) Farmer Friends will be provided with special opportunity for upgradation of skills through trainings, study tours and visits to SAUs / other institutes, by utilizing support available under ATMA. `6,000 per Farmer Friend per year will be shared equally by the Centre and the State to meet contingency expenditure which FF incurs towards discharge of his duties, including incidental expenses incurred on acquisition of knowledge. The States may decide on a higher fixed amount of more than `6,000/- per annum for Farmer Friend. However, contribution of the Government of India shall not exceed `3,000/- per Farmer Friend per year.

(iii) Since Farmer Friends provide a vital link between public extension system and farmers at village level, it is essential to select experienced, progressive and practicing farmers who are not looking forward to this task as an alternative means of livelihood. It has been noticed that in some States, unemployed youth farmers with little expertise have been designated as Farmer Friends. Consequently, they start looking at it as an employment opportunity in the State Government in the long run. Such employment was never envisaged under the scheme. In fact, the small sum of ` 6,000/- per annum has been provided to the Farmer Friends to meet contingent expenditure for assisting fellow farmers. It should not be perceived as remuneration.

Cafeteria Guidelines

Earmarking of Funds

The mandatory activities indicated in the Cafeteria should invariably form a part of the State Extension Work Plan. Administrative Expenditure including TA/DA, hiring of vehicles and POL and operational expenses at District / Block level shall not exceed the amount provided for in the cafeteria. Financial allocation has been made specifically for the conduct of BFAC, DFAC, and joint visits of Scientists & Extension Functionaries under the cafeteria. The States should compute their budgetary requirements for mandatory activities, cafeteria activities as per State's priorities, manpower support and emoluments as per approved norms. Any savings in administrative expenses can be diverted to other categories but not vice–versa.

Unit Cost Ceilings

The norms as laid down in these Guidelines have to be adhered to but in exceptional cases and for reasons to be recorded in writing, ATMA Governing Board (GB) may go beyond these ceilings by up to 10% without exceeding overall allocation. Similarly, IDWG can authorise relaxation of up to 15%. All such cases shall have to be reported in next year's Work Plan so that appropriate advisories can be issued. Any deviation of more than 15 % from the prescribed norms/ceilings or any activity not specified in the guidelines can be taken up by States only with the prior approval of DAC.

Support for ICT, Connectivity & Mobility

The modified ATMA Cafeteria has enhanced focus on use of Information Communication Technology (ICT). Experts of SAUs/ KVKs, BTMs and ATMs at Block will be available on mobile phone, to provide information of immediate importance to Farmer Friend, FIGs and farmers. SMS (Short Messaging Services) alerts on weather, incidence of pest and diseases and other crop related important matters are already being sent through the SMS Portal for Farmers. Basic IT infrastructure has been provided to SAMETI and ATMA under the Scheme. It has been and can be further supplemented under AGRISNET, NeGP-A, RKVY and other schemes. These equipment shall be fully utilized for extension related activities. In new SAMETIs/ ATMAs, requisite IT and other related equipment can be procured under AGRISNET / NeGP-A / RKVY schemes. The faculty members of SAMETI, officers of ATMA, BTMs and ATMs should be accessible on their cellular phones. A specific provision to ensure mobility and connectivity of the BTMs and ATMs has been kept in their emoluments.

Main Activities Included in the Cafeteria

Farm Schools

Farm Schools provide the vital link between the progressive / achiever farmers and others in a village. Such farmers should be selected broadly adhering to transparent methodology of selection. These farmers would normally be the ones who have been accepted by other farmers as achiever farmers for their success in adoption of technologies, yield difference, and income raised in agriculture and other allied sectors. Some important points related to Farm Schools have been listed in ATMA Cafeteria. Ceilings fixed on individual items shall be adhered to. In order to have a visible impact and to ensure proper monitoring, cluster approach needs to be adopted by covering various Blocks in rotation every year.

Demonstrations

It is necessary to follow a cluster approach in organization of demonstration plots within a block to have a discernible impact on the production of crops/ allied area. Further details regarding Farm School norms and selection criteria (by draw of lots after short-listing farmers based on location of their plots, size of holding, past track record etc.) have been given in the Cafeteria. The norms for organizing demonstrations should be in line with the approved cost norms of National Food Security Mission (NFSM) in case of Rice/ wheat/ pulses and coarse grains. These have been given in the cafeteria.

Formation of Commodity Interest Groups (CIGs)

(i) CIGs should be promoted / mobilized for all major commodities (Size 20-25 farmers).

(ii) FIG/CIG members should meet at least once in a month to discuss activities and to decide future course of action.

(iii) BTT and BFAC shall monitor functioning of all CIGs on a regular basis.

(iv) CIGs at village level should be federated at block level and subsequently at district level into Farmer Producer Companies in keeping with the overall structure of Farmer Producer Organisations.

(v) CIGs should maintain proper register & records (commodity/proceedings/ savings/accounts).

(vi) VI. To ensure household food and nutritional security, Farm Women's Food Security Groups (FSGs) @ at least 3 per block are to be formed each year. These FSGs are to be provided support for training, publication and access to inputs @ ` 10000 per group. These FSGs should serve as "Model Food Security Hubs" through establishing kitchen garden, backyard poultry, goatery, animal husbandry & dairying, mushroom cultivation, etc.

Induction Training & Refresher Courses for Extension Workers

An Induction Course with an average duration of (6 days course + travel) needs to be organized at a cost of `1000 per day per participant as and when the BTMs & ATMs are recruited. Besides, a 3 day refresher course each year on (4 including travel) at the same cost as given above on transfer of skills in specific areas (based on crop predominance) and generalized knowledge in common crops, for extension functionaries under ATMA scheme should also be organized.

Joint Visits by Scientists and Extension Functionaries

In order to ensure proper mobility for field visits of the scientists accompanied by extension workers, funds have been provided for meeting the expenditure towards hiring of vehicles or POL. The number of field visits should increase gradually from about three visits per month to one visit per week in the last two years.

Incentive for Exemplary Extension Work

High quality services of extension workers need to be incentivized by providing cash incentives and awards. While selecting the person for award, three factors should be broadly considered. Firstly, percentage increases in productivity in a demonstration plot, secondly percentage reduction in gap between productivity in lab conditions and productivity in the field and thirdly income per unit area.

Main Activities Included in Innovative Technology Dissemination (ITD) component

Display Boards

Once the desired number of extension personnel are in place, it is necessary to ensure their outreach to the farmers on a periodical basis. For this purpose, Display boards are to be put up in all inhabited villages (5.92 lakh) in the country. From ATMA funds, at least 45% of the villages need to be covered. These display boards shall indicate the name of the extension worker assigned to the Panchayat and his/her contact number. These Board will also contain broad details of main schemes (including their major components, eligibility, subsidy pattern etc.) applicable in that area. Besides Panchayat headquarters, the Display Boards can be alternatively put up in prominent places in various villages such as Fair Price Shops, Milk Cooperative Societies, PACs etc. All the display boards shall be in position by the penultimate year of the Plan period to ensure that the farmers know about contact details and visit roster of the extension workers to the designated villages for extension support to farmers. Visits of extension workers will also be monitored through Global Positioning System (GPS) as they would be carrying hand-held device during their village visit.

Pico/ Ultra-Light Portable Projector + Low Cost Films

Pico projectors (or alternatively ultra-light portable projectors) have been found to be very appropriate in rural areas and these are easier to operate without any laptop. 2 Pico Projectors per block are to be provided to field functionaries for disseminating best agricultural practices during the 12th Plan. These will be used by ATMA functionaries including BTMs & ATMs during their visits to villages. The low cost films would focus on specific themes and preferably directed by farmers themselves so as to have greater acceptability among the audience. Video need not be of broadcast quality but will have to be in High Definition Video formats such as .mp4 which can be easily played on laptops or projectors. These films will also be made available on the internet for display through e-Panchayats and Common Service Centres and also for direct use. This task of showing agriculture related films and success stories will be performed by the ATMs. Existing films in the DAC, ICAR, SAUs, States and non-Government entities should also be used after dubbing in regional language.

Use of Hand Held Devices

Use of hand-held devices for on the spot data entry and subsequent updation through voice recognition has been pilot tested with the collaboration of IIT, Chennai. Farm level planning and farmer empowerment using these technologies are being attempted on a pilot basis in two districts in Tamil Nadu. About 450 districts (i.e. 70% of all districts under ATMA) are to be covered by the end of 12th Plan period. At least 20000 extension workers (average of about 45 units per district) in 400 districts are to be covered during the XII Five Year Plan. Estimated cost of a hand-held device is approximately Rs. 20000 with three years' warranty per set at the time of formulation of NMAET. Additionally, General Packet Radio Service (GPRS) charges of aprox. Rs. 5000 per annum per set (for about 2 GB monthly 3G data connection) is required. Backend data processing, contingency charges on recurring expenses and capital & recurring cost of an authentication device (biometric or magnetic reader –to link it to National Population Register or to Unique ID) are also required @ 25% of the hardware/software cost. In most places backend data processing cost can be met from NeGP-A and AGRISNET resources and number of devices can be increased to that extent. Considering paucity of funds, it is proposed to downscale this component to 50 % of the proposed numbers from ATMA funds.

Involvement of Agri - Clinics &Agri-Business Centres (ACABC) & DAESI Trainees

Supplementary Manpower through ACABC

Assuming a growth rate of 5% in the trained agri-preneurs from 2014-15 onwards and a success rate which is progressively increasing, there will be approximately a total of 23331 ACABC ventures (including 10743 ventures established during 11th Plan). These trained candidates should be involved in delivering extension services to the farmers.

Supplementary Manpower through DAESI

Another avenue for involvement of private entrepreneurs in extension related activities is Diploma in Agricultural Extension Services for Input Dealers (DAESI).

Setting up of Community Radio Stations (CRSs)

(i) Community Radio Stations (CRS) have to be promoted in a big way to expand the reach of localised technologies to the farmers located within a radius of 20 to 50 Kilometers.

(ii) A total amount of Rs. 6.5 million (instead of existing Rs. 5.4 million) is proposed to be provided as assistance for this purpose. This enhancement is being suggested because the original norms were finalised in the year 2008.

Some NGOs and KVKs may also use innovative technologies meeting minimum system requirement or contribution from some private players to reduce the capital cost. At the same time, due to local constraints and nonavailability of paid advertisements, content creation cost may increase further. Therefore, a ceiling of Rs. 6.5 million of financial assistance for setting up the Community Radio Station including capital cost and content creation cost for first 3 years has now been included. Subject to compliance of minimum technical specification for setting up of Community Radio Station, break-up of Rs. 6.5 million may be permitted to be changed from capital cost to recurring cost or recurring cost from one year to another. During the XII Five Year Plan, 1/3rd of the districts, on an average should have a Community Radio Station per State. The amount earmarked for this purpose shall be Rs. 6 million per district subject to a minimum of Rs. 4.0 million per new CRS per year provided further that this would be the outer limit on aggregate amount for that year subject to break up of year wise of costs approved for that CRS.

(iii) Funding to private institutions along with Government and Quasi-government organizations for setting up Community Radio Stations (CRSs) can be considered.

(iv) Community based organizations, Agencies/NGOs registered under Societies Registration Act, 1860 or any other such Act and recognized by the Central Government /State Government and serving in Agriculture and allied areas including SAUs and /KVKs are eligible for funding. Registration at the time of application should be at least three years old.

(v) The willing Organizations should have basic infrastructure and facilities in the form of a room of about 400 sq. feet/ electricity/ necessary manpower to run and operate the CRSs.

(vi) ATMA MC of the District concerned may select Suitable proposal/s; recommend them to the Nodal Officer/ Commissioner of Agriculture of the concerned State for onward transmission to DAC through the competent .the IDWG having representation of SAU/ICAR Institutes/KVKs concerned so that the proposal are not referred again to ICAR by I&B Ministry which causes a considerable delay in granting approvals.

(vii) The PD, ATMA would regularly review the performance of CRSs along with other activities with the BTMs. In addition, ATMA GB would review the performance in detail with regard to content creation, involvement of local community, suitability to local conditions, release/ utilization of funds for/by CRS and convergence & synergy with SAU/ KVKs. Detailed Guidelines for Funding of Community Radio Stations are available at http://agricoop.nic.in/radiocguidelines.pdf

Convergence

Farmers' skill trainings and field extension as contained in all 4 Sub Missions of NMAET [Viz. Sub Mission on Agricultural Extension (SMAE), SubMission on Seed and Planting Material (SMSP), Sub Mission on Agricultural Mechanization (SMAM) and Sub Mission on Plant Protection and Plant Quarantine (SMPP)] will be converged with similar farmer-related activities going on through ATMA. Thus, for instance, Seed Village programmes under SMSP, capacity building of farmers through institutions identified by the State Government under SMAM and pest monitoring, Farmer Field Schools & Integrated Pest Management (IPM) trainings to farmers under SMPP will only be carried out through the district level institutions of ATMA and Block Technology Teams. Mutually synergetic linkages will be established among various activities instead of unilaterally mandating that all such farmer-centric activities shall be carried out through ATMA. For instance, half day's training given under Seed Village Programme shall also be made part of Farm Schools as, in any case, training on seed technologies form a part of 6 critical stages during which farmers are trained under Farm Schools conducted under ATMA. Similarly, the Assistant Technology Managers recruited under ATMA shall also double up as Pest Surveillance Scouts. This convergence should be institutionalized by ensuring that State Extension Work Plan [which emanates from Strategic Research and Extension Plan (SREP)] covers field level training & extension components for all modes of Mission. SREP is an ideal platform to provide convergence from the conceptual level and prioritization point of view. IDWG will further underline such a convergent approach at the State level.

A single ATMA Governing Board headed by the District Magistrate will provide commonality in approach & implementation and avoid duplication. ATMA Governing Board shall act as an over-arching umbrella at District level to oversee all extension related activities in other Missions like National Horticulture Mission (NHM), National Food Security Mission (NFSM), RashtriyaKrishiVikasYojana as also the other SubMissions under NMAET. Such a holistic approach will avoid duplication of efforts and promote more extensive & inclusive coverage of beneficiaries. The Integrated ICT platforms (such as Farmers' Portal, State Agriculture Portals, Central Agriculture Portal) will also provide requisite impetus to implementation in the field level.

Convergence with other Farmer Centric Schemes of DAC

With the given man-power support, ATMAs will also look after the work related to RKVY, NFSM, National Project on Soil Health and Fertility Management etc. as mandated under respective schemes. There should be full convergence of extension related work being carried out under different programmes/schemes. The field level extension workers under these different programmes/schemes should work in conjunction with the dedicated manpower being provided under this Scheme under the umbrella of BTT or ATMA, as the case may be. While these extension related workers & consultants under other schemes/programmes can continue to act as experts in their respective fields, they should also double up as multifunctional extension workers in the jurisdiction to be assigned to them by the BTT/ATMA. Budget for extension related components in different schemes and programmes of DAC shall be dovetailed at district level through ATMA. Once all the extension related workers start working in unison, they shall be fully responsible for achieving convergence & synergy in extension related work under RKVY, NFSM, National Project on Soil Health & Fertility Management to achieve complementarities and check duplication of efforts and resources.

Convergence with Research System

ATMA scheme provides for active involvement of Research System/ Research Agencies at different levels of implementation. State Agricultural Universities (SAUs) and KrishiVigyanKendras (KVKs) have to be fully involved not only in preparation of SREP and Extension Work Plans but also in implementation of various programmes in the field. They will be represented in all bodies, namely, ATMA GB and ATMA MC in districts, SLSC/ IDWGs at State level, BTT - BFAC Meetings at Block level. In addition, each KVK scientist may be made in charge of one or more Blocks within the district. The KVK Scientist will technically advise the BTT and will also be actively involved in preparation of BAPs, especially with regard to research related issues/gaps and strategies. He will also take feedback from his colleagues in the KVK in respect of their respective areas of expertise. A joint circular on convergence between Agricultural Research and Extension signed by the Secretary (DARE) & DG, ICAR and the Secretary (A&C) was issued to all concerned on 27th January, 2011.

(i) Zonal Project Directors, State Agriculture Commissioners/Directors and directors (Extension) of the SAUs concerned shall together take a quarterly with KVKs and ATMAs.

(ii) Interface meetings between PD, ATMA and PC, KVK should alternately take place in these two offices. However, PC, KVK should invariably attend ATMA GB and MC meetings.

Joint Visits by Scientists and Extension Functionaries

Quite often it has been seen that scientists of the Universities and ICAR Institutions refrain from making field visits due to inadequate provision of mobility in their budget. In order to ensure proper mobility of the scientists accompanied by extension workers, funds have been provisioned in the 12th Plan scheme for meeting the expenditure towards hiring of vehicles or POL expenses for field visits of scientists along with extension functionaries.It is expected that the number of field visits would increase gradually from about three visits per month to 1visit per week in the last two years.

Convergence with Development Departments

Necessary convergence with all line departments is to be ensured through their involvement in the process of preparation of SREP and Work Plans at Block, District & State levels. Work Plans to be submitted to SLSC for funding under the Scheme should explicitly specify activities to be supported from resources of other schemes and those proposed under ATMA Programme. Necessary convergence is to be ensured through integration of Comprehensive District Action Plans (CDAPs) and District Agriculture Action Plans (DAAPs) – all of which will form an integral part of State Agriculture Plan to be approved by State Level Sanctioning Committee (SLSC) under Rashtriya Krishi Vikas Yojana (RKVY). Further, within approved allocations of Scheme, the State Level Sanctioning Committee (SLSC) may also approve inter component changes as per need during the course of implementation of the Work Plan, within ambit of Guidelines, which should be reported to DAC immediately. Active involvement of Panchayati Raj Institutions (PRIs) in the selection of beneficiaries for various farmer oriented activities, including selection of Farmer Friend should be ensured.

Convergence with & Involvement of Non-Governmental Sector

In order to ensure promotion of multi-agency extension strategies, and to implement scheme activities in Service Provider or Public-Private Partnership (PPP) mode, at least 10% of scheme allocation on recurring activities at district level is to be incurred through Non-Governmental Sector viz. NGOs, FOs, PRIs, Cooperatives, Para-Extension Workers, Agri-preneurs, Input Suppliers, Corporate Sector etc., in either of the two modes listed below in sub-para (b) and (c). Since a lot of manpower and infrastructure has been provided under the Scheme for providing extension services, the State Government should fix an upper limit on extension services in Service Provider or PPP mode. This percentage limit should be clearly spelt out in the SEWP.

Networking

State level bodies/ officers viz. State Nodal Officer/ State Coordinator/ Gender Coordinator will ensure networking of all ATMAs so as to foster information sharing (success stories, best practices, research/ extension issues, application of innovative technologies & strategies, etc.). All District ATMAs shall establish their own portals to share information regarding their activities/ innovations/ successes to the outside world. This Portal shall also have links to related websites both at State and National level.

Monitoring and Evaluation (M&E)

Activities of the scheme shall be monitored and evaluated at periodic intervals through a specific mechanism generated at different levels – Block, District, State & National Level. M&E will be conducted through BFAC and BTT (Block Level) & ATMA GB (District Level). To achieve necessary convergence, SLSC set up in the State to consider and approve C-DAPs under RKVY funding will also consider and approve SEWP and carry out periodic monitoring. The IDWG under the chairmanship of APC/Principal

Secretary (Agriculture) may continue with the day to day monitoring to ensure that the Extension Reforms are executed in line with the broad policy framework.

Cumulative Monthly Progress Reports (MPRs) for each district are to be uploaded by the Project Director, ATMA in a webbased interface available at http://extensionreforms.dacnet.nic.in. by day 5 of the following month. After ensuring that all the districts have entered their data, the SNO will validate this data for the entire State by day 10 of the following month. No paper reports are to be submitted and all data is to be entered electronically only w.e.f. April 1, 2014. Non-adherence to MIS schedule described above will result in no further release of funds to the State. Effective M&E will be ensured through regular field visits of Inter Disciplinary Teams in project areas, reports, interfaces, conferences etc. The scheme also provides for third party Monitoring & Evaluation (M&E) which will be carried out by Government of India. DAC will also organize concurrent M&E including impact evaluation as needed as part of its scheme "Extension Support to Central Institutes/ DOE". Besides, all activities of the scheme would continue to be reviewed on quarterly basis in meetings held at National Level.

Progress of extension work done by ATMs at Block Level, Farm Schools, Demonstration Plots, Trainings, Exposure Visits and Farmer Friends will be closely monitored using ICT infrastructure and by regular monitoring & meetings by BTT at block level, ATMA at district level and SLSC at State level. A web-based interface has been provided for the purpose.

Impact Assessment Studies of extension work done by Farm Schools, CIGs & FFs under ATMA shall be got conducted by expert agencies and corrective action shall be taken timely to attain objectives of revitalization of Agriculture Extension System.

References

1. National Institute of Agricultural Extension Management (MANAGE), Process Change in Agricultural Extension: Experiences under ITD Component of NATP, 2004.
2. Technology Dissemination Unit and MANAGE, Project Completion Report, Innovations in Technology Dissemination Component of the National Agricultural Technology Project, MANAGE, 2004.
3. Guidelines for the Centrally Sponsored Scheme "National Mission on Agricultural Extension & Technology (NMAET)" to be implemented during the XII Plan, 2014, Department of Agriculture, Co-operation & Farmers Welfare, Ministry of Agriculture & Farmers Welfare, Government of India, New Delhi.

Innovative Extension Approaches

06 Introduction and Implementation of Unified Agricultural Extension System in Uganda

- Mr. Augustine M. Nyamwegyendaho

History of Agricultural Extension in Uganda

Agricultural Extension in Uganda has undergone numerous changes since independence. In the colonial and immediate post- colonial era, it was mainly regulatory and carried out by chiefs, who enforced the production of certain crops (e.g. coffee, cotton), which were for foreign exchange earnings. In the 50's, the Government recruited and trained workers in different sub- sector of agriculture, these were to support the chiefs. Until mid-1991, four ministries were principally involved in agriculture, namely:

- Ministry of Agriculture,
- Ministry of Animal Industry and Fisheries,
- Ministry of Environmental Protection, and
- Ministry of Commerce, Cooperatives and Marketing.

Each of these Ministries had extension workers, distributed on the basis of administrative levels (District, County and Sub-county). These were responsible for advising farmers in the disciplines of the agricultural sub- sectors of each Ministry.

In addition, various donor funded projects adopted the commodity approach and had parallel extension workers employed specially for a commodity. Examples include the Coffee Rehabilitation Project, Emergency Cotton Production Project and Diary Development project. Some projects attempted a multi – sectoral approach but were limited to covering specified geographical areas e.g. Agricultural Development Project, South West Regional Agricultural Rehabilitation Project. In the absence of a United Nations Extension structure and approach, each project had to evolve its own extension system and linkages with research and farmers.

The fragmentation of extension service into various ministries and departments had adverse effects on extension service delivery, some of which were;

- Parallel extension delivery systems and lack of a single line of command in management and organization.

- Duplication and overlapping of extension messages delivered to farmers, some of which were conflicting.
- Poor personnel utilization and in effective supervision.
- Wastage of farmer's time and other resources because of the visits by several extension agents especially if a farmer had different enterprise / commodities.
- Dilution of extension efforts, besides extension work, the extension agents were also required to deal with administrative matters and other assignments such as selling inputs, recovery of loans etc.
- Low staff motivation and morale due to inadequate logistics and funds.
- Poor coordination among the different ministries and agencies.
- Weak Research – Extension –Farmer –Market linkages.

As a result, there was low extension contact with farmers and consequently low production and productivity.

To address the above short comings, the government undertook the restructuring of the agricultural sector to strengthen agricultural support services.

In 1987, an action plan for the Agricultural Policy Agenda was developed as p[art of the economic recovery programme. As part of this action plan, Government of Uganda (GOU) and the International Development Association (IDA) agreed to set up two technical working groups -9a research and 9b extension where strategies were developed for the two working groups led to the preparation of investment programmes for effective sustainable research and extension services. The recommendations of the two working groups led to the preparation of Investment porgrammes for research and extension by the Task Force, established by the Agricultural Policy Committee (APC). This resulted into creation of the National Agricultural Research Organization(NARO) and the Unified Agricultural Extension Service (UAES).

Introduction of Unified Agricultural Extension System

The Agricultural Extension Project (AEP) was conceived to support, strengthen and expand the implementation of the Unified Agricultural Extension Service which started under the Head start Programme of Agricultural Research and Extension (HARE).

The Agricultural Extension Project was prepared during 1991 and became effective on the 18th June, 1993. It established the basic principles concerning the new extension approach and methodology after taking into account the lessons learnt from the past experiences in Uganda as well as the on –going experiences in other African countries. It employed the T & V system of management of extension which was adapted to Ugandan condition.

Key Features

The following were some of the key features of the extension programme

- Unified Extension Approach.
- Single line of command.
- Systematic and regular training programmes for staff.
- Regular and programmed schedule of visits.
- Focus on farmers' group with special attention on women and youth.
- Decentralized participatory bottom –up planning process.
- Strengthening of framers' extension, research and marketing linkages.
- Concentration of effects on selected enterprises, messages and impact points.
- Closer and focused supervision of extension staff at all levels for providing guidance and advice.
- Promoting partnership with stakeholders e.g. NGOs, University, school and other related agencies in planning and implementation of the programme.
- Scientific monitoring and evaluation of the progress of the programme.

The project therefore supported and strengthened the implementation of the Unified Extension Approach for crop, livestock and fisheries in the project areas. It also trained staff and farmers in addition to providing logistical support.

Objectives of the Project

The broad objectives of the projectwere the address to urgent needs of disease control improvements in yield and build capacity to deliver extension services as well as to specifically support an effective extension service.

Specially, Agricultural Extension Project aimed at improving;

- Efficiency in the delivery of extension services in 16 selected districts;
- Training capacity and skills of extension agents and farmers;
- Adoption of improved techniques by framers; and
- Efficiency in management services of Ministry of Agriculture, Animal Industry and Fishers.

Components of the Project

The project had four components, namely;

- Strengthening of the delivery of extension services;
- Improvements in training capacity and skills;

- Strengthening of Management Systems in MAAIF;
- Studies.

Design, Organization and Management

- **National Level**

  At national level, the project was fully integrated into the regular Organizational framework of MAAIF. The Permanent Secretary assisted by Director of Agricultural Extension (DAE), Director Crop Resources and Director of Animal Resource were responsible for the overall management of the project. However, the Director of Agricultural Extension was responsible for the day today management of the project assisted by the commissioners, Zonal Extension Officers and subject matter specialists (headquarters) from different sub-sectors.

- **District Level**

  The post of the District Extension Coordinator (DEC) was introduced by MAAIF to coordinate extension activities at district level. He/she was assisted by a team SMSs.

- **County Level**

  Similarly, the county Extension Coordinator was made responsible to coordinate at county level.

- **Circle Level**

  A new operational area calledcircle was put in place. It is a geographical operational area of the field Extension Worker covering between 1000 to 3000farm families.

  This corresponds to two to several parishes depending on population density, availability of staff and geographical spread. Several modifications were made in the original design of the project.

Strengthening the Delivery of Extension Services

- **Methodology**

  The Unified Extension Approach was characterized by programme planning based on farming systems involving farmers, extension workers and researchers. A front – line extension worker responsible for transfer of technology to group of farmers in a manner that encompasses a farming systems approach. The methodology focused on a group approach to transfer technology to farmers through demonstrations, group meetings, and on –farm trails. In addition, field days were held to show the achievements of farmers to the community. Inter and intra-district study tours also formed an important educational means. Mass media were used to disseminate educational messages as well as creating awareness of new developments to the farming community and the general public. Bottom–up planning was emphasized through systematic client consultation by the FEW and other extension staff at all levels.

Extension Activities

The extension method used for transfer of technology and motivating farmers for adoption included visits; group Identification/formation training, result demonstrations; group meetings; field Days/Study Tours; Radio and TV; and Literature.

Interventions to Strengthen Extension Delivery

In order to strengthen the delivery of extension services, the following interventions were introduced.

- Farmer – to – Farmer Seed and Stocking materials Multiplication;
- The village level Participatory Approach (VLPA);
- Information Dissemination Strategy;
- Pilot demonstrations; and
- Benefit – cost Analysis of Enterprises.

Partnership with Other Institutions and Services

The MAAIF collaborated with various institutions and NGOs. The institutions included Makerere University, NARO, and Management Training and Advisory Centre (MTAC), National Institute of Extension Management (MANAGE, India) and Agricultural and Rural Development Foundation (ARDF, Philippines). The extension programme also collaborated with NGOs such as CARE, World Vision, Uganda National Farmers Association (UNFA) and SG2000.

Training Component

The training component was incorporated to build capacity among extension workers and farmers. The following types of training were provided.

- Orientation Courses,
- Pre-seasonal Planning Workshops,
- Monthly Training Programmes (MT),
- Technical Workshops (TWS),
- Inservice Training Courses,
- Re- training of FEWs,
- Management Training Porgrammes,
- Framers Training,
- Long Term Training, and
- Study Tours, Visits, Workshops and Courses.

Strengthening Management Systems

Under the, management component, six committees were established to strengthen the management systems. They include; Policy committee, Technical Committee, Technical Advisory Committee, Procurement Committee.

Finance committee and Farmer – Research – Extension – linkage committee

In order to strengthen the above committee; a desk officer was appointed to closely work with the Director of Agricultural Extension. In addition, six zonal teams, eleven lead persons as well as eight Subject Matter Specialists were put in place.

To build the local capacity, the following consultants were hired on long – term namely; Extension Advisor, Extension Training Specialist, Procurement Specialist, Finance Officer and Civil Works Engineer.

Beside the above consultants a number of short term consultants were recruited by the project.

Studies

The studies conducted to improve the implementation and management of AEP included the following;

- Baseline Survey;
- Farmers Follow –up Survey;
- Extension Staff Surveys;
- Monitoring and Evaluation Surveys;
- Mid- term Review;
- Beneficiary Assessment of the Agricultural Extension Project;
- Man-power Status, Training Needs, Training Plan and Personnel Specifications;
- Rationalizing the roles and the responsibilities of the Various Training Institutes, Colleges and center under MAAIF and the Development of their Curricula;
- Evaluation of monthly Training programmes; and
- Design of a Follow –on project.

Benefits and Results

- The project facilitated the development and organization of an intergraded extension system which was implemented in 33 districts of Uganda. This resulted in improved extension services delivery as 75 per cent of the farmers reported to have contacts with the extension staff.

- The project significantly contributed to capacity building through human resource development by enhancing skills and capabilities of 1000 staff and 500,000 farm families. Besides, 71 SMSs and Lecturers were trained at post-graduate level, 7 undergraduate and 510 FEWs were retrained. In addition, 58 in service training programmes together with 49 study tours were undertaken.
- The diffusion of improved technologies in the districts under the project helped to generate substantial increases in agricultural production which improved the nutrition and standard of living of number of rural families. The extension services reached both directly and indirectly over 900,000 farm families.
- Women farmers benefitted from improved production and processing technologies in addition to labour and energy saving devices. Thus gender concern received special focus in the project.
- The awareness of the recommended practices by 60 per cent of crop farmers, 40 per cent of Livestock farmers and100 per cent of fish farmers was attributed to AEP. Among the farmers who were aware, 75 per cent of crop farmers, 50 per cent of Livestock farmers, and 100 per cent of fish farmers adopted the recommendations.
- The crop yield was reported to have increased by 20 to 50 per cent and milk rose from 1 to 2 liters per day in local cattle and 5 to 10 liters in exotics.

Experiences and Lessons Learned

As a result of adopting the Unified Agricultural Extension Approach, the following were experienced/leant;

- Extension contact with farmers markedly improved as a result of providing transport facilities and allowances under the project.
- Demonstrations, field days and study visits as well as farmer training generated increased level of awareness of recommended technologies and practices.
- Te research–extension–framer linkages were strengthened through regular interactions among extension staff, research scientists and farmers at various levels and for these joints activities not only accelerated the pace of technology generation and transfer but also institutionalized the linkage mechanisms.
- The low extension coverage is mainly attributed to low staff to framers ratio (1500-3000) farm families to each extension staff). This is further accelerated by bicycle as being the predominant mode of transport for the FEWs.
- Through the project, about 8000 active framer's groups were mobilized by extension staff in the project areas which enhanced extension contact.
- However, some of the groups disintegrated due to the retrenchment of staff inadequate knowledge of group dynamics, farmers' expectations of material incentives, civil strife, and lack of access to credit and agricultural inputs.

- The interactive training programs provided to extension workers through monthly training as well as short, medium and long term training, and study tours significantly contributed to human resource development.

  Their knowledge and skills to address farmer's problem on crops, improved and fisheries in an integrated manner. Besides, the effectiveness of farmers training programs was evidenced from the rate of adoption of improved technologies.
- Despite the achievements recorded, the adoption rates were not as wide spread as expected. Some of the factors which contributed to this situation were lack of access to credit, scarcity or high cost of agricultural inputs and implements, adaptability and profitability of some technologies and the retrenchment of extension staff in the wake of implementation of the decentralization of the programme.
- Generally, those 4 farmers who adopted the recommended production practices recorded increased yields and consequently incomes even when influences of market prices were discounted. Increases in yields were high for those farmers who had adopted more impact points.
- Fish farming as an income generating activity as well as a source of protein greatly gained popularity amongst the farmers. Availability of fish fry and poor fish pond management were the major constraints.
- Mass media, which is another method of complementing the delivery of extension messages was not fully utilized under the project. This would have contributed to an enhanced multiplier effect through awareness of extension messages among the farm families.
- Most extension staff were satisfied with their jobs since they were technically confident on jobs due to regular training and programme work schedules. The FEWs who undergone retraining were more competent and confident in handling the various agricultural disciplines. However, 50% of the FEWs in the project districts have so far been retrained. This therefore, suggested a need to emphasize retraining in the follow – on programmes.
- On progrmme sustainability, most farmers observed that cost sharing was not a new phenomenon in agricultural extension as it was already contributed both in kind and cash, towards extension services as long as a direct benefits were expected or assured. For example, farmers who increased their income as a result of extension effort were already cost-sharing in the areas of training and study tours.
- In as much as the project focused on increased farm production, concerns of agribusiness and farm management, which are essential for a profit–oriented farming, were not adequately addressed initially.
- The first phase of the project focused on low cost technologies.
- However, there is need for emphasis, among the farmers who had already adopted low cost technologies, to moves to intermediate and cost intensive technologies, to enhance their incomes. Furthermore, there was clear evidence of diffusion of recommended

technologies from contact group farmers to non-group farmers. This spread effect needs to be stepped up by specifically targeting for a multi-media approaches.

- The initial expectation of strengthening the management capacity of the entire ministry was rather ambitious. As a result, the entire ministry expected support from AEP resources which were limited and this constrained its operations to some extent by spreading its resources thinly.
- Inadequate and irregular flow of funds partly constrained timely implementation of planned field activities during 1995-1996. This was due to limited counterpart funding and expansion of the programme to four additional districts. In addition, the financial management unit had inadequate staff who depended on manual accounting and reporting system.
- The project depended on programme approach, therefore, it was expected that the ministry support staff will be available for project operations. However, with civil service reform, most of the support staff were retrenched and some activities in accounts section consequently suffered.
- The assigned monitoring and evaluation Officers in the districts were also assigned normal extension work as additional responsibility. This affected their performance.
- The project has collaborated with Maker ere University for developing the curriculum for an integrated diploma course for FEWs and for the training of SMSs to masters degree level. This has enabled the extension staff to do their research on home-based extension problems and other areas of interest.
- The 1995 constitution abolished, the level of a county and thus the span of control by the DEC became too wide for effective supervision. Besides, the titles of ZECs, CECs and SMSs were simply operational titles which were not institutionalized although the positions of DECs were institutionalized in few districts.
- The AEP in away pioneered the programme approach in Agricultural sector effective, promotes team spirit and performance, equitable sharing of available resources and opportunities. This contrast to project Nevertheless, it requires considerable efforts and sacrifices in its establishment.
- One of the barriers the Unified Extension Systems faced is the traditional alligence to parent departments and sub- sectors. In the process of working together coupled with management training, this could be solved.
- These implementation experiences and lessons learned have been used as a basis for designing the proposed NAEP.

07 Village Level Participatory Approach (VLPA) for Sustainable Development

- Prof. S. V. Reddy and Prof. M. Surya Mani

Introduction

Since independence, there are many approaches adopted in strengthen the delivery of agriculture services in India. in all these approaches, the emphasis has been to organize extension and research without giving due attention to the organization of client system. it has to be noted that involvement of client system in planning, managing, implementing, monitoring and evaluation of programmes is important for improving the adoption rates, incomes and welfare of rural population (Farmers), Agricultural Extension services in India have undergone a lot of metamorphosis over the years, from commodity oriented to farming systems approach. the increased involvement and participation of all the village residents strengthens the delivery of extension and makes it responsive to farmers needs and priorities. Furthermore, experience has shown that farmers problems are multidisciplinary and need to be addressed in a holistic manner. It is against this background, the Village Level Participatory Approach is being proposed to strengthen the agricultural extension services in India.

What Is VIPA

It is a bottom-up participatory process in which all the village residents and other stakeholders are involved in identifying, priority and analyzing problems, preparing action plans, implementing monitoring and evaluation through village implementation committees (elected by Gram Sabha) and self help groups (special interest groups).

Objectives

- To facilitate the village residents to identify, analyze their problems, to prepare action plans and to motivate villagers to take responsibility for the development of their villages involving various socio interest groups.

- To aid mobilization of resources from within and outside the village to implement the action plan and to put pressure on different partners (government services, NGOs etc) to deliver quality services in a timely fashion.
- To encourage and train the village residents to implement the action plan and monitor and evaluate the efforts.
- To strengthen agricultural extension and other services delivery in the villages through community mobilization and organization of self–help groups and associations.

Key Features

- All village residents (Gram Sabha) are involved.
- Adopts bottom up and systematic approach.
- Integrated multi-sector plans addressing all concerns (broad based) of rural development.
- Holistic micro plans addressing all issues of agricultural development.
- Villagers are trained on priority settings, identification of beneficiaries, decision making and setting agenda for action.
- Establishment of village committee by Gram Sabha to oversee implementation, monitoring and evaluation of the plans.
- Empowering village residents to demand for timely services.
- Establishing linkages with the existing institutions (panchayats, cooperatives, block/mandals).
- Promotion of partnership among various stake holders and ensuring their commitment.
- Rationalization, integration and harmonization of resources.

Principles and Considerations

- Involvement of Grama Sabha
- Education and training
- Triangulation
- Emphasis on village residents leadership role
- Holistic development
- Addressing equity concerns
- Self-help, mutual help and cooperation
- Programme vs project approach
- Rational use of resources.

Methods and Tools

- Collection of secondary and primary data.
- Participatory methods
    - Focused group discussions
    - Village meetings
    - Mapping
    - Transect
    - Venn diagram,
    - Well being analysis,
    - Problem trees,
    - Objective trees,
    - Prioritizing tools,
    - Semi structured interviews etc.

Phases and Process of VLPA

- Selection of a Village.
- Information and approval of the villagers.
- Secondary data collection and rapport building.
- Training of trainers.
- *Diagnosis*: Problems, opportunities and resources through village mapping, transect venn diagram and semi structured interviews.
- *Problem analysis*: Prioritization and finding causes and effects.
- *Search for solutions*: To understand the perception and motivation of village residents to solve priority problems and to identify villager` proposals to solve the problem.
- *Planning*: Translates objectives into activities/actions for implementation within a given time frame and budget. The action plans are prepared jointly by intervening agencies, expets and villagers which are presented for approval and adoption by Grama Sabha.
- *Implementation arrangements*: Grama Sabha elects village implementation committees (VICs) sector wise. Balance representation is given by including panchayat, ward members, women, youth, SC, ST and BCs. In addition to it, Village Development Committee (VDC) is constituted drawing chairpersons of VICs, Sarpanch (Panchayat President), Village Administrative Officer (VAO) /Village Development Office (VDO), who will oversee the implementation of plans, mobilize villagers and resources. Formation of VICs and VDC facilitates communication and dialogue within community, outside and intervening agencies.

- Formation of self-help and special interest groups to spearhead for economic development and livelihood improvement of farmers. Different special interest group like cotton, dairy, poultry, women, agricultural labourers etc can be formed and they can be federated at village level into farmers clubs, organizations etc.
- These committees are expected to meet on a designated date every month to review the progress and set agenda for action.
- Monitoring and Evaluation: Continuous process of VLPA. The committees are trained on this important function.
- It is hoped that members of VICs and VDC will eventually be trained and registered to take-up role of Village Based Organizations (VBOs) eg: registered farmers associations to promote self-reliant villages.
- Wherever farmers groups are available they can be strengthened and additional groups can be formed. The Village Implementation Committee responsible for Agriculture Development can evolve out of the groups and can be federated at cluster/block level to form Farmers Associations.

PHASES OF VLPA

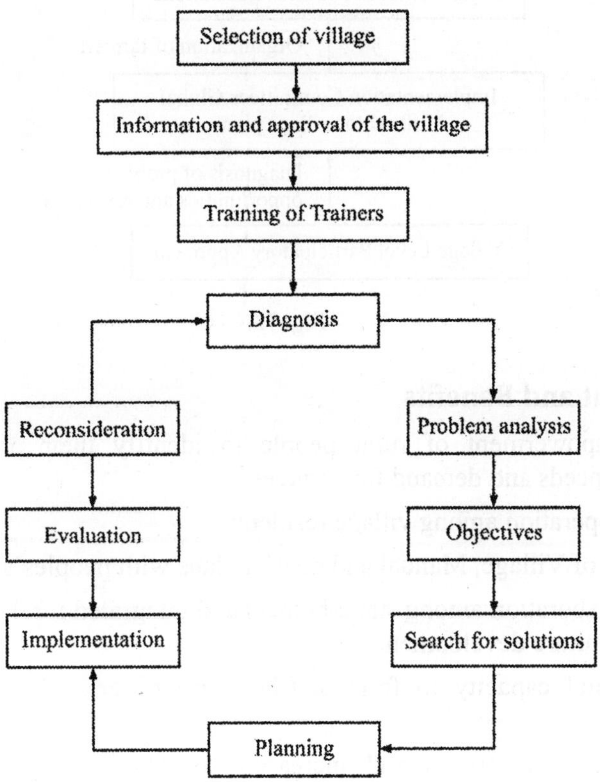

Fig. 7.1

THE PROCESS OF VLPA

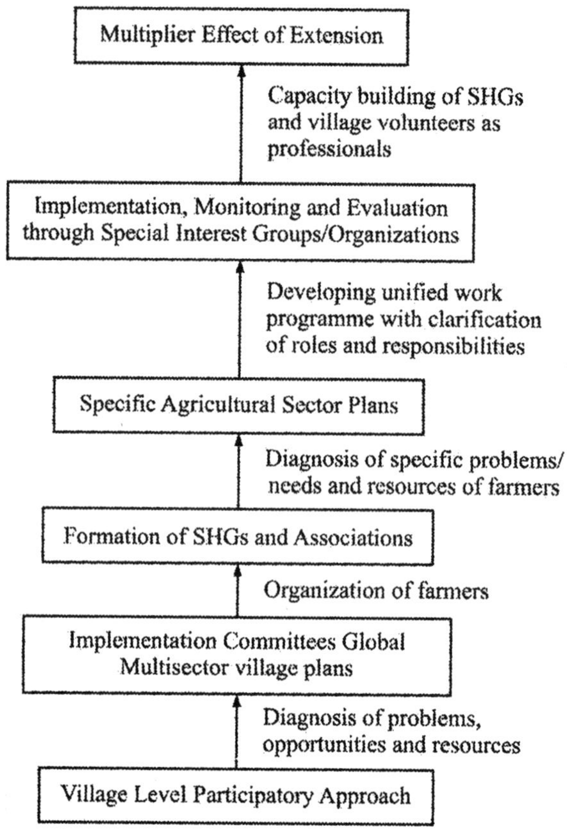

Fig. 7.2

Expected Output and Benefits

- Increased empowerment of rural people to identify their problems, determine development needs and demand for services.
- Improved cooperation among village residents.
- Development of Village, Mandal and district plans with peoples's participation.
- Increased collaboration among stake-holders and integrated rural development which includes agriculture development.
- Developed rural capacity to form self-help groups and identify viable/fundable projects.
- Optimum use of resources and increased contribution from villagers towards developmental activities.

- Capacity building of the village residents, development staff and other stakeholders in participatory planning, implementation, monitoring and evaluation.
- Strengthened agricultural extension and other services.
- Enhanced capacity of farmers to choose different services providers of extension.
- Strengthen farmer-research-extension and marketing linkages.

Integration of VLPA in Extension Services

This methodology can be used to strengthen extension services as it is a bottom-up planning methodology to support rural development in general and agricultural extension in particular. the advent of VLPA implies additional responsibilities for all cadres of extension staff as shown below.

- Developing a detailed micro plan for agricultural extension indicating what is to be done based on the global village action plan. The extension worker should extract broad agricultural needs emerging out of VLPA and prepare a detailed work programme by involving interdisciplinary team of experts in agricultural sector. the efforts also include consultations with village implementation committee responsible for agricultural development and special interest farmers groups with responsibilities of each actors, schedules of visits, target groups/interventions etc. duly specified. client consultation methodology with focused group discussions and other participatory tools can be adopted for this purpose.
- Motivating and assisting VICs to organize more special interest groups and also forming of farmers clubs/organizations which can spearhead and demand for services. Ofcourse, the mechanics of organization should be left to choice of the villagers.
- Motivating farmer`s and VICs to promote farmer to farmer extension and to arrange the input services through self help groups by mutual help and cooperation.
- Linking up the farmers groups/organizations with relevant institutions for access to credit, input and technology.
- Continuing to focus on the methods and techniques of transfer of technology (demonstrations, field visits, media etc).
- The block/district level officers to focus on the popularizing VLPA in more villages and also to coordinate, monitor and supervise the activities.
- Working out the specific strategies for capacity building of committees (VICs) and groups.

The Role of VLPA in Strengthening Extension Delivery

VLPA strengthens the delivery of extension services in the following ways:

- **Education**

  The parties involved in the VLPA benefit through learning. The village residents learn how to identify, analyze their problems and make action plans to address these problems. In this way, the educational element in VLPA facilitates better understanding between the village residents and the extension staff and leads to faster adoption of technologies and practices.

- **Involvement**

  Village residents involvement in every step in planning makes them to recognize the plan as their own. This creates a sense of commitment to implement it.

- **Multiplier effect**

  Village residents who are involved in the planning process act as disciples to all corners of the village/block. This leads to a higher multiplier effect and accelerated pace of adoption of recommended practices.

- **Empowerment**

  The village residents gain more self confidence in making decisions to determine their programmes, as well as to demand for services.

- **Participation**

  The village residents are actively involved in decision making and execution of their plans. This generates higher levels of motivation. Enthusiasm and change of attitudes. this helps the village residents as a community to develop a vision and aspirations for progress.

- **Equity**

  The participation of all the socio-interest groups with different socio-economic status and resource structure ensures equity. No sub group is marginalized in terms of resource utilization and benefit.

- **Partnership**

  VLPA promotes partnerships among NGOs, Private Sector Government Agencies, and other Stakeholders by involving them right from planning up to implementation stage. This strengthens the extension and research service and rationalize resource use.

Sustainability

VLPA sustainability entails institutional, financial and environmental aspects:-

(a) VLPA aims at human resource development to plan, implement and maintain the activities on a sustainability basis.

(b) By the mere fact that the planning is participatory, there will be high commitment in implementation of planned activities.

(c) Cost-sharing. It is hoped that when services are demand driven, adoption of improved practices/technologies will be high and thus higher productivity and incomes. This will empower the rural communities to cost share some of the activities.

(d) Integration of VLPA in extension implies that implementation of action plans will be guided by the extension services and farmers.

(e) Collaboration with NGOs involved in rural development, donors and other stakeholders will maximize use of the resources available from the different sources since plans are prepared using convergence approach.

(f) As farming becomes commercialized, demand driven extension services will be cost effective.

(g) With decentralization the local authorities Village Implementation Committees Farmers groups/clubs/Associations will readily supervise staff to ascertain implementation status.

(h) Use of village residents as trainers/volunteer leaders would reduce on the cost of implementing VLPA; particularly during expansion activities.

Conclusion

This innovative approach was extensively used with success for strengthening agricultural extension delivery services in Africa and Latin America. This was adopted to Indian conditions and tried in 17 villages of Bhoothpur mandal in Mahaboobnagar District and in 25 villages in Venkatachalam mandal of Nellore District of Andhra Pradesh by PRDIS-an NGO based in Hyderabad. Besides several NGOs have adopted the approach through grants from Council for People`s Participation and Rural Technology (CAPART), GOI. Some of the Extension Researchers also used this concept. During this process about 200 villages were involved and about 20000 villagers and other stakeholders were trained in participatory planning, monitoring and evaluation. All stakeholders are enthusiastic and share positive signals that VLPA holds the key towards rural transformation through agricultural and rural development. The VLPA can serve as a launching pad for implementation of programmes of central and state governments with peoples participation and commitment. It can strengthen democratic institutions like panchayats in rural reconstruction process. Furthermore, it can strengthen extension delivery through bottom-up planning process and farmer to farmer extension.

08 Farmer's Producers Association for Sustainable Development of Farm Families in India

- Dr. K. Narayana Gowda

The biggest challenge Indian agriculture going to face is how to ensure sustainable productivity, production, profitability and equity. The GDP from agriculture today is 14 percent which was 58 percent during 1950. There are many challenges and compulsions Indian agriculture encounters with and most important ones are: continuous decline in percapita holding and fragmentation of holdings, indiscriminate increase in cost of production, high cost and non-availability of labour, increased nuclear families over exploitation of natural resources, climate impact, multiplicity of agencies engaged in transfer of technology, drying up of public extension system, poor linkages, variety of organizations engaged in providing credit, insurance, critical inputs, increasing post harvest losses, under and unharvest wastages, marketing issues, fast decline in profit margin, farmers losing interest in farming, younger generation matching alliance are not forth coming for the prospective bridegrooms engaged in farming. Migrating to urban areas and many villages are becoming old age homes. There is increasing social disharmony in villages due to various issues like land disputes, political factors, communal factions and so on. Further, many of the farm families are located in inaccessible areas/regions like hillyareas, mountains, valleys, islands and so on, reaching them is a biggest challenge.

Cooperative movement in India was started with high hopes over a century back, but today there is no state in the country where cooperative is healthy except in dairy sector. The villagers were fairly organized in the yesteryears, partly because of joint family systems it was easy to bring them on a common platform, but today they are divided on various factors and how to bring them together for extension education purpose to a common place in harmonious way is a real challenge.

It is high time to evolve sustainable solutions to address aforesaid issues. There are many alternative options to address emerging challenges they are: Promoting Farmer's interest groups, Farmer's producers associations, Farmers producer companies, FPO's and so on. Having had closely observed all these initiatives, it is the Farmer's producers association at the grass root level and federations at the higher level is the most appropriate option for all-round sustainable development of farmers by addressing end to end issues.

Farmers Producers Association **(FPA):** It is a voluntarily established registered autonomous body, to address end to end issues in a given enterprise or groups of closely nit enterprises for the purpose of ensuring sustainable productivity profit, social harmony and equity. The size and spread of an enterprise vary from enterprise to enterprise as well as location to location keeping optimum productivity and profitability without affecting natural resources.

Increases in opportunities for speedy transfer of technologies are: more number of Literates in rural areas, availability and access to Information and Communication Technologies. There are many success stories before us which can be replicated to all types of enterprises. The Rural Biofuel Growers Association where in farmers share from the consumer is 75 percent which was less than 33 percent and while in jack case Jack producers Association farmers share is 65 percent which was 18 percent before the establishment of these associations by Rural Bioresource complex project implemented by UAS, Bengaluru during 2005-10, there after nurtured by KVK Bengaluru Rural District in Karnataka is a message not only to the developing countries but also for under developed countries in the world (Narayana Gowda,2011). Similar experiences are documented in Baramati KVK in Poona district and Ramakrishna Mission KVK in Ranchi which are the ray of hopes for addressing end to end issues (Srivastava 2011) through the promotion of FPAs.

The establishment of FPAs will help to bring all the farmers in one platform by their own initiative at a shortest possible time. Thereafter, platform can be used to discuss any innovation even complex technologies with the organised group. Relevant printed information such as leaflets, folders, bulletins, books may be provided to the office bearers for further empowerment, discussion and use. It is easy to arrange exposure visits to the organized groups and timely reaching through available ICT tools. The FPAs will address ten of the backward and forward linkages which will speed up the rate of diffusion and adoption of innovations realising sustained profit in the given enterprise.

Promotions of FPAs will not only bring in enhanced productivity and profitability, creates additional employment opportunities for under employed and unemployed youth, retains the farm youth by rebuilding confidence in rural area above all improves social harmony in rural areas.

Establishment of FPAs is a difficult proposition, left to the people on their own, they hardly come together for such initiative but constant persuasion and follow up besides need based financial support and encouragement will bring in place such organizations. This will also helps to regulate how much to produce to meet domestic requirement, value addition and processing avenues besides export opportunities. The growth of Indian agriculture is suffering partly because of inadequate and reliable data base information generation. Promotion of FPAs will help to generate data base information for every major farm produces across the country. Therefore the future of sustainable development of agriculture, food security and profitable farming goes with promotion and establishment of enterprise based FPA's in India.

References

1. Narayana Gowda. K, 2011-Project based Intervention of Organizing Farmers. Edited Book; Future Agricultural Extension, Kokate K.D et.al, Westville Publishing House, New Delhi.
2. Srivastava, J.N.L, 2011. Farmer's organizations for Technology Dissemination. Edited Book; Future Agricultural Extension, Kokate K D et.al, West ville publishing House, New Delhi.

09 Institutional Development in Agricultural Extension for Good Governance – An International Perspective

- *Prof. S.V. Reddy*

Preamble

In my journey with Agricultural Extension for about 50 years, I had an opportunity to study closely and implement different Extension models and methods in India and different countries. I would like to briefly deal in this paper the lessons learnt and best practises in Agricultural Extension Models vis a vis good governance.

Agricultural Extension

A new direction for agricultural extension is needed that captures the institutional reforms towards market-oriented privatizing innovations and non-market decentralizing reforms. But a new direction must accommodate the very many paradigm shifts taking place in the world; the green revolution, for instance, has shifted to the evergreen revolution; the commodity approach has moved to an integrated systems approach; mono-disciplinary to multi and interdisciplinary approach. Thrusts in technology research have shifted to eco-technology, and biotechnology, which are congruent with productivity, profitability, equality and sustainability. The supply–driven approach has become demand-driven. Farm employment has shifted to off-farm and non-farm employment. The definition of food security is evolving; there is increasing importance on economic access, on self-reliance rather than self-sufficiency, Extension knowledge skills, self-learning, natural resource management information technology, equality concerns nutrition, health awareness decent work and environment protection.

In current times, the main global forces of change that are affecting or are likely to affect the existing structure, mandate and practices of national agriculture extension systems in developing countries are globalization and market liberalization; privatization;

commercialization and agribusiness; democratization and participation; environment and equality concerns, disasters and emergencies; information technology breakthroughs; rural poverty, hunger and vulnerability; the HIV/aids epidemic; sustainable development; echo technology/biotechnology and genetic engineering; market led extension diversification and value addition and integrated, multidisciplinary and holistic development.

Thus, with the above backdrop of changes, keeping in view the futuristic needs direction and vision, many countries as well as bilateral and multilateral donors have seen a need to reform National Agricultural Systems in response to the global forces of change. Today coping up with changes and the need for private-public partnership promotion in Agricultural extension delivery with increasing involvement of community based organizations (CBOs), NGOs and Farmer organizations, each country is evolving New Extension Models based on its resources.

Need for Good Governance

God governance is epitomized by predicable, open and enlightened policy making, a bureaucracy with professional ethos acting for furtherance of the public good with rule of law, transparency process, accountability with active participation of civil society and stakeholders in public affairs.

Institutional Development

The possibilities for good governance depend on Institutional structures and the economics and credible human resources availability for ensuring it happens. Towards this end, many countries over a period of time brought out Institutional reforms with changing role of extension mostly directed towards.

- Broad based extension role
- Market reforms
- Public-Private Partnership
- Innovations in extension delivery such as cost sharing, contracting, out sourcing of Extension Services and use of ICT
- Farmer organizations leading the Extension Services
- Participatory transparent result oriented management
- Decentralization of Extension Services

Extension Reform Strategies in Different Countries

Market Reforms

	Public	Private
Pubic	Revision of public sector extension via downsizing and some cost recovery (Canada, Israel, Thailand, USA)	Cost recovery (free-based) system (OMECD countries, previously in Mexico) China (need based as incentive)
Private	Pluralism, Partnerships power sharing (Chile, Estonia, Hungary, Venezuela, S. Korea, Taiwan) Pluralism and partnership (India, Kenya, Uganda, Ethiopia, Liberia, Malawi)	Privatization (total) Commercialization (The Netherlands, New Zealand, England & Wales)
	Non-market reforms	
	Decentralization to lower tires of government (Colombia, Indonesia, Mexico, The Philippines, Malawi, Uganda, India & Others)	Transfer (Delegation) of responsibility to other entities (Bolivia, to farmer organizations: mixed with farmer-led NGO programmes; Peru, extension developed to NGOs) West Africa, family farm advise

The countries mentioned above are only illustration of strategies of Extension models employed. They can be classified as partnership, facilitation, collaborative, private extension models. However the association of strategy for a comity is not exclusive in some of the countries mentioned, several extension reforms have taken place and participatory extension methodology also are being incorporated. In others words, the now Agricultural Technology Management Agency (ATMA). Similarly in Uganda the reforms started with T&V system, unified Extension System, National Agricultural Advisory and Development Services (NAADS) and now unified Extension with Certain features of NAADS mostly to encourage Private Extension Services and Local communities as partners in Extension.

Indicators for Good Governance

The task of development is the art of generating change. The following indicators can able to provide guidance to Good Governance in Extension models.

- **Result Oriented Management (ROM)**

  Result oriented Management (ROM) instead of activity Oriented management. The ROM envisages setting up realistic objectives and tasks (using participatory bottom-up planning strategy) together with the strategies for implementation. With the

ROM, each functionary at the grass-root level clear cut work programme with responsibilities, obligations, time tables, activities and results with the constraints and commitments of each side clearly understood. The managers at higher level are supposed to organize themselves (making a list of positive as well as reactive tasks). Besides, they should also organize other people by delegation of authority commensurate with responsibility, train them and be available for advice. The result oriented management can improve the efficiency of the personnel in the organization and at the same time will be in a position to shown the results setting objectives and targets leading to good governance.

- **Professionalism**

  The Public and Private Extension Providers should update their technical, managerial and communication skills periodically, prepare work programmes with bottom up planning involving all stakeholders from village level preferably using village level participatory approach (VLPA) and implement them professionally. The credibility of Extension workers in terms of expertness and Trust worthiness is of permanent important. There should be regular mechanism for capacity building of extension providers and farmers with a planned scheduled period for updating of knowledge. Linking with Research, Farmers schools use of ICT will also help in this direction.

- **Single Line of Authority**

  It is important Extension operates under single line of authority from National / State/District to grass root level. Similarly, various Agricultural development schemes under operation can also be merged and be placed under single line of authority for implementation. Convergence is most important aspect.

- **Programmed Visits**

  It is essential to have a work programme developed at various levels of administration and Management where in each functionary has a plan of work with clearly defined objectives, goals and results. Such a plan should incorporate regular visit schedules to different villages and Farmers/Groups. The schedule should be known to the farmers/groups/local bodies so that they can participate in implementation supervision, monitoring and evaluation. The schedule can be revised monthly/bimonthly/quarterly.

- **Concentration Efforts**

  There is tendency on the part of Extension workers to set ambitious targets and not in a position to achieve with the resource availability. Besides too much of paper work and on extension work also remains a barrier in achieving the set targets. It is important to prioritize and focus attention on selected targets to complete them with due concentration.

- **Research and Development**

  In Extension work a considerable amount should be earmarked for research and developed R&D as it is an important aspect for technological progress and quality assurance. There is need for accelerated progress in research in Extension to cope up with the changes.

- **Participatory Programme Planning, Supervision, Monitoring and Budgeting**

  Participatory Programme planning and budgeting using the village levels participatory approach and Zero based budget involving all stakeholders, farmers, elected representative NGOs, CBOs Governments Officials experts etc, is a right procedure for planning Extension programmes. The Extension should be farmer centric and driven by local organizations/community based organizations (CBOs).

  The supervision should be focused on guidance than fault finding as typically done in government services. The target is results and methodology is guidance. Hence team visits and sprit should be encouraged.

- **Equity Concerns**

  Important consideration to be kept in view while planning and implementation, the vulnerable section of society namely women, youth, disabled scheduled castes and poor. This is important section of the target population which needs special attention and guidance to address the equity concerns. The extension models should accommodate this aspect.

- **Private and Public Extension Interface**

  The present day context, extension models should have a good interface of private and public extension services. Contracting or outsourcing the private extension providers is very important. This interface should have an MOU with clean cut roles and responsibility.

- **Motivation and Incentives**

  There is need to develop standards based on the objectives and results to be achieved. The Extension services have to set standards against which the performance can be judged. Similarly, performance indicators also should be incorporated in work programme of individuals. The indicators and standards should reflect quality of performance. Based on this, the incentive schemes have to be inbuilt in to the system.

10 Institutional Innovations in Agri-Extension for Inclusive Agricultural Growth:

Drivers and the Operational Framework

- BS Sontakki & R.Venkattakumar

Factors affecting inclusive agricultural growth

Agricultural contribution to GDP is 14.5% (2010-11 in 2004-05 prices) and has declining trend. Notwithstanding the declining trend of agriculture and allied activities to GDP, it is critical from the income distribution perspectives as agriculture accounts for 58% employment in the country (Census, 2001). Hence, growth in agriculture and allied sectors remains a 'necessary condition' for inclusive growth in terms of reducing poverty, development of rural economy and enhancing farm income (Economic Survey, 2011-12). Indian agriculture with two-third of rainfed area remains vulnerable to vagaries of monsoon, besides facing occurrence of drought and flood in many parts of the country. Climate change will aggravate these risks and may considerably affect food security through direct and indirect effects on crop, soils, livestock, fisheries and pests. Building climate resilience, therefore, is critical.

Given the growth rate of population, the rise in demand for food is a natural concomitant. Further, the rise in income levels and change in tastes and preferences of people have also contributed to the increased demand for diverse food products. An estimation of food consumption expenditure in the country during 1987-88 to 2009-10 clearly reveals the shift in expenditure towards milk and milk products, egg, fish, meat and vegetables both in rural and urban areas and the share of consumption of cereals in the total food basket has gone down (Table 10.1).

Table 10.1 Share of expenditure to total food expenditure

Items	Rural		Urban	
	87-88	**09-10**	**87-88**	**09-10**
Cereal	41.1	29.1	26.6	22.4
Pulses and products	6.3	6.9	6.0	6.6
Milk and products	13.4	16.0	16.8	19.2

Table 10.1 *Contd...*

Egg, fish and Meat	5.2	6.5	6.4	6.6
Vegetables	8.1	11.6	9.4	10.6
Sugar	4.5	4.5	4.3	3.7
Total	100.0	100.0	100.0	100.0

Hence, the country has to step-up efforts for increasing production of milk and other dairy products, egg, poultry, fish, meat etc. Similarly, a thrust on horticulture products is required for enhancing *per capita* availability of food items as well as ensuring nutritional security. Dairying has become an important secondary source of income for millions of rural families and has assumed an important role in providing employment and income-generating opportunities.

As far as the utilization of inputs is concerned, seed quality is estimated to account for 20-05% of productivity. It is therefore important that farmers have access to quality seeds. Though, the country has been witnessing considerable progress in farm mechanization, its spread across the country still remains uneven. Indian farmers are mostly small and marginal with small and fragmented landholdings. This poses challenge in terms of adoption of farm mechanization as well as generating productive income from farm operation. Pooling of many landholdings may yield better economies of scale for which land laws for leasing with sufficient safeguards in place should be considered. Current farm power availability hovers around 1.7 kw/ha which is much lower than Korea (1.7 kw/ha), Japan (14kw/ha) and the USA (6 kw/ha). Supporting and franchising rural entrepreneurs for establishing custom-hiring or farm-service centres will help extending farm mechanization to so far excluded farmers' category. Adequate availability of feed and fodder for livestock is vital for increasing milk production and sustaining the genetic improvement programme. Green fodder shortage in the country is estimated at about 34%. Similarly, given the vagaries of monsoon, augmenting irrigation potential is key to sustained growth in agriculture. There is a need to aim at making the extension system farmer-driven as well as accountable to farmers by providing new institutional arrangements for technology dissemination.

Agricultural credit and insurance plays an important role in improving agricultural production and productivity and mitigating distress of farmers. Access of small and marginal farmers to formal sources of agricultural credit is still inadequate, though the flow of agricultural credit has increased. Providing financial support to farmers in the event of failure of crops as a result of natural calamities, pest and diseases outbreak, providing insurance protection against adverse weather conditions such as deficit rainfall, floods, high or low temperature and humidity that are deemed to adversely affect crop production.

Given the compositional shift in food-basket of a common household and its impact on consumption demand, improved supply response is critical for ensuring price stability in food items. Extension programmes and guidance to farmers regarding inputs usage and alternate cropping pattern based on soil analysis need to be undertaken and intensified.

Setting-up special markets for specific crops in states/ regions/ areas producing specific crops would facilitate supply of superior commodities to the consumers. Mandy governance is an area of concern. A greater number of traders must be allowed as agents in the mandis. Anyone who gets better price outside the Agricultural Produce Marketing Committee (APMC) or at its farm gate should be allowed to do so. Perishables could be taken out of the ambit of the APMC act. The recent episodes of vegetables and fruits inflation have exposed flaws in supply chains. The government regulated mandis sometimes prevent retailers from integrating their enterprises with those of farmers. In view of these, perishables may have to be exempted from market regulations. Considering significant invest gaps in post-harvest infrastructure of agricultural produce, organized trade in agriculture should be encouraged. Setting-up of modern storage facilities need to be addressed.

Another area for improvement is the generation of real-time market intelligence and also agricultural marketing reforms. Setting-up of efficient supply chains is essential not only for ensuring adequate supplies of essential items at reasonable prices but also producers get adequately compensated. Linking farmers to market is therefore very important. Concurrently, with increasing income and population, demand for processed food is likely to increase. Hence, investment in food processing, cold chains, handling and packaging of processed food needs encouragement. Addressing the welfare of agricultural producers and consumers simultaneously is a real challenge. MSP signals the floor prices for the produce which in turn, has the potential of increasing prices. However, inability of a large number of small and marginal farmers to directly access the agricultural market puts a question mark on increases in MSP actually benefitting such farmers.

For facilitating inclusive growth in agriculture, there is a need to address the challenges of agricultural sector through comprehensive and coordinated efforts directed at improving farm production and productivity of food grains and high value crops, developing rural infrastructure, renewing thrust on irrigation sector, strengthening market infrastructure and supporting investment in R&D with due emphasis on environmental considerations.

Challenges in achieving profitability in agriculture and inclusive growth (Srinivasan, 2007)

- Profitability refers to positive returns to working capital and capital invested in various productive assets including land. In case of capital assets, profitability should ensure return of capital and also return to capital at rate equal to or exceeding the prevalent market rate of interest. The farm family must be in a position to generate a total net income that is sufficient to meet the expenses of the household and invest in future. This indicates the farm viability. Inclusive growth refers to viability of even the smallest and most marginal rural, whereas sustainability is the ability to generate positive net returns over a period of time, protecting the productive capacity of capital assets.

- A major concern is the relatively low growth in agriculture GDP as compared to overall GDP without significant reduction of labour force engaged in agriculture. It has widened the *per capita* income of labour engaged in agriculture and non-agricultural sectors. A large number of rural households share a shrinking part of the national pie. Sixty percent of rural households owning less than 1 ha, 12% households are landless and 28% owning more than 1 ha (National Commission of Framers, 2006). Such an uneven distribution of income and assets has social and political ramification for future.
- The dry land crops show much lower returns than irrigated crops. This clearly brings-out the difficulties faced by dry land farmers in generating viable incomes.
- A number of hidden costs that make agriculture are rarely recognized and computed such as failure in infrastructure, delay in receipt of subsidy, delay in delivery of services, lack of quality inputs, under-financing by banks and procedural complexities.
- The policy on land tenancy and ownership has pushed up the cost of leasing and has driven the market underground. Without effective grievance redressal mechanism in case of disputes, the land policy has made the relationship between the owner and tenant extortionate from either side.
- Failure to focus on market has impacted the profitability of agriculture and made the farmers dependent on subsidized input supply rather on the market for remunerative prices. Through a combination of state and credit policy, overall production growth was targeted rather than farm income. Further, the state intervention in agriculture has tended to benefit irrigated and larger holdings as these are better placed to use the inputs offered.
- The subsidies given by benevolent state seeking to reduce input cost has led to distortion in the market and encouraged the mis-use and misapplication of inputs. The review over the last four decades of the farm policy and the policy of inputs leads one to feel that farmers' security has become casualty to the interests of food security.
- The farm-level risks such as technology, credit availability and absence of policy continuity impacts farm profitability. The absence of market-based risk mitigation mechanism has made it difficult for farmers to use friendly and reliable insurance products. The state action in terms of risk mitigation is more evident in calamity relief rather than systematic management of risk.
- Pending application for farm power connection are fairly large and availability of power is erratic.
- The high cost of borrowing pushes up the capital costs and reduces profitability.

- Delayed supply of inputs and supply of poor quality inputs.
- Productivity of major crops of the country is much lower than world average, since farm viability has not been made an objective of technology development.
- The technology transfer mechanism is weak with a significant portion of farmers don't use extension workers as the credible source of information.
- Lack of aggregation mechanisms for produce and lack of marketing mechanism have prevented the small farmers from realizing the best prices for their outputs.
- The policy of keeping private sector out of procurement and marketing operations has led to continuous exploitation by the existing intermediaries in the chain.
- Export-import policies are used by the government in case of domestic shortages and cool down the market to maintain consumer prices at lower levels to the disadvantage of farmers. With subsidies in large quantities available to the farmers abroad (as much as 40 to 60% of their income), competing with them in global market by poor Indian farmers is a big question.

Agenda to be addressed (Srinivasan, 2007)

- Focus on farm profitability; shift focus from inputs to outputs and markets.
- Clear articulation about where consumer protection must end and farm protection must start.
- Public investment in rural must not only be enhanced but be efficient; implication is to reduce cost of agricultural operations and widening the markets. Technology also must concentrate on cost reduction and income enhancement at farm level.
- A dispute redressal mechanism that would address the issues relating to contract and corporate farming is necessary in the context of increasing coverage of such arrangements.
- Mechanism must be there to compensate farmers for failure of technology and inputs
- Income generation is an entrepreneurial activity. Hence, less state and more private involvement could provide suitable conditions for enhancing farm incomes. More private sector initiatives must be welcomed in agriculture with necessary safeguards to prevent exploitation of farming community.

Evolution of innovations in agri-extension (Rivera and Sulaiman, 2009)

The 'green revolution era' with improved technologies that are suitable for resource-rich farmers and privileged areas of the country saw 'diffusion of innovation' (Rogers, 1962) (technology transfer/ linear model) as the model for technology dissemination. Innovation was defined as a new technology developed by scientists, transferred by extension personnel and adopted by scientists. This model viewed agricultural research as central source of all agricultural innovations. The non-relevance of green revolution approach in the late 80s of 20th century led to the evolution of participatory methodologies and farming systems approach which had the concept of understanding and strengthening farmers' capacity to develop knowledge to solve their problems. The assumption was that farmers had considerable indigenous knowledge as well as acquired knowledge through experience and their ability to use and improve this knowledge could be strengthened through research and extension. This model, though had received expectations from all stakeholders, failed to create impact in technology management. Agricultural knowledge information system (AKIS) emphasized the significance of multiple information sources and set of systems that support the information sources and interaction among all the stakeholders. This model highlighted the need for strengthening the capacity of different systems (Research, extension and education) and the linkage mechanism among these systems. The concept of innovation was broadened further to include the outcomes of interaction among the diverse actors required to address a particular problem. In this scenario, the role of extension was identified as facilitating the process of reflective action, learning and decision making by the stakeholders. More recently, the agricultural innovation system concept has been applied to agriculture. Its significance is that it recognizes that innovation is not a research-driven process that simply relies on technology transfer. Rather, innovation is seen as a process of generating and accessing knowledge and putting it in to use. The main focus is on strengthening the capacity of different actors in agricultural development to create, diffuse and use knowledge or in other words, strengthening attitudes and skills to enable innovation. By this evolution, the role of extension would be setting agenda, organizing producers, strengthening capacity, coalition building of stakeholders, promoting platform for information sharing, experimenting, learning from experiences etc.

Extension innovations - retrospect (Rivera and Sulaiman, 2009)

- Public sector extension in both developed and developing countries is undergoing major reforms.
- The presence of more actors in extending services has ensured different kinds of support. However, much of the extension provision still revolves around dissemination of technical messages and problem-solving advice at the farm level.
- While pluralism in extension is considered desirable, privatization of public sector extension as a reform measure tends to lead to one-sector system. Small and marginal farmers can benefit from these arrangements, only when they are organized in to groups and supported by the state to access quality advice.
- Extension in the public sector needs to strengthen the capacity of small farmers to access, adapt and use knowledge. While the innovation systems framework emphasizes the importance of better interaction and knowledge flows for innovation, extension planning and implementation continue to be based on the research-extension-farmer linkage paradigm.

- Linking farmers to market is most important and the extension agencies need to expertise themselves to facilitate the linking farmers with market.
- Ideally extension within agricultural innovation system should act as a bridging organization, linking together the different aspects of knowledge, expertise and skills available in different organizations, so that the capacity to access, adapt and apply the knowledge can be enhanced.
- The extension reform initiatives have not yet fully addressed the issue of either broadening the mandate or building the capacity of extension personnel to perform widened roles. A related issue is the need to strengthen the technical, administrative and financial ability of decentralized units at different levels to develop, implement and evaluate programmes suitable for local contexts.
- There is no formula for reforming public sector agricultural extension systems, but the approach may be a 'best fit' one which insists on individual country analysis before attempting extension changes or development. However, the reforms and developments pertaining to extension is important for strengthening the capacity of agricultural innovation systems to deal with the rapidly evolving environment.
- The emerging extension institutions are expected to respond to the issues such as health, population, sustainable agriculture and the environment and not just the productivity and profitability concerns of linking farmers to markets and policy makers may find themselves called upon to consider, in addition to extension's commitment to agricultural advancements, its role in the development of rural economies, social equity and the protection of environment.
- Playing such wider roles requires large-scale restructuring and institutional innovations.

Emerging and evolving institutional innovations in agri-extension

Providing broad-based services to the farming community and playing a central in agricultural innovation systems in order to integrate the various components effectively are the major responsibilities expected by the contemporary agricultural extension service providers. There are many such institutional innovations are emerging with a way to provide either broad-based services to the majority small and marginal farmers of the country or playing the integration role among the major components of the agricultural innovation systems or both.

Producer cooperatives, one of the popular institutional arrangements, strive to cater to the location-specific demands of the stakeholders especially the farmers, in terms of arranging agricultural inputs including credit, market facilities, etc and thus provide scale of economies in addition to multiply the voice of the small and marginal producers. Amul model of Gujarat in milk has shown the entire world not only how an organized cooperative system can bring vibrancy to the livelihood pattern of the milk producers but also how the producers can diversify their activities through established business models. However, there are by and large inadequacies in the cooperative system with respect to efficiency, strategies in profit-making ways and eliminating the political intervention. Hence, there was an attempt in 2002 through an amendment in Companies Act, 1956, by which primary producers are allowed to establish a 'producers company' by their own with a minimum start-up capital. Here, the provision is such that only primary producers can be the members and the non-producers cannot become a member in the company and in such a way that there was an attempt to eliminate the political intervention. The producer company provisions can encourage the company to act with the goodness of the cooperatives and the vibrancy of the private limited companies. After the amendment in 2002, it has been reported that there are about 150 producer companies are established all over the country in agriculture and allied activities and activities pertaining to rural development.

The private sector started realizing the scope and opportunities in taking the services to the rural and agricultural stakeholders. In the agricultural arena there are many models pertaining to contract, corporate and cooperative farming such as the like of APNLBP for public-private partnership (Andhra Pradesh Netherlands Biotechnology Programme) (presently Agri-Biotech Foundation), PEPSI Co. for contract farming in tomato, chillies, potato, Marico in Safflower and Reliance Industries, TATA, AVT Private Limited for corporate farming in tea/ cashew etc. The private sector also understood the potential of Information and Communication Technologies (ICTs) in providing the knowledge access and market intelligence to the farming community and thus established their business around such services to their rural customers. The best model for such institutional arrangement has been the ITC e-Choupal model that has been under operation in many states of the country. Apart from this, Hariyali Kisan Bazars, Tata Kisan Sagar, Mahindra Krishi Vihars, i-kisan of Nagarjuna fertilizers are the successful models in this category. The change in the consumption pattern of the economically improved citizens of the country, especially the increasing urban population provides opportunities again for the private sector to target them with fresh, packaged, value-added and ready to eat/ serve commodities. Such private players (More, Reliance Fresh, Spencer's, Wal-Mart etc.) when procure the primary produce from the farmers, either intentionally or indirectly tend to facilitate many services to them and thereby support the livelihood.

Agri-clinics and Agribusiness Centres Scheme (ACABC) has been under operation by Directorate of Extension, Ministry of Agriculture to provide employment as well as entrepreneurship opportunities to unemployed agricultural graduates by training them in agri-business avenues, so that such entrepreneurs can extend broad-based services to the farming community. So far, there are about 10000 agripreneurs successfully established agribusiness ventures and serve the farming community in more than 30 agribusiness arenas. Similarly, there is an another programme by Directorate of Extension, Ministry of Agriculture through MANAGE, Hyderabad is to train the input dealers in agri-extension services and thus this programme on 'Diploma in Agricultural Extension Services for Input Dealers (DAESI)' provide opportunity for the farmers and rural customers to avail the services from input dealers.

Besides, many NGOs like DHAN, PRADHAN, MYRADA, BAIF, SKDRDP etc place themselves in a central position and arrange to integrate many public as well as the private and voluntary service provides for the betterment of their target farmers. Some of the mainstream media organizations like E-TV, Deccan Development Society, *Adike Patrike,* The Hindu etc attempt to provide knowledge access to the farmers and also try to integrate the extension service providers. Krishi Vigyan Kendras (KVKs) being the main outreach arm of Indian Council of Agricultural Research (ICAR), have a critical role in spearheading the agricultural extension programmes and service delivery at district level and below form. As science-based extension system, KVKs are expected to provide capacity building, on-farm technology validation, demonstration and dissemination of agro technologies and other diagnostic services for sustainable agricultural and rural development. Growth of KVKs has been phenomenal from one KVK in 1974 to about 630 in 2012. The mandate of KVKs evolved from vocational training to on-farm testing, technology assessment and refinement and now KVKs are considered as knowledge and resource centres and thus poised to extend broad-based services.

Convergence of all the emerging institutional innovations at various levels with appropriate platform is needed to facilitate efficient services and avoid duplication efforts as well scaling-up of successful models. Agricultural Technology Management Agency (ATMA), an institutional innovation in the agri-extension of the frontline system itself, has been conceived formulated and implemented in all districts of the country. It aims at convergence of not only all the agricultural and rural development schemes but also the various services providers in a single platform at village, block, district and state level with adequate and appropriate participation of the target beneficiaries either directly or through their representatives.

At this juncture, studying, deliberating and analysing the various models of institutional innovations in terms of providing the needed services and integrating the related stakeholders for the benefit of the farmers-the target beneficiaries of all such innovations is very much important for arriving at strategies for expansion and sustenance of efforts, scaling-up innovation models and refinement of such models if any. As discussed earlier, no model will be a blanket recommendation for entire country. But to arrive at 'best-fit' and 'location-specific' models, it is necessary to discuss and deliberate such innovations in order to recommend implacable strategies.

Innovations in agri-extension-guiding principles for formulating a framework (Sulaiman, 2011)

The framework (Sulaiman *et al* 2011) for strengthening and reforming extension to be a strong component in the national agricultural innovation system should have the elements such as,

- A broad scope for service provision beyond technology transfer
- Extensive use of partnerships to fulfill an expanded mandate
- To have a learning-based approach
- Preparations for having negotiations with a wide range of stakeholders for developing workable and effective service arrangements
- Ready to represent and address the client's interests at the management level, so that the programme remains accountable to its clients

Needed Shift in extension roles (Sulaiman and Hall, 2004)

Aspects of Extension	Shift from	Shift to
Content of extension	Technology dissemination, improving farm productivity, forming farmers groups, providing services, market information	Supporting rural livelihoods, improving farm and non-farm income, building independent and farmer-operated organizations, enabling farmers to access services from other agencies, market development
Monitoring and evaluation	Input and output targets	Continuous learning
Planning and implementation strategy	Doing it alone	Through partnerships
Sources of innovation	Centrally generated information for wide implementation	Locally evolved with diverse approaches and multiple partners
Role of technical research	Technology development	Source of technical expertise and supporting adaptive research
Approaches	Uniform and fixed	Evolving and diverse
Capacity development of staff	Training	Learning by doing, facilitated experimentation
Capacity development of extension system	Personnel and infrastructure	Development of linkages and networks
Policy approach	Prescriptive/ blue prints	Facilitating evolution of locally relevant approaches
Introducing new working practices	Staff training	Changing organizational culture through action learning
Paradigm	Transfer of technology	Innovation system

Indicators for Monitoring Institutional Innovations (Sulaiman, 2011)

Output indicators	Outcome indicators
• Farmer groups/ producer associations, sustenance and maintenance of records • Formation of new markets, marketing and price realization • Training organized • New inputs and technologies distributed, purchased, used • Access to credit and repayment • Development of value-added products • Infrastructure developed and capacity utilized • Partnerships, new working arrangements, new areas of collaboration, quality of interactions • Reforms promoted, changes in guidelines related to funding and collaboration	• Increase in income, production, productivity, additional employment created • Sustenance of the arrangement, continuance, expansion and impact • Enhanced capacity for collaboration and continuance of good practices, new partnerships formed, other institutional changes generated • New funding generated • Ability to respond to new demands • Governance mechanisms; how different stakeholder views are expressed and quality of response

Suggested framework for institutional innovations in agri-extension (Sulaiman, 2011)

Phase/ activity of institutional innovation/ approach	Framework activities
Planning-phase	Conduct individual consultations, workshops, sample surveys
	Identify key partners
	Develop a shared vision for the programme
Institutional human development	Recruit experts with skills that are demanded
	Negotiate to get right kind of staff on any mode including deputation
	Identify and contract consultants
	Conduct training, exposure visits and case analyses
	Conduct an organizational and management review
Technical support	Identify best technologies, refine or adapt them to local conditions, do adaptive research
	Make available best and efficient inputs in time either directly or brokering arrangements with other suppliers
	Recruit qualified technical staff and train them so that they remain up to date

Contd...

Phase/ activity of institutional innovation/ approach	Framework activities
Credit and financial support	Understand the financial/ credit landscape
	Negotiate with financing agencies
	Guarantee transactions and set up revolving funds
	Organize producers for group lending
	Influence policies for mainstream credit operations
Organizational development	Form producer organizations
	Enhance skills through appropriate capacity building
Market development	Analyze the market chain
	Negotiate with different actors in the value-chain
	Create new markets, if needed
	Develop new products

Conclusion

Extension can and should expand its role, given its significance for the larger agricultural innovation system. The guiding principles of institutional innovations provide an opportunity for expanding the role of extension by raising questions on the nature of extension's tasks, recognizing the need for new expertise, facilitating a review of extension's current interactions and highlighting the importance of institutional innovations. These tasks are important for developing and sustaining a capacity for innovation, which should be main focus of investing in this kind of approach. Before designing the programme and operational strategy for investment, it would be better to undertake an institutional diagnosis to understand the range of organizations within the agricultural innovation systems, their expertise and activities and their pattern of interaction. The scope of specific extension investment and the priorities will vary in relation to the national, district and local situations.

References

1. Rasheed Sulaiman V. 2011. Extension-Plus: New Roles for Extension and Advisory Services. In: Agricultural Innovation Systems: An Investment Sourcebook.
2. William M Rivera and V Rasheed Sulaiman. 2009. Extension: object of reform and engine for innovation. Outlook on agriculture. 38 (3): 267-273.
3. Srinivasan N. 2007. Profitability in agriculture and inclusive growth. National Symposium on Farm Credit for Inclusive Growth. College of Agricultural Banking. Pune. P (18-22).
4. Economic Survey, 2011-12. Agriculture and Food. P (179-201). http://indiabudget.nic.in

11 Agro Eco System Analysis – Transfer of Technology Prospective

- Dr. P. Gidda Reddy

Development of Agriculture continues to be critical not only for economic growth bu also poverty reduction in India. But, it is suffering from sustainable production anc marketing problems. One such problem is the climate and it's variability in a giver locality/zone/regional. We depend on the climate/environment, the natural resources o land, water, sunlight and biological organisms for agricultural production. But in the process of agriculture development modern/chemical technologies were introduced which interacted with the environment, adversely and sometimes to such an extent that natura resources essential to agriculture are harmed or destroyed. The green revolution has beer highly successful in raising agricultural productivity where India become close to cerea grain self-sufficiency. The narrow emphasis on cereals in best endowed although crucia in productivity terms, has largely ignored both environmental and socio-economic heterogeneity. As a consequence, there has been mismatch of agricultural developmen creating wedge between endowed vs less endowed farmers and also Agriculture become vulnerable to biotic and abiotic stresses because of climate variability.

Though India is basically an agricultural economy its contribution to GDP is 13.6 percent only but, 56 percent of rural population is employed in this primary sector. The country needs 300 plus million tons of food grains by 2020 at 4% growth rate in the current 5 year plan period. Hence all out efforts are required to improve productivity and production without harming the environment to keep in check on climate change and global warmingon sustainable basis. The sustainable agriculture development requires a radically different approach, one that is holistic and also more sensitive to the complexities of agro-ecological and socio-economic conditions prevailing locality/zone/reign. The pay offs come from the breeding of specifically adapted varieties and the design of inputs and techniques specially tailored to the needs of specific agro ecosystems. The target is to generate more location specific technologies each fitting particular ecological, social and economic situations.

The major concerns/challenges, the Indian agricultural are:

- Shrinking resources base
- Climate change (biotic & abiotic stress)
- Growing population
- Progressive farm biased technology a markets
- Degradations of natural resources like land & water
- Production sustainability and equitability

To overcome, the challenges the agricultural development requires different approaches which can balance both the human and agricultural ecology. The new approaches should fit into achieve equilibrium between or among various ecosystems contributing to agriculture development.

Approaches

- Farming systems approach (FSA)
- Integrated Rural Development (IRD)
- Agro ecosystem Analysis (AESA)

Farming Systems Approach (FSA)

Farming systems approach is characterized by its focus on the small farm as the basic system for development and by the involvement of the farmers at all stages to derive more income by having combinations of enterprises (Agriculture+other enterprises) to improve their livelihoods.

- Focus on the small farm as the basic unit for on-farm research and development.
- Close involvement of the farmers at all stages in the on-farm research and development Process.

Integrated Rural Development (IRD)

The second is the integrated Rural Development (IRD) approach which is even more holistic in scope, focusing on enterprises along with agriculture which encompasses off-farm activities like tailoring, handicrafts, dyeing and provision of health, education, etc. In practice IRD projects are commonly seen as a means of improving coordination and better working relations between different government agencies.

- More holistic in scope
- Focusing on the projects that go beyond involving agriculture i.e. non-farm activities

Drawbacks

- Divergence of views of various disciplines and talents
- Disagreement on worthwhile practical action

Agro-ecosystem Analysis (AESA)

The third and latest approach is Agro ecosystem Analysis (AESA). This differs from FSA and IRD in two important respects. First, it can deal with all levels in the hierarchy of agro ecosystem, from field/farm, village and watershed, to the region and nation. Second, it provides a technique of analysis and packages of technology that focus not only on productivity, but also explicitly, on other indicators of performance i.e. Stability, sustainability and equitability. However, it may not be an alternative to FSA or IRD, but is offered as an approach that can be used within the framework of FSA or IRD and indeed in any multidisciplinary agricultural development program, at whatever level of intervention. The AESA is based on the disciplines of agricultural ecology and human ecology.

What is Agro-ecosystem Analysis?

"Agro-ecosystem Analysis is a methodology of analyzing the agro-ecosystem in system's perspective with a view to improve the overall performance".

Concept of Agro ecosystem?

Agriculture ecosystem is a spatially and functionally coherent unit of agricultural activity which includes the living & non-living components involved in that unit as well as their interactions. Agro ecosystems can be manipulated to improve production and to produce more sustainably with fewer negative environmental or social impacts and fewer external inputs.

Difference between Agro Ecosystem& Natural Ecology

Five ways of differences:

- Monoculture
- Crops generally planted in rows
- Simplification of biodiversity
- Plough which exposes soil to erosion
- Use of genetically modified &artificially selected crops.

Sustainable Agro Ecosystem

- Maintain their natural resources
- Rely on minimum artificial inputs from outside the farm system
- Manage pests & diseases through internal regulating mechanism
- Recover from the disturbances caused internally or externally.

Sustainable Agro Ecosystem Design

- Combining the different components of the farm system i.e. plants, animals, soil, water, climate and people so that they balance each other to have greatest possible synergetic effects.

- Reducing the use of off-farm, external and non-renewable inputs.
- Relying mainly on resources within the agro ecosystem by replacing external inputs.
- Improving the match between cropping patterns and the productive potential and environmental constraints of climate and soil health.

Merits of Agro-ecosystem Analysis

- Deals with all levels of hierarchies - from field/farm, village and watershed, toregion and nation.
- Provides techniques of analysis and packages of technology that focus not only on productivity, but also stability, sustainability and equitability and on the trade-off between them.
- Can also be used as an approach within theframework FSA or IRD at different levels.
- It is flexible and has wide focus.

Concepts of AESA

- System
- System hierarchy
- System properties

System

System is defined as an assemblage of elements contained within a boundary such that the elements within the boundary have strong functional relationships with each other, but limited, weak or non-existent relationships with the elements in other assemblages; the combined outcome of which is to produce a distinctive behavior of the assemblages such that it tends to responds to stimuli as a whole, even if the stimulus is only applied to one part".

System Hierarchy

In the process of agricultural development, the systems are modified for the purpose of food and fiber production. Agricultural ecology provides the bridge between two hierarchies, linking the natural ecology with the multiplicity of disciplines with in the gambit of agriculture (agrl.economics, rural sociologies, soil science, agronomy, horticulture etc.) Human ecology provides the bridge between the agriculture hierarchies and the hierarchy of social systems (family, kin, group, rural institutions, etc.). Each level in the hierarchy (natural ecology, multiple disciplines and social/human ecology) has to be analyzed in its own right and this is consequently an important feature of Agro ecosystem analysis.

The natural ecosystem undergoes into an agro ecosystem after a number of significant changes. The system is also simplified by the elimination of natural flora & fauna.

However, at the same time, the system is made more complex because of the intervention of human management and activities leading to agricultural development.

e.g. in the wet lands when rice is cultivated under puddled conditions, the diversity of natural flora & fauna is reduced or limited to crop, pest and weeds. The basic process of ecology, such as the competition between the rice crop, weed, pests and predation of the pests by natural enemies remain, but now the human intervention like cultivation, application of pesticides and harvesting activities create more complex system in the process of agricultural development. So, each of this hierarchy needs to be analyzed to make the system more productive, stable, sustainable and equitable.

System properties

Agro ecosystems are increasingly complex. The various eco-sociological processes that tie people, crops, weeds, animals, micro-organisms, soil, and water together into a functioning, on-going ecosystem are so intricate that they can never be fully described, nor can they be fully comprehended. Simplification is a practical necessity essential for effectively communicating the results of analysis to agricultural practitioners. The dilemma is how to simplify without losing the essence of key relationships in the agro ecosystem as a whole. One approach to simplification is system properties (also called agro ecosystem properties in this essay), which combine large numbers of agro ecosystem processes into single, highly-aggregated measures of performance that suggest how well an agro ecosystem is meeting human objectives (Gypmantasiri et al, 1980; Conway, 1985; Rerkasem & Rambo, 1988). The system complexity in terms of its dynamic consequence can be captured by the four (4) system properties which together, describe the essential behavior of agro ecosystems. These are productivity, stability, sustainability and equilibility.

Properties of Agro-ecosystem

Productivity is the Yield or net income per unit of resource.

Stability is the degree to which productivity remains constant in spite of normal scale fluctuations in environment.

Sustainability is ability of system to maintain productivity when subjected to stress and perturbations.

(a) Stress is a regular, sometimes continuous, relatively small and predictable disturbance such as salinity etc.

(b) Perturbation is an irregular, infrequent, relatively large and unpredictable disturbance such as rare drought or flood or a new pest.

Equitability is how evenly the products of an agro-ecosystem are distributed.

The system properties they derive from the system as a whole rather than from any one of its parts. The productivity of a wet crop agro ecosystem is not determined simply by the yield potential of the particular variety that is cultivated. The yield that actually occurs depends upon the irrigation, nutrition and threats the crop experiences at each successive stage of growth, and how farmers manage the crop. The crop productivity is therefore a consequence of the functioning of the total interaction between the agricultural-environmental-social systems.

With irrigation, ecosystem in place, the productivity increases because yields per hectare are higher, because more crops can be grown each year and possibly also because the improved water supply provides an opportunity to grow crops of higher value. If the irrigation system is reliable, stability also increases as farmers are liberated from the vagaries of rainfall. These gains are only sustainable, however, if the irrigated agriculture does not encounter serious problems such as salinization, a pest or disease that arrives on the scene and is prohibitively expensive to treat, or administrative problems in the irrigation system that cause its performance eventually to decline. If fields near the main canal receive a better water supply than fields at the end of secondary canals, there may be considerable variation in production from one farm to another farm. Equitability is less than it was without irrigation, when production was uniformly low.

It is important for anyone who contemplates a new form of agricultural technology such as irrigation/weedicide application to consider all significant consequences, positive and negative. He can then decide whether the technology is really attractive in balancing system properties. Moreover, being alerted to the negative consequences, the extensions can try to minimize the detrimental effects of negative consequences. In the process of agricultural development, the experience shows that it is inevitable some degree of tradeoff between the different system properties. The new technology whenever introduced may have the immediate effect of increased productivity but at the expense of either one or more properties, as productivity increases stability suffers on one hand and sustainability and equilibrate and the other and sometimes this kind of tradeoffs between all the properties. Thus the agricultural development typically involves a progression of change in the relative value of these system properties.

Table 11.1 Agricultural development as a function of agro ecosystem properties

Type of system	Productivity	Stability	Sustainability	Equitability
A. Shifting cultivation	Low	Low	High	High
B. Traditional cropping system	Medium	Medium	High	Medium
C. Improved	High	Low	Low	Low
D. Improved	High	High	Low	Medium
E. Ideal (best land)	High	Medium	High	High
F. Ideal (marginal land)	Medium	High	High	High

Source: Conway (1985)

There are some complications in assessing agro ecosystem properties. It explains how each of the system properties for assessing production which has a combination of factors. This is primarily because there are so many dimensions to production, but it is also because the properties can be very different at different levels of an agro ecosystem hierarchy and under different environmental and social conditions (large, medium and small farm holders).

Productivity Dimensions

The system properties like productivity should be simple and unequivocal, but, in fact it is highly multidimensional because agro ecosystems have a variety of products for a variety of uses. The significance of these different measures is that production of a single agro ecosystem may be relatively high for one measure and relatively low for another. A cotton crop is high for monetary value, but low in food values. A millet crop is relatively high in its production of some food values but generally low in monetary value. Comparison of the production of different agro ecosystems is therefore meaningful only when the unit of production is explicitly defined. Monetary value is the most universal measure of agro ecosystem production, but no single measure—not even monetary value is of universal significance.

It may be desirable to express the efficiency of production with respect to very specific inputs. *For e.g*: production per unit of water input may be a primary concern in irrigation systems; production per unit of nutrient input may be the major concern with regard to fertilizer costs or where nutrient leaching loss is a problem; production per unit of weedicide application may be the concern if loss is due to weed menace. Electrical energy inputs may be more important than other energy inputs if electricity generation is expensive and time consuming.

Stability and Sustainability Dimensions

Stability a major/focal issue in productivity that result from a from variety of fluctuations in an agro ecosystem's physical and social environment: variations in rainfall, periodic pest attacks, price fluctuations, etc. Stability is assessed in terms of the fluctuation of production on a long-term average or the fluctuation of production about a long-term trend. Stability derives from productivity; stability is multidimensional in exactly the same respects. A given agro ecosystem can be relatively stable with regard to some measures of productivity and low with regard to others. For e.g. millet production for subsistence can be considered stable as long as yields are consistent for consumption but, the same crop may be considered unstable if grown for a market economy with low prices as compared to commercial crops.

Sustainability concerns whether a given level of productivity can be maintained over time. The multiple dimensions of sustainability derive in large part from the fact that it may be necessary to improve soil health with successive crops to maintain yields at the same level. For eg; if increasing fertilizer inputs are required to sustain production per hectare at a given level, the production per hectare may be sustainable even though production per unit cost is not. If weed problems require increasing labor inputs, production per hectare may be sustained while production per unit of labor is not and hence in cost terms it is not sustainable.

Equitability Dimensions

Equitability is most commonly measured in terms of the fairness of distribution of agricultural products or income. A low coefficient of variation for the distribution among households indicates a high degree of equitability. Equitability may be assessed with respect to the distribution of agricultural products or with respect to access to inputs such as land, capital or technology. Equitability of production and equitability in access to inputs are often closely linked, but not always. *For e.g*: vegetable farmers (small holders) may have equal land holdings (i.e. high equitability for land inputs) but very different incomes (i.e. low equitability for production), because some may have assess for high-paying urban markets while others must accept local market prices for a distress sale during the harvest season (production glut). Moreover, different measures of productivity can lead to different measures of equitability, in part because different kinds of agricultural products may be shared differently. *For e.g*: everyone in a community may have equal access to indigenous technical knowledge and have similar quantities of land, their opportunities for subsistence food production may be correspondingly similar, and equitability for food production is high. However, if they vary in their access to credit, modern technology, or commercial markets, equitability in income terms may be low.

Relations between Agro Ecosystem Properties

The production system for a given area which includes cropping pattern, farming system and other aspects of agricultural development may be particularly interested in the trades-off between different agro ecosystem properties. Improvements in one system property (e.g. productivity or stability) should not be at the expense of other properties so as to balance the agro ecosystem. But in reality at the field level, the agro ecosystems properties function in such a way that 'If productivity increases, sustainability declines', but it is not easy to discern a pattern. Some highly productive agricultural technology systems are quite stable while others are not, and some low-productivity agricultural technology systems are stable while others are not. For eg: in high nutrient intensive cultivation, the yields are reliable in some areas but are not reliable in other areas because of pests and disease problems. Consistent relationships between productivity and the other system properties are equally elusive, but exploring those relationships can nonetheless provide some insights into agro ecosystem design and reduce the inconsistent relationship between the properties of agro ecosystem.

Productivity, Stability and Sustainability

There are numerous ways that high levels of productivity can have a positive impact on stability and sustainability. For eg: higher productivity may be attained by increasing the yield in bad years (i.e. Life saving irrigation to reduce the impact of drought, or pesticides to reduce the impact of pest attacks), thereby making harvests more even from year to year, increasing the stability. Higher productivity can be associated with higher sustainability when a more productive crop like groundnut (legume crop) contributes for the maintenance of soil health (fixes o_2). Higher productivity can also be associated with higher stability or sustainability if it leads to higher income that give a farmer the capacity to deal with periodic problems that threaten production. In general, any attributes that increase 'fallbacks' and other adaptive mechanisms in an agro ecosystem can increase both its stability and/or sustainability (Jodha & Mascarenhas, 1983).

There could be times that productivity can be negatively associated with stability or sustainability. For eg:, higher productivity can be associated with lower stability if the higher production is achieved through high-yielding varieties that are more vulnerable than local varieties to fluctuating environmental threats such as droughts and pest attacks. Higher productivity can be associated with lower sustainability if production is at the expense of soil exploitation (e.g. by reduction in soil organic matter or exploiting/excessive use of soil nutrients), if the production is due to chemical inputs leading to major alterations in the ecosystem that eventually result in low productivity (e.g. excessive nutrients leading to soil salinity or pesticides leading to the loss of natural enemies and the emergence of secondary pests), or if higher production is a consequence of a high labor cost that place a strain on the social and economic conditions of ecosystem participating in the agricultural production.

In pest control, the use of chemical pesticides can increase stability, by eliminating even the smallest pest losses. In the process, the local crop varieties having pest resistant may be ignored/ discarded, and the pesticides may eradicate the beneficial insects along with the pests. If an insect resistant to insecticides should suddenly appear, the crop damage may be more serious than it would have been without pesticides, because beneficial insects are no longer present to keep the pest population in check. Even if the development of pesticide resistance is gradual, it eventually maybe necessary to increase the frequency of pesticide application to a point where the crop must be discontinued due to excessive pesticide costs making the crop productivity on sustainable. Therefore, it is very important to analyze the relationships between the properties in a given eco system.

Productivity and equitability: In a given eco system, the productivity and equitability are related as higher productivity can contribute to greater equitability. Considering the example of irrigation, an improved water supply increases productivity because it increases yields or provides an opportunity to grow commercial/high-value crops. With the increase in productivity there is greater stability for all if the water is well distributed as per crop demands and greater stability can lead to greater equitability because of good crop harvest and income as compared to rainfed agriculture where losses were more severe due to drought but with irrigation, their yields can be more equal.

However, greater productivity and stability can also lead to lower equitability. If the overall supply of irrigation water is not sufficient to provide reliable irrigation to all farmers in the area, only some farmers close to source of irrigation may receive the benefit. This will hamper the productivity of the area but will also increase the spread of incomes leading to low equitability.

In view of the above, if the relationship or analyzed for combination of possibilities, it is to conclude that both positive and negative relationships or possible between each of the properties. As properties of agro ecosystem fiction, the system properties are endpoints of complex ecosystem processes that can lead to both positive and negative relationships. The positive or negative dominance depends upon how the agro ecosystem is organized and the circumstances under which it is functioning.

Assumptions in AESA

- It is not necessary to know everything about an agro-ecosystem in order to produce a realistic and useful analysis.
- Understanding the behavior and important properties of an agro-ecosystem requires knowledge of only a few key functional relationships.
- Producing significant improvements in the performance of an agro-ecosystem requires changes in only a few key management decisions.
- Identification and understanding of these key relationships and decisions requires only a limited number of appropriate key questions which are to be defined and answered.

Approach and analysis of AESA

The AESA was mainly based on agro-ecological situations recognizing various components like climate, topography, Vegetation as major influencing factors. The emphases was on analysis of agro-ecological conditions and develop balanced and suitable extension programs directed squarely to solve the major problems limiting the agricultural growth in the locality/region. Through this process, technologies are being evolved/refined for each of the major commodities Viz: agricultural crops, horticulture crops, live stocks etc. package of practices for respective commodities are specific to the location/region to take care in a better way compared to conventional approach which lacks sustainability and equilibility.

In AESA, the crop/commodity situation in which it is grown is taken as the basis for resynthesizing the blanket recommendations rather than an area as a whole. The technologies are delineated with the agronomic factors like sowing time, previous crop, source of irrigation (canal, tank and well) soil borne problems, etc. It has also been observed that in many cases the productivity of improved technologies has not been found to be significantly higher across various locations. This kind of experience has often created an impression that the improved technologies does not appear to be promising across various farmers field conditions as they lack stability and sustainability. This calls for change in extension strategy to be built using AESA approach which takes care of various micro level agronomical and social factors to refine the technologies for a stable, sustainable and equitable production practices.

Pattern Analysis

The basic ingredient of AESA is the pattern analysis. Pattern analysis is the most important aspect of an agro-ecosystem analysis and is to be done using various PRA tools. Pattern analysis is done to reveal the key functional relationships of all or some of the system properties. Analysis of patters also helps in understanding the system properties. Pattern analysis is to be done with reference to space, time, flow and decisions in an agro-ecosystem. It is to be noted that the patters of space, time and flow are discipline neutral while pattern analysis of decision is discipline specific.

Space

Analysis of agro-ecosystem from spatial dimension is essential part of AESA and the same can be done with the help of maps and transects. While the maps and transacts helps in identification of physical location of various components of agro-ecosystem while the tools of overlays unravel potentially important functional relationships among various components.

Time

The pattern of time is an important component of AESA that would help in unraveling the behavior patterns, processes and relationship of components in an agro-ecosystem with that of system properties over a specific period of time. Various tools of pattern analysis of time include timeline, trends and changes, seasonality etc. which could be done in the form of simple graphs, charts and diagrams (pictorial and non-pictorial) etc. The seasonal changes represented in the form of crop calendars and cropping sequences overlaid with labour, credit and price peaks would be useful in identifying the availability of resources and timing of operations to optimize the performance system properties i.e productivity, stability, sustainability and equitability, while the timelines and trends and changes could be used to reveal the time lags that cause instability to system properties.

Flow

Patterns of flows are to be mainly drawn on transformations of energy, materials, money, information etc. among various elements and components in the agro-ecosystem in the form of simple and conventional flow diagrams. The pattern of flows in agro-ecosystem reveals the cause and effects of various elements and process on system properties. The pattern of flow also unravels fathoms of stabilizing or destabilizing feedback loops. Construction of tables, matrices, bars or histograms and regression graphs also indicate important relationships among various components in an agro-ecosystem.

Decisions

Decision is the last in the series of pattern analysis. Decision patterns are all about construction of decision tree or matrix to indicate goals and objectives of decision maker vis-à-vis constraints and choices. To have a better comprehension on patter analysis, the summary of the same is presented in Table 11.2.

Summary Table of Pattern Analysis

Type of pattern	Sub Pattern	Purpose	Technique	Remarks
Space		1. Physical location of components in AESA 2. Identification of system boundaries. 3. Identification of problems associated with ESA which affect system properties 4. Identification of micro forming situations 5. Identification of functional relationships between various components	1. Maps 2. Transects 3. Transects 4. Transects 5. Overlays	
Time	Seasonal changes Long term changes Time line	To understand what kind of changes happening to AESA Identification of periods in a year where timing of operations and availability of resources are crucial for stability and productivity Trends in productivity, stability and sustainability To understand process of socio-economic and cultural situation of AESA	Simple graph Simple graphs and pictorial diagrams Time line framework	1. Crop colanders 2. Labor peaks 3. Credit peaks 4. Occurrence of pests and diseases 5. Fodder availability 6. Fire wood availability 7. Gender analysis 8. Lively hood or activity Fluctuation in prices, roduction, climate and demographic
Flow		1. To reveal the functional relationships of resources, components and inputs and outputs. 2. To understand the major cause and effect relationship of stabilizing and destabilizing factors	Flow diagram, Tables, matrices, bar or histograms	Flow of energy, material, money and information

Contd....

Type of pattern	Sub pattern	Purpose	Technique	Remarks
Decision		To understand the process of human management of AESA (house hold to community)		
	Choices made at a given AESA	To reveal the goals of decision maker Analysis of constraints and choices available in AESA	Matrix ranking Decision tree	
	Spheres of influence of decision maker	Identification of decision makers in AESA hierarchy Extent of decision makers participation in decision making	Pictorial diagrams	

Examples of Agro Ecosystems

- Conservation Agriculture
- Forest farming
- Mixed crop-animal farming
- Parma culture
- Silvi pastorial systems
- Agro forestry
- No-tillage/zero tillage agriculture
- System of rice intensification (SRI)
- Organic Agriculture

Agro Ecosystem Analysis Tools

- Maps & transects: soil topography, land use, problems and opportunities
- Seasonal calendars: for climate, crop sequence, livestock, non-farm activities, labors requirements, capital requirements, income, and monthly prices.
- Long term graphs: showing prices, yield ,acreages, population trends etc
- Bar diagrams: showing sources of farm income, expenses on different types of production inputs.

- Flow diagrams: showing production and marketing chains, flow of income
- Decision trees: depicting choice points , key factors
- Venn diagrams: depicting overlapping institution affecting decision making, linking of institutions, and distances between/amongst institutions.

Conclusions

Agriculture being complex, diverse and risk prone in small farm production systems especially in lesser developed and developing countries, there is an increased need for participatory approaches like the one agro-ecosystem analysis discussed above. For planning people centric development projects there is increased need for AESA as it has the following ingredients:

- Extension/TOT based on AESA has better contextual basis and has greater chances of being acted up on
- The AESA analysis form the base line for any kind of research and developmental projects
- It can be also used as a tool for monitoring and evaluation purposes
- It provides better farmer to extensionist and scientist to extensionist interaction
- It is more flexible and accommodative
- It ensures equitability in the adoption and usage of technology
- It also ensures technology stability and sustainability in different locations/regions

Note: The concept and explanation described by Conway (1985) and Gerald G. Marten (1988) were kept in mind while writing this note on Agro ecosystem Analysis for local situations and the same is acknowledged.

12 Public – Private Partnership in Agricultural Extension Management – Potential, Experiences and Lessons

- *Dr. P. Chandrashekara*

Public sector extension is a State responsibility that has undergone several transformations since independence. Initially, the focus of extension was on human and community development. But there has been a steady progression toward technology transfer, within the policy framework of food security. The most significant recent development was the introduction of Training and Visit (T&V) extension management system, starting in the mid-seventies. T&V extension was well suited to the rapid dissemination of broad-based crop management practices for the high yielding wheat and rice varieties released since the mid-sixties. Given this focus on disseminating Green Revolution Technology for major cereal crops, extension activities have been largely carried out by State Department of Agriculture (DOA). Other line departments like Animal Husbandry (DAH), Horticulture (DOH) and Fisheries (DOF), have primarily focused on the provision of subsidized inputs and services to farmers, with little attention and few resources being allocated to extension.

By the early 1990's, with the completion of the third National Agricultural extension Project (NAEP), the important contributions that the T&V extension approach had made to agricultural development were duly recognized. But it was also realized that it needed to be overhauled in meeting the needs of farmers in the 21st century. It was recognized that extension should begin to broad-base its programmes by utilising a faring systems approach. For example, attention should be paid to the needs of farmers in rainfed areas and to diversify extension programmes into livestock, horticulture and other high value commodities that are capable of increasing farm income. A realization has also dawned that issues like financial sustainability, lack of farmer participation in programme planning and the weak links with research are serious constraints facing the current extension system.

Introduction of Extension Reforms is a major intervention in overhauling the extension system through process and institutional reforms mechanism. Extension Reforms focus on Reforming public sector extension, Promoting private sector, Augmenting media & information technology, Mainstreaming gender and Capacity building. Commitment to promote public-private partnership in agriculture extension management is demonstrated by reserving minimum 10 percent of the funds through public-private partnership.

System Constraints: Based on background studies carried out in the preparation of NATP project, more serious constraints identified are summarized as follows:

- Multiplicity of Technology Transfer System
- Narrow Focus of the Agricultural Extension System
- Lack of Farmer Focus and Feedback
- Inadequate Technical Capacity within the Extension
- Need for Intensifying Farmer Training
- Weak Research-Extension Linkages
- Poor communication capacity
- Inadequate operating resources and financial sustainability
- Some of the problems in public extension system identified by the practitioners in the field of Agricultural Extension are; Public extension services are widely viewed as supply driven rather than demand driven; Commercialization of agriculture gave rise to specialized client and demand for location specific extension services which are not catered by public extension system; Public extension deals with a large area, large population and diverse cropping pattern; Extension services provided are general in nature rather than specific and intensive; high cost, low impact of extension programmes; growing conflicts between farmer's interest and policy goals; poor motivation of staff and conflicting roles are observed in public extension; Insufficient face to face contact between extension worker and farmer; Inadequate funds for operational purpose; Majority of the extension services are curative in nature; Inadequate technical qualifications of Village level workers; Incomplete extension services; Inadequate internal organization structure; Inefficiency of extension personnel; Inappropriateness or irrelevance of extension content; Dilution of impact etc.

This public sector monopoly came under increasing threat in the 1980's as many started questioning the desirability of this situation on economic and efficiency ground. Increasing restraints on government finances and emergence of new extension arrangements offered by the private and voluntary sector have accelerated the process of limiting the role of government in extension.

Need for pluralism in Agriculture Extension: It is becoming increasingly evident that public extension by itself can no longer respond to the multifarious demands of farming systems. There is need for reappraisal of the capacity of existing agricultural extension systems to address, effectively, contemporary and future needs of the farming community. Public funding for sustaining the vast extension infrastructure is also under considerable strain. Meanwhile, in response to market demand the existing public extension network is inexorably being complemented, supplemented and in some instances replaced by private extension. As the nature and scope of agricultural extension undergoes fundamental changes, the outlook is for a whole new policy mix nurturing a plurality of institutions.

National Agriculture Policy on Public-Private Partnership in Agriculture Extension Management: The references made on public-private partnership in National Agriculture Policy are;

- The involvement of co-operatives and private sector will be encouraged for development of animal husbandry, poultry and dairy.
- Role of KVK's, farmer's organisations, co-operatives, corporate sectors and para – technicians in agricultural extension will be encouraged for organizing demand driven production systems. The government will endeavor to move towards a regime of financial sustainability of extension services through affecting in a phased manner, a mere realistic cost recovery of extension services and inputs, while simultaneously safeguarding the interests of the poor and the vulnerable groups.
- Development production and distribution of improved varieties of seeds and planting materials and strengthening and expansion of seed and plant certification system with private sector participation will receive a high priority.
- Protection to plant varieties through a sui generis legislation will be granted to encourage research and breeding of new varieties particularly in the private sector in the line with India's obligations under TRIPS agreement.
- A conducive climate will be created through a favourable price and trade regime to promote farmers own investments as also investments by industries producing inputs for agriculture will also be encouraged more particularly in areas like Agricultural Research, Human Resource Development, Post – Harvest Management and Marketing.
- Collaboration between the producer co-operatives and the corporate sector will be encouraged to promote agro-processing industry. The small farmers agro-business consortium will be energized to cater to the needs of farmer entrepreneurs and promote public and private investment in Agri-business.
- The private sector participation will be promoted through contract farming and land leasing arrangements to allow accelerated technology transfer, capital inflow and assured markets for crop production especially of oil seeds, cotton and horticultural crops.
- The government will provide a true support for the promotion of co-operative form of enterprise and ensure greater autonomy and operational freedom to them to improve their functioning.

Policy framework for Agriculture Extension on Public-Private Partnership in Agriculture Extension Management

- **Promotion of Community – based private extension services:** Group approach is the cornerstone of the restructured extension mechanism. A major component of extension services will be the mobilization of the community into farmers groups – FIGs, FOs and SHGs. Farmers' Organisations will be linked with Panchayats through existing statutory institutional arrangements such as the Land Management Committees, Development Committees etc. FOs will be supported directly through public funds and will be involved in the planning, implementation, monitoring and feedback of programmes. FOs at the village level would be federated at higher levels. Representatives of FOs would be members of decision making bodies such as ATMAs, Block level Farmer Advisory Committees, Watershed Associations. Ultimate aim is for FOs to internalize extension services for its members and provide backward (inputs, credit, technology) and forward linkages (post-harvest facilities, markets, value addition) in a vertically integrated arrangement.
- **Promotion of NGOs based private extension services:** Strength of NGOs is in their ability to mobilize communities into Farmers Organisations / Farmer Interest Groups / Watershed Associations / Market Associations. As such NGOs complement the public extension effort in several centrally sponsored programmes. Also extension services are contracted out and out-sourced to NGOs at the Block level in some states. In such cases the NGOs substitute for public extension. Public funds are used to support NGOs and are usually met from the provision of administrative expenses built into the Project Costs. NGOs are also supported directly by the central government in undertaking extension work. Significant number of KVK's are operated by NGOs. A systematic training, capacity building and technical backstopping mechanism, supported through public funds is to be developed for NGOs involved in providing extension services.
- **Promotion of para-professional based private extension:** Para-extension workers normally supplement public extension in a relatively cost-effective manner and overcome constraints of absentee public extension functionaries (Gopals for AI services, Mitra Kisan for agri-services such as soil testing etc.). Under the new policy agenda para-extension workers at grass-root level will be supported through publicly funded training and capacity building and payment of honorarium in the early years. The honorarium will be routed through the Farmer Organisations / Farmer Groups serviced by the para-extension workers to ensure accountability to the client group. Once the para-worker is able to demonstrate his/her usefulness to the client group the honorarium provided through public funds will be phased out and the client group would take on the onus of paying for the services of the para extension worker. The public extension machinery will also assist para-workers in procuring loans from credit institutions for equipment, mobility and linkages with

SMSs in line departments and SAUs. There will be an element of partial / full cost recovery of services provided by para-workers who must ultimately become economically viable units except in the case of vulnerable clients where the state may continue the targeted subsidy.

- **Competitive Agriculture extension Grant Fund:** Similar to the Competitive Agriculture Research Grant Fund set up in ICAR and several state governments, wherein both public & private sector research institutions compete for funds to address specific research problems, it is proposed to set up a Competitive Agriculture Extension Grant Fund. Resources under this fund could be accessed through a competitive bidding process. Contracting out extension services to private sector, community-based organisations or NGOs in selected geographical areas (e.g. A village, cluster of villages, block) would be done through a transparent, laid out procedure under this fund. This would also imply a strict monitoring and evaluation process.
- **Linkage of performance with funding for public sector:** In a manner similar to the private extension agencies who must compete with one another to access funds and whose subsequent eligibility to compete for funds will depend upon their performance as indicated by an independent impact evaluation, it is proposed that on a pilot basis Public extension agencies also be made to compete with private extension agencies for operational funds under Competitive Agriculture Extension Grant Fund (CAEGF).
- **Contracting out extension support services:** Wherever possible extension services in whole or in part could be contracted out for greater cost effectiveness. This applies, in addition, to administrative services such as security, mobility, computer and secretarial services, participatory planning to NGOs (being done in watershed management), staff training to a University / Institute, monitoring to a Farmer organisations / IIM / other institutions.
- **Training Institutes and SAUs to train private extension functionaries:** Facilities of public training institutions and SAUs would be available to NGOs and private extension agents.
- **Private Information Shops / Kiosks:** The ultimate aim is to promote private information shops / kiosks franchised out to private sector especially unemployed rural educated youth, in the manner of PCOs/STD shops. Private sector will be encouraged to establish information shops at block / mandal / village level. A major programme for development of software will need to be mounted so that information shops could have access to suitable material. Electronic connectivity and access to e-mail would put the franchisees in contact with district KVKs, line departments, markets and the resources of information. Such information could be dispensed to farmers, farmers groups on payment. Credit facilities for purchase of equipment for

setting up such information shops would be permissible under the micro-credit programme for agriculture and allied activities.

- **Capacity Building for use of IT:** Application of IT is constrained by lack of or inadequacy of complementary inputs (equipment, power, etc), appropriate organizational and institutional structures, information management and skills development. A major training programme for developing capacity for it usage will be promoted. Training Institutes will run suitable courses for the purpose.
- **Cost-cutting mechanisms for extension services:** Cost effectiveness may be improved by relying on fewer but better qualified (graduate or post-graduate) field advisers who interact directly with researchers for subject matter advice and then multiply their impact in the field by working with farmer groups rather than individual contact farmers. Cost cutting mechanisms, including the exploitation of mass media, encouragement of NGO and private sector involvement in extension, or needs-based coverage.
- **Privatization of agro-services:** An environment in which private investment in technology generation and transfer is more attractive will be created. Product diversification both horizontal and vertical shall be promoted to not only improve profitability sustainability and more efficient use of production resources but also to encourage greater involvement of the private sector. Where opportunities exist to contract out publicly-funded services, or to transfer costs to the corporate sector or to users themselves, these opportunities should be exploited-for instance for diversification into higher-value or export crops, or to develop new commercial inputs or machines. Privatization of selected "private goods" and agro services wherever a competitive market exists, such as AI services, soil testing, fertilizer advice, farm improvement plans or breeding plans would be undertaken. Wherever feasible contract farming through the involvement of private sector would be taken up, particularly, in the area of high value / export oriented agriculture.
- **Towards a realistic cost recovery of agro-services:** Wherever farmers have capacity to pay for public services, which are in the nature of private goods, realistic cost of such services should be recovered. However, provision is made for targeted subsidies to protect the vulnerable class of users.
- **Co-financing of public extension:** Co-financing of public extension services by farmers and farmers' associations to reduce pressure on public finances and to improve the accountability and responsiveness of extension to farmers.

Scope for Public-private partnership in Agriculture Extension Management: Historically public extension played dominant role in extension delivery mechanism. However, commercialization of agriculture facilitated the emergence of private sector. Sustainability of partnership is ensured by equal partners. Hence, strengthening of private sector is pre-requisite for promotion of public-private partnership in agriculture extension management. Post Green revolution period witnessed the emergence of strong private sector which was confined to few sector like marketing of inputs like seeds, fertilizers, pesticides, machineries etc. At present, private sector is not only diversified its activities from marketing of inputs to extension advisory, value addition and agriculture trade. Horizontal expansion of private sector increases through partnership with public extension where vertical expansion of public extension increases through partnership with private sector.

Potential private extension service providers who could be the partners in public-private partnership in agriculture extension management are as follows:

1. Agripreneurs: Unemployed Agriculture Graduates/ Diploma holders
2. Farmers Organisations
3. Input Dealers
4. Agri-Business Companies
5. Non Governmental Organisations (NGOs)
6. Print and Electronic Media
7. Private Banks
8. Funding Agencies

1. **Agripreneurs - Unemployed Agriculture Graduates/Diploma Holders:** Unemployed Agriculture Graduates/Diploma Holders are available in plenty due to lack of adequate job opportunities in public sector. Central sector scheme of Agri-Clinics and Agri-Business Centres Scheme train them in Agri-Business, provide start up loans and facilitate in establishment of Agri-Clinics and Agri-Business Centres. This ensures strengthened professionalized extension. Recent past has witnessed emergence of large number of Agri-Clinics and Agri-Business Centres throughout the country i.e., 19,500 in 32 categories of Agri-ventures. Studies have proved the impact of such centres impact on income of farmers. There are opportunities exist to channelize central / state government programmes through agripreneurs, networking Agri-Business companies with agripreneurs and effectively utilize Agri-Clinics and Agri-Business Centres in implementing ATMA activities. Channelizing 10 percent of ATMA funds through private-public partnership is a positive steps in promoting agripreneurship development.

2. **Farmers Organisations:** Group led Extension efforts resulted in formation farmers organisations / commodity interest groups creating a favourable condition for partnership. Pro-active farmers organisations namely Maha Grapes, Maha Mango,, Sugar Co-operatives, United plantation association of South India (UPASI), Rythu Mitra Groups, Commodity Interest Groups promoted by ATMA;s prepared the necessary condition for partnership. It is relatively easy for Agri-

Business Companies to partner with homogenous commodity interest groups with facilitation of ATMA. Partnership with farmers organisations are expected to enhance farmer to farmer extension.

3. **Input Dealers:** Information consultancy pattern studies have repeatedly proved the importance of input dealers. Transforming input dealer into para-professional and partnering in extension delivery mechanism has great potential in the present scenario. Diploma in Agriculture Extension for Input Dealers (DAESI) promoted by MANAGE is a significant step in this direction. There is great potential for networking input dealers and channelizing public extension services. Transformation of DAESI programme into a Central Sector Scheme is a welcome development.

4. **Agri-Business Companies:** Traditionally Agri-Business Companies are confined to marketing of inputs. Recent time has seen the diversification of Agri-Business Company activities. Dhanuka experiences in Hoshangabad of Madhya Pradesh, ATMA initiatives in Sangrur in Punjab, Ratnagiri of Maharashtra, Chittoor of Andhra Pradesh are few evidences to prove the potential of Agri-Business companies in agricultural extension in partnership mode. Policy framework on Public-Private Partnership in Agriculture Extension Management is expected to yield huge results in Agriculture extension. Mandatory spending under Corporate Social Responsibility is expected to boost PPP.

5. **Non-Governmental Organizations (NGOs):** Effective Non-Governmental Organizations exercise strong influence in their service area. NGO's could be used as change agents in agriculture extension. NGO's are proved to be effective in the fields of social mobilization, gender mainstreaming, livelihood improvement etc., could be used in agricultural extension work. However, the challenge is in selection of right NGO's to partner in agriculture extension activities.

6. **Media:** Increase in the literacy rate and wider coverage by print and electronic media, there is great scope for use of mass media / ICT in agricultural extension. There is scope to increase quality and quantity of Agricultural information through mass media. As majority of print, electronic media and ICT initiatives are undertaken / controlled by private sector, there is scope to partner such players in agricultural extension.

7. **Private Banks:** It is mandatory for private sector banks to push credit under priority sector lending for agriculture. There is good scope for private sector banks to support private-public partnership in agricultural extension management through client specific products.

Public-Private Partnership in Agriculture Extension Management-experiences

I. Public Led Partnership – Experiences

- **Agri-Clinics and Agri-Business Centres Scheme - Public fund – private delivery:** Agriculture and allied graduates are trained under Agri-Clinics and Agri-Business Centres Scheme in Agri-Business, supported with start up loan and also provided with handholding support in establishing Agri-ventures under central sector scheme of Agri-Clinics and Agri-Business Centres. Under the scheme, more than 46,000 agriculture and allied graduates have been trained, 19,500 have established 32 categories of Agri-ventures in different parts of the country. Recent MANAGE study indicates that agripreneurs are earning decent monthly income and providing value added extension services resulting in increase of yield and income of farmers. Attempts are already made to channelize central sectors schemes like National Horticulture Mission, Rastriya Gramin Bandaran Yojan etc., Agriculture Insurance Corporation is channelizing its programmes partially through Agripreneurs. There are examples of state governments making use of Agripreneurs in implementation of their schemes. Agribusiness companies have come out with specific programmes to involve Agripreneurs in their activities. Separate provisions are made under Extension Reforms to popularize the scheme and implementation of their programmes through Agripreneurs.
- **Diploma in Agricultural extension Services for Input Dealers (DAESI) - Public-Private fund, private delivery through public facilitation:** MANAGE sponsored DAESI programme aims at converting input dealers into para professionals in Agriculture extension. One year Diploma programme through distance mode, on self financing pattern could able to cover 3,200 input dealers and good number of input dealers are in pipeline. There is greater scope for central / state governments, Agri-Business companies to partner in this programme for further promotion as DAESI has been covered under centrally sponsored scheme.
- **Public-Private Partnership in Agriculture Extension Management - Hoshangabad Model – Public-Private fund and delivery:** The pilot project implemented during 2001-03 in Hoshangabad district of Madhya Pradesh is first of its kind in which 18 activities were implemented jointly by Dhauka Group and Department of agriculture at district level through MoU. The lessons available from the project is useful for upscaling the public-private partnership in agriculture extension management in other parts of the country.
- **Public-Private Partnership – ATMA Experiences:** A large number of ATMAs have taken initiatives to develop partnership with the private sector like processing industry, farmers organizations, cooperatives, corporate bodies etc. in different areas. This partnership has facilitated dissemination of technologies, supply of quality inputs (seed, fertilizers, micro-nutrients, bio -fertilizers, pesticides and bio-

pesticides and other technological tools) and marketing of farmers produce. The examples of market interventions through public private partnership undertaken by ATMAs are summarized below. These examples of NATP pilot phase provide good PPP Models.

- **ATMA-Chittoor:** identified and enrolled three private entities, namely, Food World, Agri-Horticulture Society and WORD organization. ATMA charged a registration fee of Rs. 50/- from each of these firms/societies. ATMA has entered into an agreement with Poultry Association, Food World and Vennar Organic Fertilizers for maize buy-back with support price; direct buying of mango without middlemen; and quality component.

II. Private Led Partnership Models

- **SBI, Dabur tie-up to finance farmers**: State Bank of India has signed an agreement with Dabur India for financing farmers for production of medicinal and aromatic crops under contract farming arrangements. The Memorandum of Understanding constitutes a broad understanding between SBI and Dabur India, manufacturer of Ayurveda medicines, to finance farmers for cultivation of medicinal and aromatic crops in Uttaranchal.
- **McCain in Punjab:** McCain Foods, the world's largest producer of French fries and potato specialties, is currently busy guiding the farmers in Punjab on developing and growing high-quality potato seed varieties. McCain agronomists are coordinating with the farmers and laying stress on mouth-watering French fries varieties.
- **Andhra Bank in pact with NCMSL for loans against farm produce:** Andhra Bank entered into an MoU with the National Collateral Management Services Ltd (NCMSL), promoted by the National Commodity & Derivatives Exchange of India (NCDEX).
- **Food Bazaar in tie-up with NAFED:** Pantaloon's food and grocery retail outlet, Food Bazaar, has entered into a sourcing tie up with National Agricultural Cooperative Marketing Federation (NAFED) for the supply of onions to its all outlets in order to keep the price per/kg at Rs.16/- across its outlets in the country, although the prices have shot up across the country in the open market. Food Bazaar in all its 34 outlets has made special arrangements for customers and additional quantities were being brought in.
- **Tata Kisan Kendras of Tata Chemicals:** Having Service Units in 470 locations, managing contract farming practices in 40,000 acres covering Basmati rice, wheat, mustard, soybean and potato·Services provided for sowing, fertilizer application, plant protection, custom hiring and other Agri value chain.
- **Chambal Fertilizers –Uttam Bandhan :** Serve entire value chain through FIGs and FOs including Soil testing, Advisory, Input supply and Marketing.

- **Poultry Association**: ATMA, Chittoor has contacted the District Poultry association and came to an understanding on purchase of Maize from the growers. The poultry association given a written agreement that they will purchase the maize seed at Minimum support price i.e., *@Rs.4851/-* per Quintal. The Poultry association also agreed to supply 2 MTs of poultry Manure on free of cost to the maize growers. To boost up the Maize crop, ATMA, Chittoor has supplied Maize seed on free of cost in an extent of 400 acres in Kharif, 2002 and for 1000 acres in Rabi, 2002-03, as an incentive technical support is given by the ATMA and BTT Officers for cultivation of Maize.
- **ITC – E-Chaupal:** Operates on the principle of information disintermediation. Based on the fact of difference between Farm Gate Price and Retail / Consumer Price. Creating a win-win situation for farmers and company by reducing 30-40 per cent transaction cost on procurement.
- **Contract Farming:** Cadbury (cocoa), Pepsi (potato, chillies, groundnut), Unilever (tomato, chicory, tea, and milk), ITC Limited (tobacco, wood trees and oilseeds), Cargill (seeds), domestic corporates like Ballarpur Industries Limited (BILT), JK Paper and Wimco (in eucalyptus and poplar trees), Green Agro Pack (GAP) Ltd., VST Natural Products, Global Green, Interrgarden India, Kempscity Agro Exports and Sterling Agro (all in gherkins), United Breweries (UB) (barley), Nijjer Agro (tomato), Tarai Foods (vegetables), A M Todd (mint in Punjab), McCain India (potato in Gujarat).
- **Need for policy framework for Public-private partnership in Agriculture Extension Management:** Public-private partnership needs favourable state policies. Strong private extension is pre-requisite for promotion of sustainable public-private partnership. The policy framework reassures a conducive conditions to strengthen the private extension. The present policy approach is in favour of public extension which needs to be oriented to promote public-private partnership. This policy framework aims at building the confidence among private partners, clearing the hurdles in the way of promoting partnership and to provide a clear picture of agreements to be executed between public and private partners. The framework assures level playing field for both the partners and channelize the symbiotic efforts for the benefit of farmers. Policy ensures the sustainability of partnership and also ensures cost effective, accountable, efficient and result oriented delivery mechanism in extension. The policy framework aims at efficient utilisation of existing resources of private and public sector to increase the efficiency of extension delivery mechanism to benefit the farmer. It also provides a roadmap for extension approaches to emerge in future.

❖ **Potential Private Partners in Public-Private Partnership:** Present scenario, potential private extension service providers are Agripreneurs, farmers organizations, input dealers are in the unorganized private sector whereas Agri-Business Companies, NGO's, print and electronic media, private banks and funding agencies are in the organized sector. Hence, it is important to initiate partnership with organized private sector at National, State and district level to start with followed by unorganized private sector below district level later to realize immediate benefits. Organized private sector may make use of unorganized private players in delivery of services at grassroots level resulting in unorganized players getting organized over a period of time. Each private sector has strength in their own commercial activity. To benefit the farmer holistically by assisting him in production, procuring and marketing, consortium of private players is desirable condition. Policy framework promotes various combinations of public-private partners.

❖ **Potential areas for Public-Private Partnership in Agriculture Extension Management:** Potential areas for Public-private partnership in agriculture extension management are broadly classified under;

1. Technology Dissemination
2. Sale of Inputs and Processing, Marketing of Agri products
3. Infrastructure support for production, processing and marketing
4. Different combinations of all the above three areas

However, nature of partnership solely depends upon need of the farmer and relative strength of public and private partners. It is desirable to have private partner or consortium of partners to work with public sector to benefit group of farmers in the entire process of production, processing and marketing on a project mode.

❖ **Technology Dissemination:** The following extension activities are categorized under this area;

1. Training of farmers and extension functionaries
2. Demonstrations
3. Farmer's study tours
4. Exhibitions / Kisan Ghostis
5. Field days
6. Production and dissemination of extension messages through print and electronic media
7. Award to successful farmers

8. Mobilization of farmers groups
9. Any other activity in technology dissemination

As majority of technology dissemination activities are handled by public sector, any partnership in this area is essentially public led partnership. These areas may not generate attractive revenue for private sector, but expected to boost their corporate image. Hence, to sustain partnership in this area, public sector has to consciously acknowledge the contribution of private sector in public forums.

❖ **Sale of Inputs, Processing and Marketing of Agri products**

Sale of Inputs such as seeds, fertilizers, pesticides, machineries, feeds, medicines etc.

Processing of Agriculture / Horticulture / Live Stock / Fishery products.

Marketing of Agriculture / Horticulture / Live Stock / Fishery products.

As majority input sale, processing and marketing activities are handled by private sector, any partnership in this area is essentially private led partnership. Partnership modalities should ensure agribusiness freedom of private partner with proper regulation. Regulation should ensure agribusiness freedom of private partner, prevent exploitation of farmers under public-private partnership banner to sustain the partnership.

❖ **Infrastructure support for Production, Processing and Marketing**

1. Soil, Advisory centres, Fertilizers, pesticides, water & seed testing facilities
2. Training centres
3. Demonstration farms
4. Nurseries
5. Seed production farms
6. Bio-control laboratories
7. Agro-processing units
8. Godowns
9. Cold storages
10. Veterinary hospitals
11. Artificial insemination centres
12. Custom hiring units
13. Feed mixing units
14. Seed processing units
15. Bio-fertilizer / Bio-pesticides production units
16. Agriculture Information Kiosks (FIAC etc)

17. Printing press
18. Any other infrastructure available with public sector.

Majority of the above infrastructure are established with public investment. Policy framework promotes maximum utilisation of such infrastructures with private investment and management. Private participation is expected to turn these units as profitable while delivering efficient services to farmers. To facilitate the same, existing policies on infrastructure management need to be revised to promote partnership. Partnership should ensure sharing of profit generated with private sector while protecting the interests of the farmers to sustain partnership.

- ❖ **Different combinations of all the above areas:** The nature of partnership is solely decided by need of the farmers and relative strength of public and private partners. Hence, in reality the partnership deals with one of the area / activity or combination of areas / activities. To focus the partnership efforts to yield maximum benefit to the farmers, it is necessary to design partnership models involving public and private partner / consortium of partners working with group of farmers directly growing one crop in all the stages ie. Production, processing and marketing on a project mode. To ensure this approach, public extension may provide necessary incentives to private sector wherever necessary. All the existing government schemes / programmes / incentives may be converged to ensure this support.
- ❖ **Selection of Partners:** Selection of partners may be done through transparent process, based on comparative strengths of partners through decentralized decision making process.

Steps to be followed in selection of the partner is as follows:

- ❖ Selection of the activity/programme under partnership mode
- ❖ Working out delivery mechanism under partnership mode by consulting all the stakeholders.
- ❖ Working out MoU which covers duration of implementation, roles of partners, implementation process and expected end results.
- ❖ Modification of existing finance/administrative guidelines to suit effective implementation of MoU
- ❖ Inviting expression of interest through advertisements/through communication to potential partners.
- ❖ Selection of partner/partners based on the comparative strength by group of experts
- ❖ Execution of MoU

- **Framework of MOU :** The MoU should contain
    1. The activity/programme selected by implementation under PPP mode
    2. Process of implementation
    3. Role clarity of partners
    4. Geographical area/target group covered under implementation
    5. Duration of implementation
    6. Monitoring and evaluation mechanism
    7. Joint Account
    8. Regulation
    9. Expected end results
- **Joint Account:** Joint Account is a confidence building measure under public-private partnership in agriculture extension management. Joint account is opened by Nodal Officers of private and public sectors jointly and operated together. All the funds committed by public and private sector should go to joint account immediately after signing MoU or thereafter as decided. The account is operated jointly by Nodal Officers of both the sectors. Revenue generated by activity under public private partnership should go to joint account. The profit if any should be shared among partners proportionality as agreed.
- **Certification:** Certification process aims at assessing the strength of private partner based on performance which enable the government to promote such Private Extension Service Providers. The certificates are issued to Private Extension Service Providers on yearly basis by an expert committee. The performance of the Private Extension Service Provider is assessed based on the monitoring and evaluation reports through participatory monitoring and evaluation process. The Private Extension Service Providers may be classified into five groups based on their performance.

    A Grade Private Extension Service Provider - Excellent

    B Grade Private Extension Service Provider - Very good

    C Grade Private Extension Service Provider - Good

    D Grade Private Extension Service Provider - Satisfactory

    E Grade Private Extension Service Provider - Poor

The MoU with the E-grade Private Extension Service Provider shall not be renewed and are not eligible to enter into partnership for next two years. D grade Private Extension Service Provider shall be given one year time to improve their performance. A,B, and C grade Private Extension Service Provider are allowed to continue. The government shall prefer better-graded Private Extension Service Provider for future partnership mode extension delivery systems. The expert committee should also assess the overall impact of partnership mode of extension delivery system in terms of cost saving to government, accountability and efficiency of extension services.

- **Regulation:** The activities of Private Extension Service Provider shall be under the close observation of expert group. Activities of Private Extension Service Provider such as non co-operation, indiscriminate promotion of commercial interest, suppression of Agriculture information meant for dissemination, sale of spurious inputs, cheating, use a government name for anti- farmer activities would be taken note of and a penalty equivalent to the damage would be levied on the Private Extension Service Provider as found guilty after enquiry. These activities of Private Extension Service Provider shall be taken into consideration while renewing MoU. On the other hand, good work of Private Extension Service Provider activities will be recognized through good grade during certification.

A caution deposit may be considered at the time of opening joint account by both the partners. In case of violation, the caution deposit should go to the partner who suffer loss due to violation of MoU by the other partner. The violation may be decided by expert group consisting of private and public partners situated at state and national level so as to handle all such grievances at different state / central programmes under public-private partnership.

- **Participatory Monitoring and Evaluation Mechanism:** Participatory Monitoring and Evaluation mechanism ensures that the ultimate beneficiary i.e. the farmer, monitor and evaluates the Public-Private Partnership mode extension delivery system. Here, the Private Extension Service Provider and department Nodal Officers jointly identify representatives of farmers for the purpose. Selected members are empowered through training to monitor and evaluate the programme with the help of designed tools and techniques. Such reports collected at the end of each season will reach expert committee which would be used in evaluating the performance of Private Extension Service Provider.

An expert group may be constituted at national and state level to handle the public-private partnership proposals on case by case. The expert group may consist of representatives of public, private sector and expert agency like MANAGE. All the MoU's at national level may be handled by expert group at national level, MoUs at state level may be cleared by expert group at state level. Decisions of the expert group may be abide by all the partners. The expert group may meet quarterly or as and when the need arises. A full time member secretary may co-ordinate the activities. At district level, ATMA may be empowered to take decisions regarding public-private partnerships with the approval of expert group at state level.

❖ Essentials in upscalling Public-Private Partnership in Agriculture Extension Management

Based on experience, essentials required for upscaling Public-Private Partnership in Agriculture Extension Management is as follows:

Proposed Model for up scaling PPP in Agricultural Extension Management

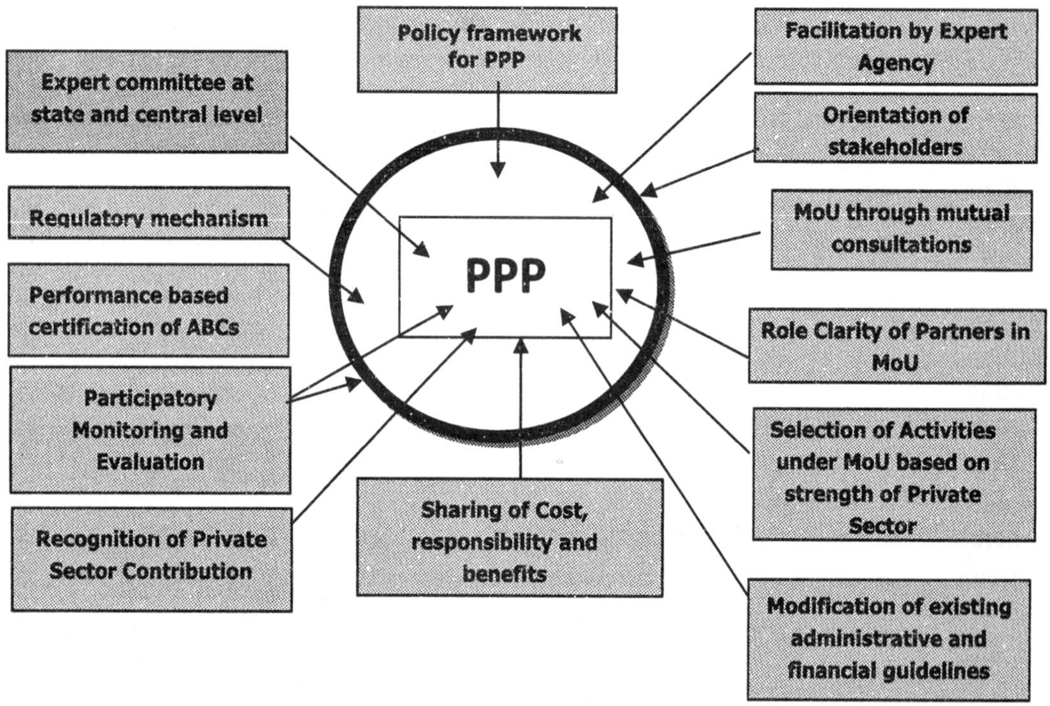

13 Market Led Extension

- Dr. M. N. Reddy and Prof. M. Surya Mani

Indian agriculture has made rapid progress in the last half century by augmenting the annual food grain production from 51 millions tones in the early fifties to 209 million tones in 1992-2000 and steered the country to a status of self sufficiency. It has been successful in keeping pace with the rising food demand of a growing population. Food grain production quadrupled in the last 50yrs while population tripled from 350 million to billion during this period. Significantly the extension system had played its role untiringly in transfer of production technologies from lab to land besides the agricultural scientists, farmers and marketing network.

Though the production has increased dramatically, not so much bothered about remunerative prices, small and marginal farmers generally prone to sell produce on "as is where is basis" due to several constraints like repayment of personal hand loans and to meet domestic needs.

With globalization of market, farmers have to transform themselves form mere producers-seller in domestic market to producers cum sellers in wider market sense to best realize the returns for his investments, risks and efforts. This to be achieved, farmers need to know answers to questions like what to produce, when to produce, how much to produce, when and where to sell, at what price and whom to sell his produce. Farmers received most of the production technologies from extension system. Extension system now needs to be oriented with knowledge and skills elated to market.

An efficient marketing system is essential for the development of the agricultural sector. The marketing system contributes greatly to commercialization of subsistence farmers. Effective linkages of productions system with marketing. Agro-processing and other value added activities would play an increasingly important role in the diversification of agriculture (MOA,GOI 2001)

The government provides much of the infrastructure required for efficient marketing. One of the most important is the information and extension services to farmers besides transport & communication facilities, public utility supply, like water, electricity, fiscal and trade administration and public storage, market and abattoir facilities (FAO, 2000).

In the changing scenario of Indian agriculture, with newly added face of marketing, the extension system is likely to undergo series of crises:

***Knowledge-skill input crisis*:** Besides the production technologies, the extensionists now, have to get equipped with market information which requires further training to the extensionists and additional funding.

***Efficacy crisis*:** Already, the extension system is under criticism. With the increased and enriched role, they have to perform multiple activities to prove their efficacy.

***Credibility crisis*:** even with all the market knowledge and efficacy in performing their role, the extension system may face the credibility crisis due to rapid and unexpected changes in the market.

***Reorganization structure crisis*:** with assumption of new roles, the organization structure may be prone to changes and the system has to adjust itself to this shock.

Challenges

* The gigantic size / mechanism of the public extension system in the country is heavily burdened with performance of multi-farious activities in the field. Extension system acts as liaison between the researcher and farmer. They are endowed with the responsibility of conveying research findings from the scientists to the farmers and feeding back the impressions of the farmers to the scientists. The new dimension of 'marketing' may overburden and become an agenda beyond their comprehension and capability. The public extension system is already under severe criticism for its inability to deliver the services. In the light of this, the challenge remains to motivate the extension personnel to learn the new knowledge and skills of marketing before assigning them marketing extension jobs to establish their credibility and facilitate significant profits for the farming community.
* Sporadic success stories of using information technology by farmers are publicized. There is an urgent need to strategically frame an information policy to make the farmers info-rich. Internet technology has percolated down up to taluq level and in some states up to village level. Search engines and the present websites furnish general information presently. Agricultural Market related information on the internet is inadequate. Hence, a whole network of skilled personnel need to be engaged in collection of current information and creation of relevant websites pertaining to / serving specific needs of farmers. Creation of websites should be mandatory in different languages to equip the farmers with information. These websites should contain information like market networks, likely price trends, current prices, demand status options for sale, storage facilities etc.

In short, Kipling's, seven servants may be employed to get answers to questions like what and how much to produce, when to produce, in what form to sell, at what price to sell, when to sell and where to sell. Information technology should be able to provide this kind of information to the farmers with 'press a button' on the computer on a continuous updated basis. Then and only then, the much talked about IT revolution would be beneficial to farmers.

* Generation of data on the market intelligence would be a huge task by itself. Departments of market already possess much of the data. Hence, establishment of linkages between agriculture line departments and Departments of Market strengthens the market-led extension.
* Extension cadre development poses a new challenge to the newly designed role. The present extension system suffers from several limitations of stationery, mobility, travel allowances, personnel development, etc. There is a dire need to upgrade these basic facilities and free the extension cadres from the shackles of the hygiene factors and enthuse them to look forward for the motivating factors like achievement, job satisfaction, recognition etc.
* Reorganization of the extension system like the position of Additional Director Extension may be redesignated as Additional Director Extension and Marketing and be made to look after the extension and marketing.

Enhanced roles of Agricultural Extension personnel in light of Market Led Extension:

- SWOT analysis of market. Strengths (Demand, high market ability, good rice etc), Weaknesses (inadequate staff, poor skills and information) Opportunities (Export, appropriate time of selling etc) and Threats (imports and perishability of the produce etc.) need to be analyzed about the markets. Accordingly, the farmers need to be made aware of this analysis for planning of their production and marketing.
- Organization of Farmer's Interest Groups (FIGs) on commodity basis and building their capabilities with regard to management of their own farm enterprise.
- Enhancing interactive and communication skills of the farmers to exchange their views with customers and other market forces for getting feedback.
- Establishing marketing and agro-processing linkages between farmer's groups, markets and private processors.
- Advice on product planning: Selection of crops to be grown and varieties suiting the land holding and marketability of produce will be stating point of agri-enterprise.
- Educating the farming community: To treat agriculture as an entrepreneurial activity and accordingly plan various phases of crop production and marketing.

- Direct marketing: farmers need to be informed about the benefits of direct marketing.
- Capacity building of FIGs in terms improved production, post harvest operations, storage and transport and marketing.
- Acquiring complete market intelligence regularly on various aspects of marketing.
- Regular usage of internet facility through computers to get updated on market besides internet.
- Organization of study tours of FIGs : To the successful farmers / FIG's for various operations with similar socio-economic and farming systems as the farmers learn more from each other.
- Production of video films of success stories of community specific farmers.
- Creating of websites of successful FIGs in the field of agribusiness management with all the information to help other Figa achieve success.

Required information to extension system and farmers

- Present agricultural scenario and land use pattern
- Suitability of land holding to various crops/enterprises
- Crops in demand in near future
- Market prices of crops in demand
- The extent if demand
- Credit facilities
- Desired qualities of the products by consumers
- Market network of the local area and the price differences in various markets
- Network of storage and warehouse facilities available
- Transport facilities
- Regular updating of market intelligence
- Production technologies like improved varieties, organic farming, usage of bio-fertilizers and bio-pesticides, IPM, INM, and right methods of harvesting etc.
- Post-harvest management like processing, grading, standardization of produce, value addition, packaging, storage, certification, etc. with reference to food grains, fruits and vegetables, eggs, poultry, fish, etc.

Flow Chart of Agriculture as an Enterprise

Rupee (credit/investment)

↓

What to produce

↓

Analysis of land holding for suitability of enterprises/crops

↓

Decision on how much to produce/ how much land holding to each enterprise/crop

↓

How to produce

↓

Post harvest technology

↓

Value addition

↓

Storage/transport

↓

When to sell

↓

Where to sell

↓

At what price to sell

↓

Selling options

↓

Rupee Investment

Paradigm shifts from Transfer of Technology to Market Led Extension

Aspects	TOT Extension	Market led extension
Purpose/objective	Transfer of technologies	Enabling farmers to get optimum returns out of enterprise
Expected results	Delivery of messages. Adoption of package of practices by most of the farmer	High returns
Farmers seen as	Progressive farmers High producer	Farmers as an entrepreneur "Agripreneuer"
Focus	Productivity / yields "Seed to seed"	Whole process as an enterprise. High returns "Money to Money".
Technology	Fixed' package recommended for an agro climatic zone covering very	Diverse baskets of practices suitable at local situations/ farming systems
Extensionist's interactions	Messages, Training, Motivating Recommendations	Joint analysis of the issues, varied choices for adoption, consultation
Linkages/Laison	Research-Extension farmers	Research – Extension – farmers extended by market linkages
Extensionist's role	Limited to delivery mode and feed back to research system	Enriched with market intelligence besides the TOT function. Establishment of marketing and agro-processing linkages between farmers groups, market and processors
Contact with farmers	Individual	FIG's/Focused groups/ SHG's
Maintenance of records	Not much importance as the focus was on production	Very important as agriculture viewed as enterprise to understand the cost benefit ratio and the profits generated.
Information Technology support	Emphasis on production technologies	Market intelligence including likely price trends, demand, position, current prices, market practices, communication network, etc besides production technologies.

Conclusions

Market – led extension system establishes its position by helping the farmers realize high returns for the produce and minimize the production costs and improve the product value and marketability.

Information technology, electronics and print media need to be harnessed to disseminate the production and market information.

Indian farmers have moved from subsistence to self – sufficiency due to advent of production technologies. In order to be successful in the liberalized market scenario they to shift their focus from 'Supply driven to market driven' and produce accordingly to the market needs and earn high returns.

References

1. Agricultural marketing in FAO: concepts policies and services, FAO, 2000 policy framework for agricultural extension, Extension Division, Dept. of Agriculture & Cooperation, Ministry of Agriculture Government of India, 2001 (Draft).

New Extension Methods and Skills

14 Farmer Field School

- Prof. S.V. Reddy & Prof. M. Surya Mani

The Farmer Field School (FFS) approach was developed by an Food and Agriculture Organisation project in South East Asia as a way for small-scale rice farmers to investigate, and learn, for themselves the skills required for, and benefits to be obtained from, adopting on practices in their paddy fields.

The first Field Schools were established in 1989 in Central Java during the pilot phase of the FAO-assisted National IPM Programme. This Programme was prompted by the devastating insecticide-induced outbreaks of brown plant hoppers (Nilaparvata lugens) that are estimated to have in 1986 destroyed 20,000 hectares of rice in Java alone. The Government of Indonesia's response was to launch an emergency training project aimed at providing 120,000 farmers with field training in IPM, focused mainly on recording on reducing the application of the pesticides that were destroying the natural insect predators of the brown plant hopper.

Since then, the approach has been replicated in a variety of settings beyond IPM. The FARM Programme (FAO/UNDP), for example, has sought to adapt the FFS approach to tackle problems related to integrated Soil Fertility Management in the Philippines, Vietnam and China. The IFAD/FAO programme in East Africa has adapted the approach for Integrated Production and pest Management (IPPM) and poultry production. The Livestock farmers field school programme by ILRI in Kenya has adapted the approach to dairy production etc.

After Asia the FFS approach has been extended to several countries in Africa and Latin American. At the same time there has been a shift from a focus on a single constraint of a single crop (IPM for rice based systems) to an emphasis on the multiple aspects of crop production and management, to cropping systems, to non crop/forest (livestock production etc) to natural resource management (Soil fertility, water conservation etc) to Socio-cultural dimensions of community life (food security &nutrition, savings, health, HIV/AIDS, literacy training, livelihoods etc).

New commodities were added and local adaptation and institutionalisation of these programms was encouraged. At present, IPM Farmer Field School programmes various levels of development, are being conducted in over 30 countries across the globe. African countries implementing the approach are among others Kenya, Uganda, Tanzania, Zimbabwe, Zambia, Malawi, Ethiopia, Ghana, Nigeria, Gambia, Egypt, Lesotho, Swaziland and Mozambique.

Concept

Farmer field schools (FFS) is described as a Platform and "School without walls" for improving decision making capacity of farming communities and stimulating local innovation for sustainable agriculture.

The FFS approach is an innovative participatory and interactive learning approach that emphasizes problem solving and discovery based learning. FFS aims to build farmers' capacity to analyze their production systems, identify problems, test possible solutions, and eventually encourage the participants to adopt the practices most suitable to their farming systems

FFS is a place where farmers undergo a field oriented, discovery based training that enable them to become field experts and be able to grow a healthy crop. FFS is recognized as an effective extension tool, which can be used for empowering the farming community, developing self-confidence, increase in social and human capital and promote better living. In FFS, there will be 25 farmers. Each FFS is facilitated by trained facilitators for the entire season. They are assisted by concerned local extension officers.

A Field School is a Group Extension Method based on adult education methods. It is a "school without walls" that teaches basic agro-ecology and management skills that make farmers experts in their own farms.

It is composed of groups of farmers who meet regularly during the course of the growing seasons to experiment as a group with new production options. Typically FS groups have 25-30 farmers. The FFS members are divided into four to five sub groups and each subgroup has a leader. Through managing the FFS group and subgroups, these appointed leaders as well as the rest of the members build up skills of group management and leadership. Furthermore, FFS encourages cohesiveness among members and develops team work. Although FFS is a time bound project activity, many FFS groups continue after the FFS learning cycle is completed for self motivated study of other subjects, development of collective marketing of agricultural produce and to establish cooperatives.

FFS does not rely mainly on information and techniques brought by extension agents and transferred to farmers. Instead, it aims to encourage farmers' systematic observation and informed decision making based on discovery based learning so that new knowledge and practices are generated by the farmers themselves. This process stimulates ownership of the learning process and ensures local adaptation. The main role of extension workers is to enhance farmers' skills in practicing new ideas, discovering their own solutions, and developing coping strategies to deal with ever changing situations. Technologies practiced use them because the FFS participants themselves set up learning sites and put technologies into practice. As a result, adoption rates are usually high among FFS members. Transfer of knowledge to neighbours is also common in FFS since learning results are based on farmers' experiences applicable to their neighbours.

FFS provides farmers with the opportunity to try out new practices on a group farm where risks are minimal, and potential losses would be shared by group members. Learning sites are usually very small in size; sufficient only to test and compare new technologies and farmers' own conventional farmers' practices. They need only to contribute a half day per week of their time to participate in FFS, while they can continue working on their own food crops using their conventional farming methods. FFS does not promote new methods in isolation from regular farmer practices; rather it provides an opportunity for the participants to test and compare alternatives in a relatively risk free environment with measurable figures for discussion and debate among participating farmers. FFS is therefore a less risky approach for subsistence farmers compared to most conventional extension methods.

FFS aims to increase the capacity of groups of farmers to test new technologies in their own fields, assess results and their relevance to their particular circumstances, and interact on a more demand driven basis with the researchers and extensionists looking to these for help where they are unable to solve a specific problem amongst themselves.

FFS Objectives

To empower farmers on decision making

The FFS approach empowers farmers in various aspects through confidence building and decision making exercises. Unlike in other extension approaches, farmers in the FFS approach are facilitated to take a lead in learning sessions under a participatory manner. Every FFS session allocates time for presentation of field observations followed by group discussion. In addition, participants in FFS are divided into subgroups and discussions among sub group members are encouraged. These exercises involving tangible field results usually provide a foundation for participants to "own" the learning process, build their confidence and personal skills, and thus become empowered in their farming activities and collaborating with other farmers in finding solutions.

To educate farmers with science based learning

FFS provides an analytical structure and season long regular interactions with the field, facilitations and other FFS members which enable participants to learn fist hand technologies in Participatory technology development and to understand the behaviour of introduced crops. The FFS experience can as well assist them to recognize misunderstandings and avoid errors in farming practices or beliefs.

To make farmers the experts and evaluators instead of possive acceptors of technology

The FFS approach recognizes community members as the experts within their particular contexts, and considers indigenous and local knowledge an important source of information to be used within the FFS learning process. Through the process, FFS members learn how to improve their own abilities to observe and analyse problems, and to develop practical and relevant solutions. The approach inspires members to learn continuously by exploring and educating themselves on issues and topics that affect their livelihoods.

To facilitate confidence building with field interaction and discovery based learning

The process of learning adheres to principles of adult education and "learning by doing". Adults tend not to learn and change behaviour by passive listening, but as a consequence of experience. Through learning by doing in a discovery based manner, group members cherish ownership over their knowledge and gain confidence in what they have learned.

To encourage experimentation with skill orientation

Knowledge is gained through practical experiments where different options are compared with each other. The trials are regularly observed and analyzed. Issues are discussed as they occur — in reality. This aspect of the approach dictates the "duration" of an FFS cycle that has to match the life cycle of the enterprise being studied.

Conservation and utilization of natural resources

The learning process within the field school programme is strongly linked to our perception of human nature. Farmers gain an understanding of how the ecology of their fields operatives and being able to manage the complex process occurring. Farmers conserve natural resources and promote natural defence mechanising and indigenous technologies in problem situations.

Core principles of the Farmer Field School Approach

Farmers as Experts: Farmers 'learn-by-doing' i.e. they carry out for themselves the various activities related to the particular farming/forestry practice they want to study and learn about. This could be related to annual crops, livestock, farming systems, mixed farming and orchards. The key thing is that farmers conduct their own field studies. Their training is based on comparison studies (of different treatments) and field studies that they, not the extension/research staff conduct.

What is learned is a process, not pieces of information. This process allows farmer to face new challenges and the ever changing dynamics of the enterprise. Education is seen as a process that takes place in the learner not in the facilitator/teacher. Therefore, at a field school, it is farmer who laid experiments gather data, discuss and analyse. They conduct research on their own imitative. A farmer who masters a process can also teach the process to other farmers also allowing them to discover for themselves. In so doing they become experts on the particular practice they are investigating.

The Field is the Learning Place: The field is the learning place usually on a host farm where an FFS is established and all learning sessions are held. Participants observe and learn from the field work instead of from textbooks and lectures from extension workers. Improved farm practices must be suitable for the local context, which is usually influenced by local ecological and socio economic conditions as well as farmers' preferences. All learning is based in the field. The cotton field, vegetable field, banana plantation, or farming system is where farmers learn. Working in small subgroups they collect data in the field, analyze the data, make action decisions based on their analyses of the data, and present their decisions to the other farmers in the field school for discussion, questioning and refinement.

Extension Workers as Facilitators not a Teachers: The role of the facilitator is crucial for successful learning and empowerment because FFS does not focus on teaching but on guiding FFS members through the learning process. To foster the learner centred process, the facilitator remains in the background, listening attentively and reflectively, asking questions and encouraging participants to explore more in the field and present their ideas. The facilitator must stimulate FFS members to think, observe, analyze and discover answer by themselves. The role of the extension worker is very much that of a facilitator rather than a conventional teacher. Once the farmers know what it is they have to do, and what it is that they can observe in he field, the extension worker takes a back seat role, only offering help and guidance when asked to do so.

Presentations during group meetings are the work of the farmers not the extension worker, with the members of each working group assuming responsibility for presenting their findings in turn to their fellow farmers. The extension worker may take part in the subsequent discussion sessions but as a contributor, rather than leaders, in arriving at an agreed consensus on what action needs to be taken at that time.

Scientists/Subject Matter Specialists Work With rather than Lecturing Farmers: The role of scientists and subject matter specialists is to provide backstopping support to the members of the FFS and in so doing to learn to work in a consultative capacity with farmers. Instead of lecturing farmers their role is that of colleagues and advisers who can be consulted for advice on solving specific problems, and who can serve as a source of new ideas and/or information on locally unknown technologies.

The Curriculum is integrated: The curriculum is integrated. Crop husbandry, animal husbandry, horticulture, land husbandry are considered together with ecology, economics, sociology and education to form a holistic approach. Problems confronted in the field are the integrating principle.

Training Follows the Seasonal Cycle: Training is related to the seasonal cycle of the practice being investigated. For annual crops this would extend from land preparation to harvesting. For fodder production would include the dry season to evaluate the quantity and quality at a time of year when livestock feeds are commonly in short supply. For tree production, and conservation measures such as hedgerows and grass strips, training would need to continue over several years for farmers to see for themselves the full range of costs and benefits.

Regular Group Meetings: Farmers meet at agreed regular intervals. For annual crops such meetings may be every 1 or 2 weeks during the cropping season. For other farm/forestry management practices the time between each meeting would depend on what specific activities need to be done, or be related to critical periods of the year when there are key issues to observe and discuss in the field.

Learning Materials are Learner Generated: Farmers generate their own learning materials, from drawings of what they observe, to the field trials themselves. These materials are always consistent with local conditions, are less expensive to develop, are controlled by the learners and can thus be discussed by the learners with others. Learners know the meaning of the materials because they have created the materials. Even illiterate farmers can prepare and fuse simple diagrams to illustrate the points they want to make.

Group Dynamics/Team Building: Training includes communication skills building, problem solving, leadership and discussion methods. Farmers require these skills. Successful activities at the community level require that farmers can apply effective leadership skills and have the ability to communicate their findings to others.

Farmer Field Schools are conducted for the purpose of creating a learning environment in which farmers can master and apply specific land management skills. The emphasis is on empowering farmers to implement their own decisions in their own fields.

Process in Conducting FFS

1. Situational analysis
2. Training of facilitators programme (ToF)
3. Establishment of FFS
4. Maintaining and Evaluation
5. Conduct of field days
6. Follow up

Situational Analysis

Identify focus area or enterprise and visit the site/enterprises with the help of secondary data. Involve all the stakeholders to suggest solutions to the identified problems. This should be included in the FFS curriculum.

Training of Trainers (ToF)

The major objective of the ToF is to develop a cadre of experts in the subject. For this purpose the participants are existing extension officers of the state department, staff from concerned fields and NGOs etc. All put together trained in a particular Research Station or department site for the entire season. During this training trainees are given experience to become master trainers. They in turn organise Farmer Field Schools in the subsequent seasons.

Establishment of Farmer Field School

Selection of right village for establishing FFS and the selection of right kind of farmers as trainees will greatly contribute to the success of the FFS. Prior to starting the FFS, at least four visits to the villages by the facilitators are considered essential. Suggested activities during such visits are given below as a guideline.

First visit: To Select FFS village

It is expected that the FFS facilitators will be establishing FFS in their own area of posting and hence should be familiar with the villages. Based on information that one may already have, select potential villages for visiting.

Some of the Criteria that may be used for selecting a village for establishing FFS are as follows.

The village with:

- Maximum pest problems
- High pesticide use and pesticide related problems
- Enough number of farmers on a particular community on which FFS is Irrigation facilities
- Should not be too close to a city or big town
- Politically sensitive villages should be avoided.
- Villages having Irrigation facilities should be preferred
- Fairly easy access.
- Where facilitator has comfortable rapport with farmer.

During the first visit to the potential village, the facilitators should:

- Meet the village Panchayat president /prominent farmer
- Explain the purpose of visit in, brief i.e. organizing a FFS in the village
- Request him to invite all the farmers (Specific crop) for general meeting ("gramsabha" meeting) the following week. Also ensure that the SHGs, farmer groups, panchayat members/groups are invited to attend this meeting and so are the women farmers. Ask for the convenient place, date and time for this general meeting.
- Request him to announce meeting to the villagers through appropriate media such as "Tom Tom".
- Collect a village profile. - All basic information such as the population of the village, number of families, number of farming families, gender, education /literacy level, castes, occupation, average land holding, etc. These will be obtainable from village penchant office. These information may also be obtained from Mandal office and

therefore, even before going to the village on the first visit, this information may be collected and taken along.

- *Other important information that should be collected are:*
  - ➢ The area under specific crop/ commodity
  - ➢ Number of farmers
  - ➢ Problems related to crop/ commodity.
  - ➢ Pesticide usage.
  - ➢ Information about irrigation facilities.
  - ➢ Other relevant information
- Request the Pinhead president/prominent farmer to kindly organize the "gramsaba" meeting the following week.
- The date, time and place of the meeting should be fixed with the village head. Also request him to announce the meeting to the villagers through appropriate media such as "Tom Tom".
- Women farmers should be asked to attend for the meeting.
- Village map may be collected from penchant office or from mandal office.

Second visit: To Identify Potential FFS Farmers

- This visit should be within a week's time from the first visit.
- Organize the general meeting as planned.
- Ensure participation of different groups (Shags, farmer groups and women farmers)
- At this meeting, discuss the important problems related to crop management in the area, description of FFS, FFS structure, procedures, nature of training etc. And the criteria for selection of the farmers.
- Discuss the important problems related to crop management in the area, and its structure, FFS process/training nature and FFS procedures. It should be emphasized that:
  - ➢ The farmers will benefit by increasing their profit through reduction in pesticide cost and increase in the yield, improved health of their family members etc.
  - ➢ There will be 20 sessions. (Approximately for cotton crop)
  - ➢ Duration of each session will be 4-5 hrs.
  - ➢ The farmers should be prepared to attend all the sessions and should be on time
  - ➢ No substitutions will be allowed
  - ➢ There will be slight refreshment in each session.
  - ➢ After completion of FFS the participating farmers will be provided with the certificates

- ➢ No honorarium will be paid to the farmers for attending the sessions
- ➢ It can also be said that the farmers selected for FFS can form in to a group of their own (Such as "IPM farmer group" or "commodity farmer club") and this may bring additional benefits in the long run
- ➢ Clarifications should be given to the farmers if asked
- ➢ The facilitator should announce that the farmers interested in joining the FFS to give their names to the village head/prominent farmer.

Third visit: To Select FFS Farmers

The Criteria for farmer selection

- As far as possible, the farmers should be typical of that area in terms of cotton cultivation including pesticide usage, land holding, ownership
- Must be a specific crop growers (what ever crop / commodity is selected for the conducting FFS)
- The farmer should be highly interested in learning IPM, willing to attend the FFS regularly, curious, motivated and cooperative
- Literate farmers are preferred but not a must.
- Should be energetic, physically fit, young (preferably less than 45 years)
- Women farmers should be given preference
- Willing to learn on their own.

The village panchayat leader/some of the prominent farmers will help the facilitator team to choose the right farmers.

After the third visit the facilitator team should visit the village 2-3 times and meet the identified farmers individually or in small groups and ascertain their suitability and willingness to join FFS

- Select 30 farmers so that at least 25 farmers will be available for FFS in case of drop outs if any
- Finalize the list of 30 farmers and intimate to the concerned farmers.
- Request the selected farmers to attend the next meeting in the following week (specify date, and time and place)

Fourth visit: FFS planning meeting

- Conduct this meeting at least two weeks before the commencement of FFS sessions.
- Conduct signs and symptoms exercise.
- Conduct meeting with all selected FFS farmers.
- At this meeting, hold a participatory discussion to identify local problems related to crop management and identify the local needs of the farmers.

- Explain FFS activities and procedures in detail.
- Discuss the farmer's practice, especially the spacing, fertilizer dose and application method, and the pesticide usage. Arrive at a consensus with the farmers what farmers practices should be followed in Farmers Practice Plot. Indicate that in IPM practice plot the crop management decisions will be taken by the FFS farmers as a group.
- Also explain the farmers what IPM practices are likely to be followed in the IPM plot
- With the consensus of the farmers, select suitable FFS field and training site (gathering place) as per the criteria.
- Finalize the inaugural day for FFS and develop plan for the inaugural session.

Criteria for FFS Field Selection

- Select minimum 2 acres (if possible three acres)
- The field should be belonging to one farmer. However, if one farmer with two or three acres of land is not available, then two farmers with contiguous land area may be selected.
- It should be easily accessible and should not be too far from the FFS village
- The field should be as close as possible to the meeting place (see the criteria for gathering place below)
- Some shady areas should be nearer to the field.
- No abnormalities in the field.
- Identify Farmer who is willing to give the land for conducting field experiments and CESA
- He should agree to carryout the farming operations in IPM plot as per CESA decisions taken by the farmers.
- He should agree to follow the schedule of farming operations as finalized by the farmers, in the farmers practice field.
- He should agree to meet all the expenditure involved for both the fields except the IPM inputs for pest management.
- He should agree to allow the farmers participants to work in that field or to organize field day.
- Training team will bear the cost of IPM input for IPM plot.
- The facilitators should enter into verbal contract with the farmer collaborator in the presence of all the farmers' participants.

Criteria for training site (gathering place) selection

- Close to the training field

- Accessible by road.
- Provision for protection from sun and rain.
- Sufficient arrangement to accommodate all the participants who will be working in-groups.
- Select the site in concurrence with farmer participants.

Learning Contract

At this planning meeting, through participatory discussion with all the selected farmers, discussions on the following should be made and the facilitators should make a verbal "learning contract" with the FFS farmers. All farmers should express their agreement.

- The FFS day of the week (avoid market day)
- The duration of each FFS session (4 - 5 hours)
- The time to start FFS session
- Norms, regulations/rules for FFS (regular attendance, on time arrival, stay throughout the session etc.) and expectations. Farmer's participation and inputs in the norms/rule setting is important.
- Group the participants into 4 groups using suitable group dynamics method and assign the names for the group
- Allow the participants o to select the leader for each group
- Finalize the responsibilities of group leaders in the learning contract

Baseline Survey

The Baseline Survey form is given (see annex). The survey should be conducted in FFS and Non-FFS villages. All 25 FFS farmers and 10 Non-FFS farmers from a neighbouring village should be interviewed. The survey should be completed immediately after the selection of FFS farmers and prior to conducting planning meeting.

FFS Records

Record keeping is an important activity. The facilitator team should maintain one comprehensive record for each FFS. The record should contain the village profile, farmer participants bio-data, session-wise activities carried out, attendance, raw data of all the observations made by different groups, list of materials purchased, amount utilized, balance amount available etc. Record format shown (see annex). The records should be stored properly. Important activities should be documented using photographs

Monitoring

The FFS will be monitored by three groups namely, ToF trained colleagues of yours, one or two staff of your department and thirdly by staff of the project. The first group will mainly assess the state of the progress and provide feedback to the project as well as to the concerned department of the FFS facilitators. In addition to monitoring, these facilitators will be running one FFS in their posting place. The second group will monitor for proper functioning of the FFS and give feedback to the project. They will make three visits evenly spread over the entire season. These monitoring officers will not interfere in the normal activities of the FFS. The third group (project staff) will make announced or flash visits to assess the progress and to provide technical guidance and support. It is to remember that the department is determined to ensure proper running of the FFS and maintaining of the training quality.

Evaluation

In order to assess the level of the IPM related knowledge before and after the FFS and to certain extent measure the impact of IPM training, pre- and post-FFS ballot box tests are conducted. The details of conducting ballot box test is given (see annex).

Mid-season review

During 8th -10th sessions of FFS, the mid session review will be conducted at a convenient place chosen by the project. All the FAX teams should visit and present the progress made so far.

Conduct field day

At the end of the season, the participants should organize a Field Day in which local community members and local policy makers are invited to the field site to be instructed by the participants. Typically, the facilitators remain in the background on this day as the participants run the programme. Graduation ceremonies in which certificates are awarded by local politicians can also be done at this time. The certificates should be prepared by the facilitator. Additional details are given (see annex).

Other General guidelines

- Farmers should be involved in planning for the FFS right from the beginning.
- All activities should be truly participatory.
- Welcome stretching: Each FFS begins with physical stretching to warm up and wake up mentally. Stretch the neck, back, shoulders and arms. (10 minutes including hello and welcomes)
- All efforts should be made to promote IPM through local media and inviting local leaders to FFS
- Must try to arrange visits by students and teachers from locals schools to the FFS
- The facilitator should maintain FFS register recording all the day's activity
- Only those persons that participated fully and achieved an acceptable level of progress during the FFS should be allowed to graduate, although other participants should be recognized for the contributions and participation.

Follow up activities

The facilitators in consultation with project office, nodal officers and the FFS farmers should plan for post FFS activities such as the follow up visit by the farmers and to encourage the farmers to start IPM clubs.

After 4 to 6 visits the farmer participants should be encouraged to take up the FFS activities, which will help them to organize and continue FFS activities. While preparing the curriculum this factor must be taken into consideration.

Reporting

The facilitators should keep the records properly. They should spend one day in a week (Friday) for preparation of reports in the prescribed format supplied by the project and send it to their headquarters and to the project. Final report of each FFS should be prepared and sent to FFS project office.

FFS FIELD GUIDE

- To help us carry out the activities smoothly within time
- To put each activity in perspective
- To ensure that farmers understand the objective of each activity
- To ensure that everyone knows their role
- For facilitator to prepare to handle any topic
- To ensure all necessary materials are available

15 Farmer Life School

- Prof. M. Surya Mani and Prof. S.V. Reddy

Farmers Life School idea originated from South-east Asian HIV program of United Nations Development Program (UNDP). The UNDP in collaboration with the Food and Agriculture Organization (FAO) agreed to develop a pilot program based on Farmer Field School approach and IPM strategies (Polo Yech, 2003). Thus, approaches for enhancing small scale livelihood in marginal rural areas should look beyond agriculture problems.

Farmer life schools are based on the learning approaches in the IPM farmer field school. In each weekly FLS a group of farmer meet in the field where a number of learning activities take place such as Human Eco-System Analysis (HESA); presentations; discussions; special topics and group dynamics. These activities assist farmers in recognizing and analyzing the inter-related elements of their lives, in much the same way as they apply their mastery of ecological concepts to their rice fields.

The discovery-learning process of FFS generates a deep understanding of ecological concepts and their practical application. This ecological approach to identifying problems and finding solutions has been applied to pests but can be easily applied to HIV/AIDS and the understanding of human behavior and development. An identification of the farmer's socio-economic ecosystem and its vulnerabilities can be adopted from the FFS.

Similarly, The Farmer Life Schools (FLS) an extension of FFS thinking has emerged after studying the power of Farmer Field Schools in empowering farmers on skills and confident to achieve higher productivity through quality education. Yet, the thinking also was further compounded with the success of capacity building and skill development programmes with SHGs and also Farmer Young Professionals. Furthermore, the Cambodia and African countries experiences in running Farmer Life Schools specially to educate farmers on different aspects of life.

In the FLS farmers discuss and identify problems, which threaten their livelihoods and analyze the factors and issues in relation to economy, education, health, social relations, environment, and culture factors that influence their daily life. A few real cases in their village are taken to study and analyze by participants. Issues addressed in FLS range from poverty, landlessness, domestic violence and the attendance of children at school to specific health problems concerned with different diseases such as dengue fever, malaria, STD, and HIV/AIDS. As a result, farmers can identify the root causes of the problem and make decisions about what actions they should take. It is important to note that the FLS is not focusing only on HIV/AIDs, but because this is such an important issue in their communities, HIV/AIDS is invariably one of the issues addressed in the FLS and is placed appropriately within the socio-economic context of their agricultural livelihoods.

The core processes within the farmer life school make linkages to ecology, group organization, and student centered learning and are applied through what is termed, “Human Eco-system Analysis" (HESA). The HESA involves groups of farmers investigating various threats and adverse factors in their lives, in the same way that farmers observe pests in their fields through Agro Eco-system Analysis (AESA). This process enables farmers to build their skills in problem identification and analysis, and then make appropriate decisions to avoid and remedy problems.

What is a Farmer Life School?

Farmer life school is aimed at building on the risk assessment knowledge that farmers already have through a holistic approaches combined with long – term outlook.

Proactive charge is implemented through emphasizing the importance of local resources and network.

Farmer Life Schools are based on a non-formal, experiential learning process, similar to the FFS. Often, the FLS is a natural follow-on activity from a previous IPM FFS. A FLS consists of a group of about 20-25 farmers who meet somewhere in the village, one morning a week for 18 weeks.

It is an open school that is created for and by farmers. The participants are villegers from the village. The facilitators can be trained graduates on FLS or Farmer Field School Methodologies. It is possible to bring experts like Agricultural Officers, Teachers, Health workers etc occasionally to answer questions unanswered by the trained facilitators.

From Farmer Field Schools to Farmer Life Schools

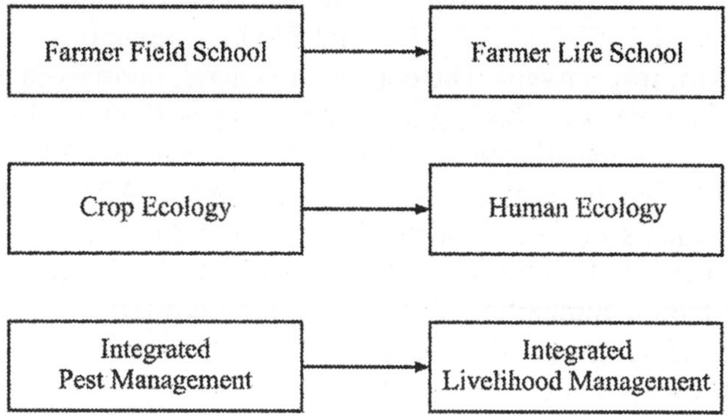

The curriculum design of the FLS follows the same general processes used in the ecology-based FFS, but with a different content. While the main learning context in an FFS is the field, the FLS is based in the community, addressing the complex range of issues related to farmers livelihoods. Each weekly meeting consists of a defined set of activities: Human Ecosystem Analysis (HESA), presentation of the HESA, a special topic and group dynamics

Aim

Empowerment of farmers to lead a quality, problem free and happy life with self help, mutual help and cooperation.

The Objectives

- To provide a free forum for discussion and problem solving
- To provide the tools for community empowerment by developing skills which can help to identify and analyze problems related to Decent Work (Freedom of association, child labour, forced labour, non-discrimination), health, environment and supply chain.
- To place prime importance on local knowledge and resources to initiate changes
- To make them aware of the developmental programmes in action and benefits as well as the importance of self help, mutual help and cooperation.
- To make the people understand the vulnerability that each individual and family member affects due to common farming problems and concerns as well as findings solutions.
- To build social capacity of the community through critical decision making and positive attitudes thus alleviating the stress of development and creating a happy and healthy society.

Farmer Trainer Orientation Course (FTOC)

Farmer trainers are the main facilitators in farmer life schools. To become farmer trainers of farmer life schools, farmers need to complete an IPM farmer field school first and then volunteer to be recruited to participate in the Farmer Trainer Orientation Course (FTOC). To run this training, the UNDP SEA Project and FAO Community IPM provide financial and technical support, respectively. The course, in general, consists of 19-25 participants and lasts at least three weeks. The age of trainees ranges from 18-50 years old. Some trainees have already had experience in running IPM training as IPM farmer trainers. The educational level of most trainees is less than high school. After dropping school they become full-time farmers in their villages. This training, therefore, provides the needed training skills to farmers to become good farmer trainers or facilitators who are capable of organizing FLS in their communities. The training provides farmers with a wide range of training skills including group organization, facilitation, introducing special topics, facilitating human ecosystem analysis processes, proposal writing, communication and management of FLS.

The Broad Areas of Learning

- Finding out real problems affecting people`s lives and livelihoods in the farm and villages through Micro Planning using various participatory tools.
- Imparting knowledge and skills for finding solutions on special topics (DW,Health, Environment etc., problems affecting lives and livelihoods.
- Identification, analysis and decisions making for solutions through participatory tools.
- How to improve the quality of ones own life by utilizing networks, value chain and resources available within and outside the community.

The Methodology

- Size of the class 25-30 (a LG)
- Participatory tools/ methods will be used for learning and identification of problems and solutions.
- Small Group Discussions and problem analysis
- One session per week in evening hours in case of open Farmer Life Schools for 4 (February to May) months duration.
- Facilitators should be choosen from same village with various backgrounds having good interpersonal and literacy skills.
- The schools can be conducted for LGs, and selected farmers, youth, women and mixed groups (mostly drawn from organized or informal groups accepted by gramasabha composed of different castes and religions)

The Resources

- A room for conducting Life School can be from school building, Panchayat office, open place like "Racha Banda" etc.
- A coordinator having good interpersonal and literacy skills
- (Farmers Field School Facilitator, School teacher retired employee etc)
- Drawing other resource persons mostly from the existing institutions of the village, cluster/mandal representing different sectors, agriculture, health, education, cooperative, sanitation, women and child welfare etc.,
- In house professionals of Implementing organization.

Steps in organizing Pilot Farmer Life Schools

- Development of broad curriculum for Farmer Life Schools
- Training of Coordinators

- Introduction and orientation about the schools to Panchayat President, Members, VOs, VDO, School Teacher, Health Worker, Agricultural Assistant and other stakeholders.
- Organizing the FLS through LGs
- Monitoring and Evaluation
- Follow-up

Expected Outcome

- Educated and empowered farmer Community that can identify, analyze and solve the problems of themselves (Pertaining to Decent Work-child labour, freedom of work choice, forced labour, non-discrimination etc., Health, Environment, Marketing and supply chain) and villages community at large.
- A socially and economically developed community leading happy life.

The Conclusion

The Farmer Life School is a process of empowerment. They are not intended to solve problems faced by the community. It makes people confident decision makers to utilize their expertise to solve their own problems and problems of farmers. If Farmer Life School participants decide to follow-up issues discussed, the knowledge and social networks established will contribute for problem solving. This will particularly help to motivate farmers to become effective decision makers of their own life, live of their families and their community networks. Above all the distress of the farmer`s unemployed youth, socially vulnerable groups can be mitigated through empowerment and confidence building process that can stimulate development and in turn improve their lives and livelihoods.

16 Psychological Counselling as part of Agricultural Extension Services

- Prof. C. Beena, G. Ravi Chandra Raju and G. Angela Sahayam

At a time when Indian economy is supposed to be gearing up to take on the world? Indian agriculture is predominantly dependant on nature. Irrigation facilities that are currently available, and do not cover the entire cultivable land. Any failure of nature, directly affects the fortunes of the farmers. Secondly, Indian agriculture is largely an unorganized sector, there is no systematic planning in cultivation, farmers work on lands of uneconomical sizes, institutional finances are not available and minimum purchase prices of the government do not in reality reach the poorest farmer. Added to this, the cost of agricultural inputs have been steadily rising over the years, farmers' margins of profits have been narrowing because the price rise in inputs is not complemented by an increase in the purchase price of the agricultural produce.

Repeated crop failures, debt hassles, lack of alternative sources of income, absence of institutional finance have left the farmers with no other solution other than ending their lives. Farming can be stressful in the best of times. Financial worries, unpredictable weather, plant pests, livestock diseases, and isolation all contribute to farmers' anxiety leading farmers to distress.

Personalized Counselling is the most Important step in Empowering Farmers

After the suicide of the farmer the families suffer a lot. Farms are confiscated due to inability to pay back high interest loans, Harassment of the family by corrupt money lenders, widows burdened with the new responsibility as the sole breadwinner, Children sometimes lose both parents to suicide forcing their education to a halt, especially if they have to work in order to provide for their needs. The problem cannot be solved through economic packages alone. And what is needed is social and counselling interventions so that the farmers realize that suicide is not the way out and they should understand that they need to develop self confidence. The future generation should have the mental strength and resilience to face life's challenge.

Even though the other interventions are happening by various other stake holders; from a psychological perspective, counselling must be seen an immediate intervention to all the crisis farmers face; especially to prevent suicides. Research study reveals that most of the time victims do not want to die; and give a lot of cues which are warnings even before they take the last step to end their lives. Counselling helps in not only understanding the issues the victims face, it also helps to identify these clues and help the victims in providing assurance, confidence and trust in their ability to solve every issue troubling them.

The aims of Counselling is gaining insight, building self-awareness, enhance self-esteem, equip the client with problem solving skills, develop knowledge, psychological education, help in acquisition of social skills, build right perspectives, cognitive changes and finally bringing in changes in one's behaviour. The essence of counselling is to accompany the victim in their journey towards holistic development.

In India there are very few population based data on prevalence of depression. In India, the point prevalence of Depression and mental health disorder is about 10 to 20 per 1000 of the population. Indians are among the world's most depressed - according to a 2011 world health organization sponsored study, around 9% of people in India reported having an extended period of depression with in their lifetime, nearly 36% suffered from what is called Major Depressive Episode (MDE). It is the second contributor to shorter lifespan for individuals. In south India a study in Chennai revealed the prevalence of depression was 15.1%.

Unfortunately, many victims do not report incidents of depression but keep it secret due to a variety of complex reasons. Some of the most common reasons are stigma, fear associated with mental illness, the severity of the situation, economic limitations, lack of knowledge about alternatives available through social services, and more.

The broader purpose of the counselling intervention is to identify suicidal ideation among farmers, which could be consequence of depression. Depression is a common mental disorder that presents with depressed mood, loss of interest or pleasure, feelings of guilt or low self worth, disturbed sleep or appetite, low energy, and poor concentration (WHO). It can affect a person's ability to work, form relationships, and destroy their quality of life. These problems can become chronic or recurrent, substantially impairing an individual's ability to cope with daily life. At its most severe, depression can lead to suicide. Depression can be assessed and addressed reliably.

Counselling plays a vital role in addressing this issue, it is in fact the first step towards enabling the farmer/victim to move on and make some hard decisions. Counselling provides both physical and psychological space to feel safe and secure. Counselling provides the opportunity to express, articulate without being judged, giving respect, love, acceptance and unconditional regard to vent, to the victim.

Counselling is a meaningful process that enables the victim to think clearly, to figure out what can work, and what cannot, gradually helping the client to seek for inner resources, for strength, and also explore outer resources to be able to assert oneself, bringing in change in the way the problem is perceived.

Counselling sessions could help the farmer to mirror themselves and help rediscover the value of being a person/ who deserves love, respect and dignity. The aims of Counselling is gaining insight, building self-awareness, awareness, enhance self-esteem, an could equip the farmers with problem solving skills, develop knowledge, psychological education, help in acquisition of social skills, build right perspectives, cognitive changes and finally bringing in changes in one's behaviour. The essence is to accompany the client in their journey towards holistic development and help in focusing on other perspectives of life and living.

Psychological Counselling Services in the Rural and Farmer Community

One of ways to address the issue of farmer's suicide is to empower the rural/farmer community in bringing about awareness, sensitization, and capacity building trainings related to psychological problems. One of the major issues to address here is the suicide. Unless we enable the community to identify the triggers the victims generally give before they take the step as a last resort, suicides cannot be prevented. Psychological services in the form of Counselling are the key to address a variety of issues in the community. Counselling can be sought to address fear, anxiety, sadness, and depression and also addictions to substance abuse.

Suicide not only affects the person his/her life, but also the lives of family members. It directly creates interpersonal problems for all family members. Suicide is a serious problem adversely affecting the social fabric of community. To reduce suicide among farmers, there is a need to increase the knowledge and awareness of the community through sensitization and awareness on psychological Counselling and suicide prevention programmes.

Additional Services through the Extension Education Workers

- Education through awareness and sensitization programme to the farmer community, local medical practitioners on important thrust areas in suicide prevention like early sign of depression and psychiatric disorders with serious suicidal risk that will help implement to prevent social disintegration, reduce social isolation.
- Financial counselling in order to prepare farmers to qualify for proper bank loans. Self help groups of farmers have proved to be very useful world over.
- Working together to solve the crisis by setting up a Crisis intervention centre by having either a helpline or counselling centre. Social support provides a buffering effect for stress. Counselling will help farmer and whole community to understand and seek help and support of friends, family, local communities.

Operational Framework

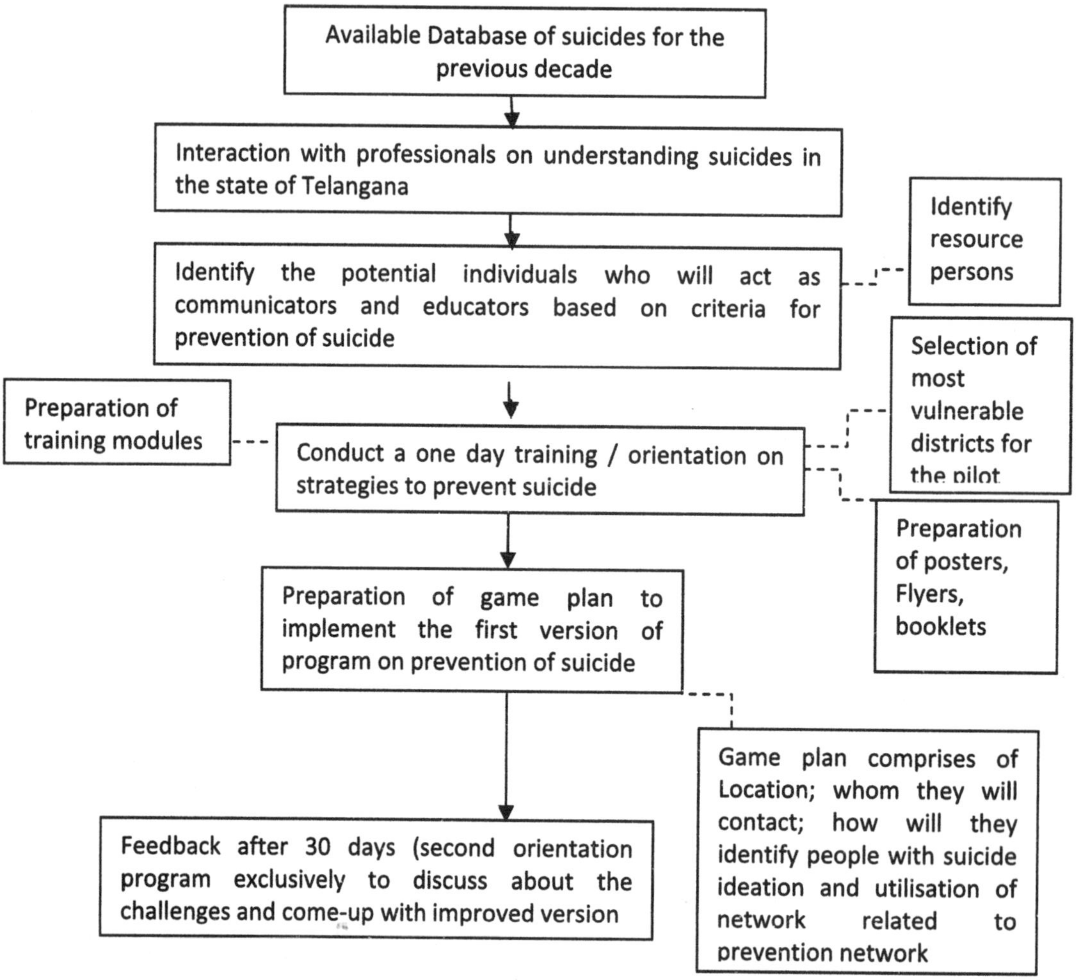

Therefore, besides the enabling provision of loan waiver intervention, the following are also necessary by the **respective state and national governments** to formulate wider safeguards to be implemented to ward off suicidal tendencies in the farmers.

1. Ensure that the banks extend the crop loans with farmer group collateral, minimum conditions and complexities
2. Provide uninterrupted at least eight hours of electricity supply
3. Implement the GO No. 421/2004 that too with an enhanced ex-gratia Rs 5 Lakhs (GO No 421 has a provision of Rs 2 Lakhs ex-gratia, however, considering the present market inflation request for enhancing to Rs 5 Lakhs) and rehabilitation

assistance to the families of more than 600 farmers who committed suicides from 2nd June, 2014 to till date.

In addition to the above processes, the **communication strategies could be designed** by the following;

1. Short films with the specific themes to build one's confidence.
2. Visual messages emphasising handholding, especially in situations of distress, trauma and depression as well as to be calm, to be peaceful, to rethink and change the decisions from ending life to beginning a new life.
3. Audio visual equipped vehicle with trained extension service workers who can extend help in greater understanding of typical stressful situations, can explain about posters, pamphlets, play the short films appeal to them primarily to live and not end life.
4. Setting up of 24/7 toll free call centre which could be accessible to farmers and make resource person available who can appeal and counsel if necessary.

Deliverables

It may be summarized that the broader purpose of the approach mentioned is to drive the multi intervention model embedded in psycho-socio and cultural milieu of a farmer while dealing with suicides. And the specific deliverables are;

- Reducing suicidal ideation in farmers who are at risk
- Bring awareness to farmer about alternative crops and other livelihood opportunities
- Bring about overall reduction in the rate of farmer suicides in the villages.

17 Personality Development for Effective Extension Services

- Prof. M. Surya Mani

Extension organizations in developing countries face the major problems of professional incompetence and lack of motivation among their employees. Further, many of the agricultural extension departments of these countries do not have a well-defined system of human resource management. Proper planning and management of human resources within extension organizations is essential to increase the capabilities, motivation, and overall effectiveness of extension personnel.

Managing relations is essentially managing people and their interests, aspirations and goals. Leaders of various corporate enterprises say that of all the resources of an enterprise managing the human resource is the most critical and challenging. Critical because, the human resources are the dynamic, life giving elements of the organizations and without their involvement and participation, the physical and financial resources remain resources and never become production and profits. Challenging because, managing human behavior is unique and unpredictable. Humans vary in attitudes, learning's, beliefs, values, perceptions and performance from one another. Also every individual varies his behavior from one situation to another. It is for this reason that managers take interest in understanding and moulding personalities of their employees.

One of the most significant developments in the field of extension in recent times is the increasing importance given to personality development. More and more attention is being paid to motivational aspects of human personality, particularly the need for self-esteem, group belonging, and self-actualization. This new awakening of humanism and humanization all over the world has in fact enlarged the scope of applying principles of personality development in extension education. The development of people, their competencies, and the process development of the total organization are the main concerns of human resource management (Pareek & Rao, 1992).

The importance of assessing personal and professional attributes for selecting productive extension personnel has been reported by several researchers (Gupta, 1963; Perumal, 1975). Assessment is essential because an unsatisfactory educational level of extension staff is one of the most serious problems of extension in countries like Bangladesh, Botswana, Kenya, Malaysia, Sudan, and Zambia (Blanckenburg, 1984). A worldwide analysis of the status of agricultural extension reveals the low level of formal education and training of field extension agents in developing countries (Swanson, Farner, & Bahal, 1990).

An efficient extension organization needs to develop the capability of responding to changes in relation to its environment. Extension organizations have to cope with changes within and outside the organization, such as changes in farm technology, communication methods, needs of farmers, rural situations, export and import of farm produce, and market economy. Personality development of its employees allows for planned changes in the organization's tasks, techniques, structure, and people. Attitudes, values, and practices of the organization are changed so that it can cope with changing situations. The employees also gain greater skills to deal with new problems.

There is a great need of soft skills for personality development programmes in extension organizations because they face complex situations due to changing agricultural scenarios. Further, extension managers have to be exposed to modern management techniques and methods. Soft skills development programmes have to suit to the needs of all-level extension managers and should be based on needs analysis. Methods such as coaching, job rotation, training sessions, games and simulated exercise are used to train extension managers. In India, a separate institutes like SAMETIs, EEIs, MANAGE etc., are established to train extension managers in soft skills, managerial skills and human relations besides extension methods and skills.

The key performance areas indicate the important roles and contributions of different categories of extension personnel. Once the roles are delineated, they can be analysed to indicate the attributes which can discriminate an effective from an ineffective role occupant. These critical attributes consist of qualities such as educational qualifications, skills, experience, physical characteristics, mental abilities, values, and attitudes needed for extension. The critical attributes needed for field-level and supervisory extension staff are necessary formal training in agriculture along with some important skills for personality development such as attitudes, motivations, presentation, leadership, team building and interpersonal skills are needed by extension professionals.

Soft skills play an important and significant role in building one's success, more particularly in professional life. The ever changing life style, hybrid – mixed cultures, emerging management styles, technology revolution essentially required refined skills set consisting of Soft skills, professional skills and life skills.

Soft skills are wide, mostly people skills, centered around one's personality, which are oral, intangible, physical, psychological, socio-cultural and non-technical. They can be acquired right from the stages of / through cuddling, caring, grooming, education, observation, adoption, teaching cum training methodology only and not by mere class-room lecture / reading. These skills are in great demand as they are associated with the management of people.

Some soft skills like

- Self improvement
- Motivation
- Leadership
- Presentation skills
- Interpersonal skills
- Positive Attitude
- Team work and team building are some of the focused techniques to improve the efficiency of extension professionals at all levels

SELF IMPROVEMENT

Development of self is the core part of personality development programme. Self improvement programs are many and varied. However, all of them start with the idea to develop self and assessing the predispositions, traits and behaviors. Based on the assessment and individual expectations for development, the development path and programmes will be identified.

Self improvement is not a single step process or an instantaneous outcome. It is a long journey. The journey demands to cross hurdles, to rejuvenate our mind, to reinforce our strength and reorient our outlook.

Simple Technique for Self Improvement and Self Growth

Nowadays, there is a growing interest in self growth and self improvement techniques. It seems that people are turning within them, to find the solution to their problems. They seek knowledge, techniques, workshops, lectures, and teachers, who can show them the way.

People are beginning to understand that self improvement and self growth can improve the quality of their lives.

The Process of Inner Change Requires Inner Work

For achieving this, one had to practice what they read. Inner change requires motivation, desire, ambition, perseverance and dedication.

In a self improvement program, it is common to encounter inner resistance that comes from our old habits and from our subconscious mind, and also resistance and opposition from the people around us.

The desire to change, build new habits and improve oneself must be strong enough to resist laziness, the desire to give up, and the fear and ridicule of opposition from family, friends or colleagues.

Identify Improvement Needs

Need is a gap between actual and expectation. The expectations can be social or personal. Our friends may expect us to score first class in the semester end examination and we are weak in certain subjects, obviously we feel improvement needs in our performance in respect of those subjects in the examination.

Corporate hirers often point out that candidates are weak in oral communication and skills. What should the candidates do to improve their skills in communication? In both cases, we will assess our strengths and weakness to go about an improvement plan.

Make an Individual SWOT

An honest evaluation of current circumstances, will enable us to work out the steps to help us move from where we are now to where we want to be.

The first step to improve oral communication skills is to have our speaking and listening skills.

Step to find how well you can speak and listen.

The most common thing in the world every individual faces is the honest and fervent desire for self-improvement, followed by inaction or giving in to temptations, followed by guilt or giving up. It happens to the best of us. It's inertia at work, mixed with a bit of laziness as well as the very human trait of giving in to desires despite all the good intentions in the world.

So how do we beat inertia and temptations?

Get moving, a bit at a time: Inertia is beat only by movement. Once we get going, momentum builds up and inertia is no longer a factor. So the key is to get started, and we do that not by trying to go from 0 to 60 in 5 seconds, but by trying to go from 0 to 5mph in a day or two. That's doable. It's all about baby steps.

Be accountable: Laziness, the second culprit, is beat by a bit of public pressure. We all get lazy from time to time (or, to be more honest, all the time), and there's nothing wrong with that. But to beat laziness, we must apply a bit of pressure, in the form of accountability. There's nothing wrong with a little pressure, as long is it's not overdone. Pressure is a motivating thing, especially when it's positive. Positive pressure includes encouragement from family or friends, an online forum, a help group in your neighborhood, or the readers of your blog.

Ignore failures — giving in to temptation is OK: We will always give in to temptation. Plan for it, accept it, move on. There's no need to beat yourself up.

Motivate yourself: Most importantly, we want to really want it. It's not enough to feel pressure to do something — we have to really desire it. I mean, really desire it, not just think it's something we should do, or that we will be a better person for doing it. If pressure gives us the push toward your goal, motivation gives you the pull.

1. **Make a date. Right now:** All the good intentions in the history of the universe mean nothing if we don't actually get started. And the only way to get started is to take action, right now. Not tomorrow, not later today, not in an hour, but, Right now! Look at the calendar, and make an appointment to create the action plan, or to take the first action. Decide what that is and make an appointment for it, right now. Second part of this step: make that appointment the most important appointment on the schedule, more important than a doctor's appointment or a meeting with the boss.

2. **Set a small, achievable goal:** Remember, inertia is a powerful force. If we haven't been exercising for a couple years, it's hard to get started. People used to the way things are, and even if they want to change, it's difficult. So don't start out trying to conquer the world. Just conquer something exceedingly small. It might sound wimpy to say, "I'm going to walk for 10 minutes" or "I'm going to do 10 pushups and 1 chinup", but those are much more likely to beat inertia than, "I'm going to exercise for 45 minutes today." Be realistic, and make it very very achievable. It's the only way to beat inertia.

3. **Commit thyself, big time:** It's this commitment that will keep us going after overcoming inertia. Sometimes we get filled up with enthusiasm, but then a few days later, that enthusiasm wanes and we submit to our old buddy laziness. So instead, make a commitment, publicly. State small, achievable goals, and tell it to as many people as we can. Call or email friends and family, tell all coworkers, join an online forum related to the goal and tell all of them. Put it on the blog. However it is done, do it, make sure people are aware of your goal, and that there's sufficient pressure to overcome laziness.

4. **Baby steps, baby:** Again, inertia is a very strong force. But this is a very important step here: the best way to change is through baby steps. One small step at a time. It is different from the above step in setting a small and achievable goal It's the same concept, but extended beyond the initial goal. It's taking things one little goal at a time, a bit at a time. For example, instead of running a marathon, all at the start, instead, preparing to run a marathon by walking 10 minutes a day. Then when it becomes a habit, increasing gradually till one is fit enough to run a marathon.

5. **Hold oneself accountable:** Committing publicly is not enough to tell people our goal. We have to make it clear that they must hold us accountable to reporting to them our progress. Then report the progress to them regularly. Daily is better than weekly. Reporting makes sure that we will think twice about being lazy and forgoing the action plan.

6. **Motivate yourself:** Accountability and commitment are two ways to put positive pressure on oneself — a form of motivation. Those are great, but we also want other types of motivation. We want to find ways to make our progress feel great ... either through rewards, or the positive way we feel about our progress, or the positive way you feel when others see how well we are doing. Find a few different ways to motivate oneself— the more the better. Incorporate these into the plan.

7. **Just keep doing it, no matter what:** Even if we encounter obstacles, falter and fall, we must get up and keep going. Just keep going. We will make mistakes and get discouraged. No matter ... just keep going. We must learn from the mistakes, and ... keep going. No matter what happens, keep going. If we are taking baby steps, we are holding ourself accountable, and we are actually doing something, we will get there.

Tips for a Simple Self Growth Technique

1. Look around and watch how people behave in various circumstances. Watch the people at home, work, at the supermarket, on the bus, train and on the street. Also watch and learn from people being interviewed on TV.
2. Watch how people talk, walk and react, and how they are treated by others.
3. Pay attention to the way people use their voice and how they react to other people's voices. Watch how one feels and reacts when people shout, or speak softly. Watch what happens when people get angry, restless and upset, and what happens to you and others, when they are calm and relaxed.
4. Analyse things that are not liked by us, our own behavior and other reasons. Be honest and impartial in the analysis.
5. When undesirable traits of character are found within oneself, conscious effort should be made to avoid these tracts.
6. Play in the mind, the mental scene of an ideal behavior and repeat in several times in a day
7. When we detect in someone a sort of behavior or character traits we like and desire to possess, try to act in a similar way. Here too, visualize several times each day a scene, where we can act and behave in that different way.
8. Think and visualize over and again in mind how we would like to act and behave. Constantly, remind yourself of the changes you desire to make, and strive to act according to them.
9. Do not be disappointed or frustrated if we do not attain fast results. It does not matter how many times we fail or forget to behave as desired. Persevere with efforts and attitude, will begin to see how we and our life can change.

MOTIVATION

The work motivation and morale of extension staff, as reported earlier, are very poor in many countries. The reasons are many. The bureaucratic structure of extension administration, lack of rewards and incentives, poor facilities, poor promotional avenues, and the low esteem given to extension are the major causes of poor motivation and morale. Extension supervisors should have the ability to motivate and lead the field extension workers so that the field agents perform more than routine jobs, and supervisors should be involved in attaining excellence in extension work. This calls for extension managers having an understanding of various theories of motivation as applicable to frontline extension agents. Therefore, a knowledge of major theories of motivation such as Maslow's hierarchy of needs theory, Herzberg's two factory theory, McClelland's need theory, theory X and theory Y, and expectancy theory of motivation is essential (Stoner & Freeman, 1992). Special training for developing motivation among field-workers has to be undertaken by the institutional management.

Motivating people means to persuade them to take part in certain activities. Motivation initiates a conscious and a purposeful action. It is motivation which arouses the action towards a goal directed and need satisfying behavior. Thus, it is a hidden force within individuals that causes them to behave in certain way. It is in-separably linked with all the developmental programmes.

Motivation is the process of channeling a person's inner drives so that he wants to accomplish why people behave as they do. It concerns those dynamic processes which produce a goal oriented behavior.

Concepts of Motivation

It is a cyclic state which starts with need and terminates with the achievement of goal. It includes need, drive, action, goal attainment and satisfaction.

The motivates may be classified as internal v/s external, conscious v/s unconscious, etc. Maslow, a famous psychologist, has classified the motives as physiological, safety, social (love and belongingness), esteem and self – actualization.

Maslow's Need- Hierarchy Theory All people have a variety of needs. At any given time, some of these needs are satisfied and others are unsatisfied. An unsatisfied need is the starting point in the motivation process. It begins the chain of events leading to behavior. When a person has an unsatisfied need, he or she attempts to identify something that will satisfy the need. This is called a goal. Once a goal has been identified, the person takes action to reach that goal and thereby satisfy the need. According to A.H. Maslow, needs are arranged in a hierarchy or a ladder of five successive categories. Physiological needs are at the lowest level, followed by security, social, esteem and self-fulfillment needs.

Physiological needs to feel free from economic threat and physical harm. These include protection from arbitrary lay – off and dismissal, disaster and avoidance of the unexpected.

Social needs are needs to associate with other people and be accepted by them; to love and be loved. These needs are variously referred to as the herd instinct", "gregariousness" and the like, but at base, they point to the fact that man finds a satisfaction in association with others and feels a real deprivation when it is not possible.

Egoistic or esteem needs are those which relate to respect and prestige. A need for dominance may be thought of as one of the egoistic needs. These are of two types: self-esteem and esteem from others.

Self fulfillment needs are needs for realizing one's potential. These include the need for realizing one's capabilities to the fullest- for accomplishing what one is capable of accomplishing. For becoming what one is capable of becoming? A musician must make music, an artist must paint. A poet must write if he is to be ultimately happy. This need is also called need for self-realization or self-actualization. This term, first coined by Kurt Goldstein, refers to the desire for self-fulfillment, namely, the tendency to become actualized in one's own potentiality.

According to Maslow, people attempt to satisfy their physical needs first. As long as these needs are unsatisfied, they dominate behavior. As they become reasonably satisfied, however, they lose their motivation al power and the next level of needs. Security needs, becomes the dominant motivational force. This process continues up the need hierarchy. As each level of needs become relatively satisfied, the next higher level becomes dominant.

Motivation Techniques

Since motivation is a complex state, there can be many factors and techniques which can motivate the behavior of the people. Hence, it is important to understand the important techniques which can help in the motivation of the people. Some important techniques are as under:

Need Based Approach

People can be motivated by knowing their needs. The various needs of the adults. Therefore, should be identified. The programme should be linked to the needs and aspirations of the learner. If the learners are interested in learning some new skill. The learning can be directed towards imparting that particular skill. The learning should fit in the recognized needs of the adults.

Novelty and Variety

Every novel thing attracts more attention of the individual and makes him more interested. So, the extensionist must introduce novelty into his messages. The teacher should resent the message in variety of ways by using various methods. To motivate the learner's proper and effective audio-visual aids can also be used.

Lecture-cu,-discussion method, demonstration along with effective audio visual aids sustain the interest of the learners and increase the enthusiasm to learn more. Mass-media play a very significant role in motivating learners, reinforcing learning and in general creating a more conducive environment. Role of projectors and documentary films can be widened for motivating people.

Personality of the Teacher

The personality of the teacher and his understanding serve as a motivation force. He should be frank, cheerful and should be sympathetic to the problems of the learners. His thoughts and recommendations should be clear to the people. The educator can add some humor to make the atmosphere interesting.

Rewards and Incentives

Rewards create interest in the people. They tend to increase the efforts and participation of the learners. Punishment should be avoided. Though sometimes punishment can also increase the performance of the people but overall it has negative effect on them.

There are different types of rewards such as material, symbolic and psychological. All the rewards have a positive effect on motivation of the client. However, the communicator must remain cautious that the rewards should not become an end in itself but it should create learning desire and adoption in the learners.

Appreciation and Praise

It is human nature that everyone likes praise and wants some praise for his achievement. However, individual differences exist in this regard also. Some learners may be praised for minor achievements because of their limited abilities but others will be motivated by praise of most worthy accomplishments related to their high ability. Praise can be used in many ways like a smile or verbal praise etc.

The teacher should be cautious while using praise. For each and every activity, learner should not be praised. Actually it should be according to the personality and experiences of the people.

Provide Feedback

Knowledge of results or performance of the learner works as motivating force in learning task. Learners continuously remain involved in the learning activity. Learning is faster and effective when the learners are provided with the knowledge of their progress. It has been experimentally established by the psychologists that knowledge of result facilitates motivation and improves content and mastery of the people. The communicator should provide proper feedback to the audience to motivate them in learning.

Ego Involvement

The ego consists of attitudes related to self. One should try to involve their ego in the job and motivate them by appealing to the ego maximization.

Diversification

The people will be motivated if the programme is in conformity to the needs of the community, ambitions of the people and suited to the environment. It must satisfy the diverse needs of the learners and should be flexible in nature. The various programmes should be relevant to the needs of the learners.

Realistic Approach

Another way to motivate is to help them realize their own expectations and aspirations. It is necessary for the teacher to guide the expectations of the learner on the real lines so that every expectations turns out be an achievement which in turn would provide motivation for sustaining the onward activity towards the goals. The activities /efforts should be such that they help in solving their actual problems and help them achieve the goals which they have set for themselves.

Rapport with the Clients

Developing good informal relations/contacts with the people helps a lot in motivating them. There should be constant dialogues and discussions with the learners about their specific problems. If possible, the clients should be approached separately and should be motivated by taking up their individual problems. In this way a good rapport with the learners can be created.

Creative Activities

The learners can be further motivated if the programmes further includes some creative activities. The programme should require some creative thinking, working and problem solving etc.

Use Competition and Cooperation

Competition and cooperation work as strong incentive for motivating individuals.

Increase Level of Aspiration

Level of aspiration means the level of performance to which one aspires for future. The motivator must see that the work should be in accordance with aspirations level.

Provide Real Life and Symbolic Models

It is fact that most of the learning in human beings is acquired through the process of observation and limitation. The motivator can influence the behavior of his clients by giving examples of real life successful and ideal persons or by providing written or verbal presentation or audio-visual information to them.

Inspite of all the techniques referred above, the place of the motivator remains supreme, his method of imparting knowledge and skills, warmth and his style of discussion and handling problems play an important role in the process of motivation in the class rooms as well as in job situations.

LEADERSHIP

All people are not born with the same potential to lead well; rather some may not have the ability to lead at all. Different personal characteristics help or hinder a person's leadership successes that require formalized programmes for developing leadership competencies.

It is leadership shown by leaders that can make impossible things possible. An effective leader understands ones strengths and limitations and teams up with other leaders having complementary strengths so that the leadership team can perform different roles for success.

Everyone in an organization can develop their leadership skills and effectiveness. Classroom style training, workshops, and associated reading and field experiences can contribute to developing leadership skills. It can inform one about what ones has in possession and what else is needed, and what is involved in leading well. Leadership development is person-oriented action and focuses on the development of mangers as leaders, such as the personal attributes, desired ways of behaving, and approach to thinking.

Leader development focuses on three main areas: (a) providing opportunities for development, (b) stimulating the ability to develop including motivation, skills and knowledge for change, (c) and providing a support context for changes to occur. The leadership development process encompasses leader development and follower development. Traditionally, leadership development has focused on developing the leadership abilities and attitudes of individuals. A leader must have qualities like integrity, loyalty, commitment, energy, decisiveness, and selflessness.

Integrity: Full devotion to the highest personal and professional standards.

Loyalty: Faithfulness to superiors, peers, and subordinates, and, above all, to the organization.

Commitment: Attachment and declaration to duty to achieve team goals, and objectives of the organization to the maximum extent.

Energy: Enthusiasm, enterprise, and drive to take initiative.

Decisiveness: Willingness to act and courage to decide.

Selflessness: Preparedness to sacrifice personal objectives and gains in the larger interest of the organization.

Need for Leadership

The socio-economic forces, effective market forces, changes in public policy and concepts of social, industrial democracy, ecological imbalances, ekistics (problem of human settlements), and ergonomics (human engineering) are changing with time.

Leadership development refers to the strategic investment in, and utilization of, the human capital within an organization. Henry Ford said, 'Burn my company, destroy everything I have, ruin all my assets just give back my men. I will create another Ford.' The gamut of leadership processes will include the social influence process and the team dynamics between the leader and each team member at the dyad levels, and the contextual climate and social network linkages between the team and other groups in the organization.

Some of the underlying concepts behind leader's development are the following.

- Managers/Administrators need development throughout work life and professional career. Thus management development is an ongoing activity.
- There always exists a gap between 'required performance level' and the 'capacity' of an individual. Management must fill the gap to provide an opportunity for improvement.
- Some personal variables (age, habits, level of motivation, state of mind, etc.) retard the growth of an individual.
- In a work situation, growth involves stresses and strains, Development can seldom take place in a completely peaceful atmosphere.
- Involvement and participation is inescapable for growth.
- There must be defined objectives and goals required to be achieved in addition to the methodology of achievement.
- Shortcoming must be identified. Feedback and counselling to junior colleagues by apprising them of their shortcoming along with relevant HRD instruments to overcome such shortcomings are essential.

Leadership Development Objectives

Based on perceptions and notions, organizations arrange leadership development programmes with the aim to achieve objectives like the following:

- To provide the organization the required number of leaders having the ability to meet present and future organizational needs.
- To instil leader capabilities among managers sense of self-dependency, achievements, and affiliation to tem members.
- To encourage leaders to keep themselves up to date and grow to meet the challenges, cope up with the changes, handle complex situations and greater problems.
- To discharge their responsibilities with improved performance.
- To sustain good performance and gain distinctive competence.

Leadership Development Techniques

The techniques lay emphasis not on the skill of leaders but on their capacity to handle complex situations, deal with tough-minded and negative persons, and solve managerial problems. The techniques generally used are:

- Creation of 'Assistant to' position – The junior colleague (the follower) can perform the activities of the senior colleague (the leader). In turn, the senior colleague discharges a teacher-like role while judging the decision-making ability, and leadership skills, traits, and other potential.
- Making member of junior boards of management – The behavioural dispositions if junior managers are observed, their views/divisions are very often implemented. This is done to instil a sense of responsibility among prospective leaders.
- Coaching and counselling – Adopt to transform the new subordinate, to make him/her job-oriented in order to accomplish the goals.
- Syndicate – team of persons of matures judgement and proven ability drawn from different functional areas exchange and share their experiences and ideas. Large numbers of new recruits are divided into small groups of 7-10 persons each. A facilitator observes and guides the group.

Process of Leadership Development – Role of Crucibles

An experiential dimension in the lives of all the leaderscrucibles and an intense, transformational experience that set them on the road to where they are now' (Pareek 2008). They identified four major types of experiential dimensions: mentoring relationships, enforced reflection, experiencing a new world, and disruption and loss.

Mentoring relationships: Mentoring have always been found to influence and develop leadership in their protégés or mentees. Some mentors attract protégés when the process of mentoring happens automatically. This is termed as 'natural mentoring'. In many cases, organizations, in a planned way, select employees to be mentored by one senior, which is termed as 'planned mentoring'. In the knowledge industry, often a younger person mentors an elder person and this is termed as 'reverse mentoring'. Mentors offer valuable advice to protégés, develop work leadership, and help them to advance in their professional career.

Enforced reflection: This crucible refers to an opportunity for both exploration and reflection, found in such events as going away to a school or to an *ashram*. 'Through these crucibles, individuals learn preparedness an alertness to the rich signals that surround them and a willingness to experiment in the internet of advancing self-knowledge and, by extension, knowledge of the world around them.'

Experiencing a new world: Leaders demonstrate a remarkable capacity not only to survive in new situations, but also to derive profound insights from them. Leaders capture insights and means to advance from their own experimental tapestry, from one's own understanding, familiarity, and experience. For Mahatma Gandhi, the crucible experience was his stay in South Africa.

Disruption and loss: Remember what Gautama Buddha asked the person who approached him following the death of a near one. Personal loss of a parent, a grown child, bankruptcy, failure in an important assignment or understanding, marriage of a girl whom one loves very much, cheating by a wife, or any similar loss or defeat can stimulate a search for greater understanding of self and relationships, a striving to explore, venturing into problem-solving and larger webs of affiliation.

A good student affected by chickenpox had to appear in a university examination from hospital. During the examination, due to mass copying, all the students were 'reported against'. One year was spoiled. This 'shocking step' of the university was the first 'crucible' for the students. But the student did not get frustrated. Instead he pursued a realistic analysis, and accepted the crucible as a simple roadblock in his career. As on date, he is an accepted academician. Again, many persons have transformed following untoward incidents, for the crucible produced negative impacts. Thus the effects of crucible can be positive as well as negative. A leader should react to crucibles in a positive manner and help the followers likewise.

Advanced Leadership Skills

These are the additional leadership skills which help a leader to effectively deal with followers and have great impact on leader's relationship with followers.

Basic Leadership Skills	Advanced Leadership Skills
Learning from Experience	Delegation Skills
Communication skills	Providing Constructive feedback
Listening Skills	Team Building and Building High Performance Team
Assertiveness	Empowerment
Effective Stress Management	Managing Performance of subordinates
Building Technical Competence	Goal Setting Skills
Managing Superiors	Leading Meeting
Building Effective Relationships	Managing conflict with Peers
Building Credibility	Negotiation Skills
Developing Planning	Problem Solving Skills
	Improving Creativity

Delegation

Entrusting your authority to others in certain areas and enabling them to act independently and under their own initiative, so that they are assuming responsibility with you for certain task. They also have the authority to react to situations and decide on a course of action, without always having to refer back to. A leader must delegate the authority but should not loose full control over the process.

Constructive Feedback

Feedback is essential to know how people and organization is processing to meet objectives. Thus feedback is to measure performance. And this becomes the essential component of leader's responsibilities.

Team Building

Team building refers to a wide range of activities, for improving team performance. This is a process of making team effective. This purpose of team building is to increase cohesiveness, mutual cooperation, and identification with the group.

Managing Performance of Subordinates

One of the major challenges faced by a leader is regarding management of performance of individual subordinates. The challenges increase when it comes to manage individuals who are not performing up to the expectations. To manage performance of individuals, a leader must be well aware about the performance discrepancy and cause of the same. A leader needs to identify the performance gap and analyze the capabilities of individuals, help them by motivation coaching and provide opportunities.

Goal Setting Skills

A leader is required to set goals for the followers. As proposed by Edwin Locke that intensions to work towards a goal are major source of work motivation. Goals have strong values. Goals give a sense of purpose as to why we are working to accomplish a given task. Goals tell an employee what needs to be done and how much effort will need to be expended. A leader must put his best efforts in goal setting for followers.

Leading Meetings

A leader's ability to lead and conduct meeting directly have impact on his participative decision making and problem solving.

Managing Conflict

Handling conflict is part of leadership responsibility. Conflict is inevitable in any group of people whether they are co-workers, family members, community members or other groups if shared common interest. Organizations are made of people. And conflict exists between people. So conflict is inevitable of organizations. The ability to manage conflicts is therefore an essential skill required to be a good leader.

Negotiation

Negotiation is a problem solving process in which two or more people voluntarily discuss their differences and attempt to reach at joint decision on their common concerns. Negotiations can help in conflict resolution also. Leader needs to negotiate with seniors for the resources, peers for the support and cooperation, and with subordinates for completion of the task. Sometime, leader is required to negotiate with outside parties also.

Problem Solving Skills

Problem solving skills are leader's creative ability to solve new and unusual, ill defined organizational problem. These skills are about being able to identify and define significant problem, gather information, analyze the situation, develop solutions, evaluate solutions, selecting the best solutions and prepare implementation plan.

Improving Creativity

A leader should be rational decision maker as well as creative decision maker. Creativity is the ability to produce novel and useful ideas. Creative ideas are different and also appropriate to the problem. Creativity makes leader look at the problem in different manner. It helps leader to develop alternatives which are not readily apparent. So creativity is about seeing things in new and different ways. Creativity has three important components: task motivation, expertise, and creativity skills. A leader should be creative in his approach and should also try to make followers creative.

PRESENTATIONS SKILL

Presentations skill are very useful in many aspects of work and life and are important in training, teaching, lecturing, and generally feeling comfortable speaking to a group of people.

A person can have the greatest idea in the world, completely different and novel, but if he can't convince enough other people, the idea goes in vain.

Developing the confidence and capability to give good presentations, and to stand up in front of an audience and speak well, are also extremely helpful competencies for self-development and social situations.

Presentations are an extremely complex and expensive form of human communication. Presentations are useful not just in offices, business, corporate etc., but also form an integral part in educational institutes, social life and day to day interactions. Almost any kind of activity involves some degree of presentation skills. Sharing your knowledge among the peers, communicating the purpose and content of any trainings or projects made during the school or college days, guiding a new colleague through the basic office procedures, reporting back to a departmental meeting, or giving the members of the board an overview of a new product, highlighting the positives of your products – in each situation, you will be giving a presentation.

The formats and purposes of presentations can be vary different, for example: oral (spoken), multimedia (using various media - visuals, audio, etc), powerpoint presentations, short impromptu presentations, long planned presentations, educational or training sessions, lectures, and simply giving a talk on a subject to a group on a voluntary basis for pleasure. Even speeches at weddings and eulogies at funerals are types of presentations.

A presentation is a means of communication which can be adapted to various speaking situations, such as talking to a group, addressing a meeting or briefing a team. To be effective, step-by-step preparation and the method and means of presenting the information should be carefully considered.

A presentation is a structured talk given to an audience to communicate one's views and information with a specific purpose. Presentations can be either technical or social. Technical presentations are related to the occupation and social presentations are related to cultural or social events.

Preparing Presentation

Preparation is the single most important part of making a successful presentation. This is the crucial foundation and you should dedicate as much time to it as possible avoiding short-cuts. Presentation should cover the important elements i.e., the objective; the subject; the audience; the place; time of day; length of talk.

Organize the Presentation Material

Regardless of whether the presentation is going to be delivered formally, such as at work or informally, for a club or perhaps a Best Man's speech. One should always aim to give a clear, well-structured delivery. That is, we should know exactly what we want to say and the order in which you want to say it.

Structuring or organizing the material clearly is vital for an effective presentation. In other words, the presentation should have the following format:

- ***Introduction***: Should grab attention, introduce topic, contain a strategy for establishing credibility, preview your speech, establish rules for questions and have a smooth transition to the main text.
- ***Main body***: Contains all topics/the entire matter organized into a logical sequence.
- ***Conclusion***: Contains signal, highlights / summary, closing statement/re-emphsis, a vote of thanks, and invite questions.

Body of the Speech

The body can be developed by appropriate choice of sequence and style.

Sequencing

Sequencing can be done in two ways for easy comprehension.

- **Logical sequences** : are essentially rational and message oriented. Emphasis should be on getting ideas into the nervous system of your receivers without great concern for their feelings.

  Logical sequences are categorized into five types.

  (i) *Chorological pattern* : In this, the points are arranged in order of their occurrence. For example, a talk on computers can trace its evolution in different periods.

  (ii) *Spatial/ directions* : Spatial arrangements refers to layout of physical things. For some topics, this pattern is suitable. A talk on the safety requirement is arranged according to the places where they are located or necessary. Thus, it explains the safety of shop floor equipment, in different workshops – Safety in fabrication shop, safety in welding shop, safety in lathe shop and so on.

  (iii) *Topical/categorical* : The most commonly followed pattern is the topical pattern, here, you divide the topic into some logical themes or categories. For example, a talk on effective presentation can be divided in to: Effective introduction, effective presentation and effective close.

  (iv) *Cause and effect*: One can use the causes first and effects last pattern, or vice versa. A present on strike in a division of your company can be presented as (a) the event, (b) causes, (c) the effects and (d) the action to be taken.

(v) Climactic : In this, ideas are arranged from least important to most important. For example, while presenting the details of an accomplished project, you may start from the basics, and slowly take the audience through the various stages of the project.

- **Psychological sequences :** These are essentially feeling oriented. The primary focus is on handling the receiver's emotional states in order to prepare him to listen to your message.

Working with Visual Aids

One should only use visual aids if they are necessary to maintain interest ad assist comprehension in your presentation. Do not visual aids just to demonstrate your technological competence – doing so may compromise the main point of the presentation – getting the messages across clearly and concisely.

If visual aids are used well they will enhance a presentation by adding impact and strengthening audience involvement, yet if they are managed badly they can ruin a presentation.

Different Ways to Improve the Presentation Skills

***Practice*:** Rehearse the presentation multiple times. While it can be difficult for those with packed schedules to spare time to practice, it's essential if you want to deliver a rousing presentation. Try to practice in the actual venue. Some acting strategists suggest rehearsing lines in various positions – standing up, sitting down, with arms open wide, on one leg, The more we mix up the position and setting, the more comfortable we feel with our speech. Do a practice run for a friend or colleague, or try recording the presentation and playing it back to evaluate which areas need work. Listening to recordings of the past talks can clue us in to bad habits we may be unaware of, as well as inspiring the age-old question: "Is that what I really sound like?"

***Transform Nervous Energy Into Enthusiasm*:** It may sound strange, but it will often down an energy drink and blast hip-hop music in earphones before presenting. Why? It pumps up and helps someone turn jitters into focused enthusiasm. Studies have shown that an enthusiastic speech can win out over an eloquent one. Make sure that as enthusiastic and energetic as possible before going on stage. Of course, individuals respond differently to caffeine overload, so know your own body before guzzling those monster energy drinks.

***Attend Other Presentations*:** If you're giving a talk as part of a conference, try to attend some of the earlier talks by other presenters. This shows respect for your fellow presenters while also giving you a chance to feel out the audience. What's the mood of the crowd? Are folks in the mood to laugh or are they a bit more stiff? Are the presentations more strategic or tactical in nature? Another speaker may also say something that you can play off of later in your own presentation.

Arrive Early: It's always best to allow yourself plenty of time to settle in before your talk. Extra time ensures you won't be late and gives you plenty of time to get adapted to your presentation space.

Adjust to Your Surroundings: The more adjusted to your environment you are, the more comfortable you'll feel. Make sure to spend some in the room where you will be delivering your presentation. If possible, practice with the microphone and lighting, make sure you understand the seating, and be aware of any distractions potentially posed by the venue.

Meet and Greet: Do your best to chat with people before your presentation. Talking with audiences makes you seem more likeable and approachable. Ask event attendees questions and take in their responses. They may even give you some inspiration to weave into your talk.

Use Positive Visualization: Plenty of studies have proven the effectiveness of positive visualization. When we imagine a positive outcome to a scenario in our mind, it's more likely to play out the way we envision.

Remember That Most Audiences are Sympathetic: One of the hardest fears to shake when speaking in public is that the audience is secretly waiting to laugh at your missteps or mistakes. Fortunately, this isn't the case in the vast majority of presentations.

The audience wants to see you succeed. In fact, many people have a fear of public speaking, so even if the audience seems indifferent, the chances are pretty good that most people listening to your presentation can relate to how nerve-racking it can be. If you start to feel nervous, remind yourself that the audience gets it, and actually wants to see you nail it.

Take Deep Breaths: The go-to advice for jitters has truth to it. When we're nervous, our muscles tighten--you may even catch yourself holding your breath. Instead, go ahead and take those deep breaths to get oxygen to your brain and relax your body.

Smile: Smiling increases endorphins, replacing anxiety with calm and making you feel good about your presentation. Smiling also exhibits confidence and enthusiasm to the crowd. And this tip works even if you're doing a webinar and people can't see you.

Work on Your Pauses: When you're nervous, it's easy to speed up your speech and end up talking too fast. Don't be afraid to slow down and use pauses in your speech. Pausing can be used to emphasize certain points and to help your talk feel more conversational. If you feel yourself losing control of your pacing, just take a nice pause and keep cool.

Don't Try to Cover Too Much Material: Yes, your presentations should be full of useful, insightful, and actionable information, but that doesn't mean you should try to condense a vast and complex topic into a 10-minute presentation.

Knowing what to include, and what to leave out, is crucial to the success of a good presentation. I'm not suggesting you skimp when it comes to data or including useful slides (some of my webinars have featured 80+ slides), but I am advocating for a rigorous editing process. If it feels too off-topic, or is only marginally relevant to your main points, leave it out. You can always use the excess material in another presentation.

***Actively Engage the Audience*:** People love to talk and make their opinions heard, but the nature of presentations can often seem like a one-sided proposition. It doesn't have to be, though.

Asking the audience what they think, inviting questions, and other means of welcoming audience participation can boost engagement and make attendees feel like a part of a conversation. It also makes you, the presenter, seem much more relatable. Consider starting with a poll or survey. Don't be put off by unexpected questions – instead, see them as an opportunity to give your audience what they want.

***Be Entertaining*:** Even if your presentation is packed with useful information, if your delivery bombs, so will your session.

Including some jokes and light-hearted slides is a great way to help the audience (and myself) feel more comfortable, especially when presenting them with a great deal of information. However, it's important to maintain a balance – after all, you're not performing a stand-up routine, and people didn't come to your presentation with the sole intention of being entertained. That said, don't be afraid to inject a little humor into your talk. If you're not sure about whether a presentation is "too much," run through it for a couple of friends and ask them to tell it to you straight.

***Admit You Don't Have All the Answers*:** Very few presenters are willing to publicly concede that they don't actually know everything because they feel it undermines their authority. However, since we all know that nobody can ever know everything about a given topic, admitting so in a presentation can actually improve your credibility.

If someone asks a question that stumps you, it's okay to admit it. This can also increase your credibility with the audience, as it demonstrates that, no matter how knowledgeable a person might be, we're all learning, all the time. Nobody expects you to be an omniscient oracle of forbidden knowledge – they just want to learn from you.

***Use a Power Stance*:** Practicing confident body language is another way to boost your pre-presentation jitters. When your body is physically demonstrating confidence, your mind will follow suit. While you don't want to be jutting out your chest in an alpha gorilla pose all afternoon (somebody enjoyed *Dawn of the Planet of the Apes* a bit too much), studies have shown that using power stances a few minutes before giving a talk (or heading to a big interview) creates a lasting sense of confidence and assurance. Whatever you do, don't sit--sitting is passive. Standing or walking a bit will help you harness those stomach bats (isn't that more appropriate than butterflies?). Before you go on stage, strike your best Power Ranger stance and hold your head high!

***Drink Water*:** Dry mouth is a common result of anxiety. Prevent cottonmouth blues by staying hydrated and drinking plenty of water before your talk. Keep a bottle of water at arm's reach while presenting in case you get dry mouth while chatting up a storm. It also provides a solid object to hurl at potential hecklers.

***Don't Fight the Fear*:** Accept your fear rather than trying to fight it. Getting yourself worked up by wondering if people will notice your nervousness will only intensify your anxiety. Remember, those jitters aren't all bad – harness that nervous energy and transform it into positive enthusiasm and you'll be golden. We salute you, O Captain! My Captain!

Dealing with questions

Many otherwise extremely competent and confident presenters will tell you that they really dread the question and answer session of a presentation

They seek ways to 'avoid' difficulty questions. But it doesn't have to be like that dealing with questions in a presentation is a skill which anyone can master.

Perhaps the most important thing to understand is that, as a general rule, if people ask you questions, even hostile ones, it's not to trip you up but because they genuinely want the answer.

INTERPERSONAL SKILLS

Building personal skills have long been the hallmark of winning individuals and organizations. Because the working environment is constantly changing, these personal skills need to be built, enhanced and refreshed on an ongoing basis.

As the Agricultural Scenario remains unpredictable, Extension organizations are being asked to deliver more with less resources.

While evidence shows that employers are making efforts to hang onto their skilled people in preparation for when the climate improves, many organizations still have a headcount freeze, resulting in people who leave not being replaced.

This brings huge challenges for the teams left behind in terms of productivity. But it also becomes an issue for extension professionals as they must continue to motivate and engage remaining employees. There's no doubting the importance of management in achieving targeted goals, and a recent report from the Institute of Leadership and Management found that 93% of managers were concerned that low levels of management skills are having an impact on their business goals.

To meet this challenges, if was found that, "if people are trained to improve interpersonal skills, there would a productivity gain, equal to a whole extra person.

Recent global research by a talent management group found that there were huge differences in productivity when comparing good and less effective employees. The majority of respondents (53%) said they weren't currently working with the best of people they had ever worked for – and said they'd be 20 to 60 percent more productive if they were working with better people boss. A quarter (26%) said they would be 41 to 60 percent more productive.

The ability to provide support without removing responsibility is the driving force behind the sense of empowerment, and extension professionals are seen as more effective when they perform supportive behaviours without taking over. While business and technical knowledge are important in extension professionals, the reasons most professional fall short, are shortcomings in their interpersonal and communication skills.

The difference between good worker and great worker is not always something one can put a finger on. It is a pleasure to work with some people ....And the secret with them in that they have good interpersonal skills.

The most important interpersonal skills required for effective performance are

1. Verbal Communication

Effective verbal communication begins with clarity. This often requires nothing more than slowing down and speaking more thoughtfully. Many people feel rushed to respond to questions and conversations immediately, but it is better to pause for a moment in consideration, especially if the question merits it. No one expects, or wants, a gun-slinging attitude in important conversations. A thoughtful person is generally taken more seriously.

The best part of this skill is the ability to stay calm, focused, polite, interested and to match the mood or emotion of the situation.

2. Non-Verbal Communication

Non-verbal communication is largely underrated and underestimated. Those who can communicate non-verbally can almost subliminally reinforce what they are saying verbally. They can also exude confidence, or any other emotion they feel, not to mention respond tactfully to a conversation without saying a single word.

Non-verbal communication is something that other people notice whether one is aware of their actions or not. Our body language is constantly speaking. Everything we do or don't do says something about us and how we are feeling. Our facial expressions (especially eye contact), our posture, our voice, our gestures with our extremities and even the way we position ourself physically in a room or amongst colleagues is constantly revealing our true attitude, for better or for worse.

3. Listening

While it may should simple, the fact of the matter is only a small percentage o people listen well. It's critical that we actually hear the person's message. Missing just a little piece can change the entire meaning or perception of that message. Listen for tone, inflection, and volume. For example," what's your problem?'' said in a soft tone would imply you want to assist someone, but said in a loud boisterous voice might indicate someone is angry.

4. Questioning

Questioning is a lost art that can serve many purposes. Questioning is something that often builds upon listening, but it is not merely a device for obtaining information.

Questioning is a great way to initiate a conversation. It demonstrates interest and can instantaneously draw someone into our desire to listen. Smart questions show that we know how to approach problems and how to get the answers we need. Fortunately, questioning can be learned more easily than other skills.

5. Manners

Good manners tend to make many other interpersonal skills come naturally. With extension becoming increasingly more global, even for small businesses, manners are more important than ever. A basic understanding of etiquette translates to other cultures and their expectations.

We are all guilty of assuming people are less intelligent if they have sub-par manners. This same judgement is reflected back on us by the people we interact with. Anyone who has visited other countries knows how sensitive its residents are to visitors' manners. Extension outreach and related interactions function in much the same way.

6. Problem Solving

A rare day would be one without problems. What makes this a skill is not necessarily how quickly one can solve a problem, but how we go about doing it. No plan is a guarantee, so there is always an element of risk. Some people can weigh risk better than others.

The key aspects of successful problem solving are being able to identify exactly what the problem is, dissecting the problem so that it is fully understood, examining all options pertaining to solutions, setting up a system of strategies and objectives to solve the problem, and finally putting this plan into effect and monitoring its progress.

7. Social Awareness

Being in tune to others' emotions is an essential interpersonal skill. This dictates how many of our other interpersonal skills should function.

Social awareness is crucial to identifying opportunities, as well. People will often unconsciously test others ability to respond to a social situation; for example, a person who is struggling professionally will be desperate for help but, naturally, wary on revealing the fact that they need it. Being able to identify something like this demonstrates that we are operating at a higher level of social awareness.

8. Self-Management

Self-management allows us to control our emotions when they are not aligned with what would be considered appropriate behavior for a given situation. This means controlling anger, hiding frustration, exuding calmness, etc. Undoubtedly there are times to show our true nature, but remaining composed is almost always the desired course of action.

9. Responsibility and Accountability

Responsibility and accountability are two reliable indicators of maturity. Saying that we are going to do something and then actually doing it is a sign of responsibility. This builds trust between oneself and those they rely on us and it encourages others to seek our counsel and assistance.

Holding our self accountable for our actions is one of the most difficult things to do, both professionally and personally. This is also a crucial element of conflict management. When conflicts arise between oneself and others, or when we have made a mistake or at fault, that is when accountability becomes difficult. Admitting to our mistakes isn't enough. We have to understand the situation fully and respond in a way that addresses the issue comprehensively.

10. Assertiveness

There is no denying the importance of being assertive. However, this is also where we are most likely to offend or come off as too aggressive. Being assertive is the only way to get our ideas onto a competitive table.

It also means standing up for what we believe it, defending our ideas with confidence, instructing others on what needs to be done, etc. we are all familiar with the fact that most people who ask for raises receive them; and yet very few of us are assertive enough to make it happen. When used tactfully, assertiveness can gain us a kind of respect that we won't be able to attain by other means.

ATTITUDE

Attitude is a hypothetical postulate, which implies a broad-based behaviour tendency in individuals. Attitudes have to be analyzed in the light of real life situations to which they are related. Attitude cannot be directly observed, but only through inference it is possible to understand.

Attitude is associated with physical neural state, verbal opinion, ideologies, and integrated philosophy of life. Attitude is the detailed direction of human behaviour. Attitude is a stage of sensitiveness and process to act.

Attitudes are evaluative statements; they refer one's feeling either favourably or unfavourably towards person, objects or events. Attitudes are nothing but how one feels about somebody or something.

Attitude is a "learned predisposition towards aspects of our environment." Attitude is a complex mental state involving beliefs, feelings, values and dispositions to act in certain ways.

Attitude is the degree of positive or negative affect associated with some psychological object i.e., may be symbol, phrase, slogan, persons, institutes etc.

The important features of attitudes are as follows:

Multiplexity: Attitudes are constituted by as number of elements. That is, there are many ingredients in the formation of attitudes. For example, family, society, environment etc.

Relations to needs: Attitudes vary in relation to the needs they serve, e.g., attitudes of a person towards sports and games may serve only entertainment needs of favourableness or unfavourableness towards something or somebody.

Valence: Valence refers to the magnitudes; could be low or high. Low valence means a person has an indifferent attitude towards something or somebody. High valence refers to a person having extremely favourable or unfavourable attitude towards something or somebody.

Related to feelings and beliefs: Attitudes are related to one's feeling and beliefs towards something or somebody.

Attitudes affects behaviour: Attitudes affect behaviour either positively or negatively because attitudes affect one's perceptions.

Attitudes undergo changes: Generally speaking, attitudes remain in a person for a longer duration. But the same can undergo changes of persons are exposed to different situation they like.

Individuals are not born with certain attitudes nor inherited. These are acquired and learned by individuals from the situation they face in their life. The formation of attitudes depend on various factors. Unlike personality, attitudes are expected to change as a function of experience. Hereditary variables may affect attitudes – but may do so indirectly. Let us discuss the factors that play a significant role in formation of attitudes.

Psychological factors: The attitude of a person is determined by psychological factors like his ideas, values, beliefs, perception etc. All these have a complex role in determining a person's attitudes. For example, if a person perceives that all the workers are lazy, he is likely to develop a negative attitude towards his workers.

Family: Family plays a significant role in the primary stage of attitudes held by individuals. Initially, a person develops certain attitudes from his parents, brothers, sisters and elders in the family. There is a high degree of relationships between parent and children in attitudes found in them.

Society: Society plays an important role in forming attitudes of an individual. The culture, the tradition, the language etc. All of which influence a person's attitudes. It is the society, the tradition and the culture which teaches an individual what is acceptable and what is not acceptable.

Political factors: Political factors such as ideologies of political parties, political leaders and political stability affect the attitudes of the people.

Economic factors: A person's attitudes also depends on issues such as his salary, status, work as such etc.

Change of attitudes: Changing attitudes is a long – term process that takes into account the whole man, his home life, goals, hopes and concept of himself. Change of attitudes is person-centred rather than work-centred.

Attitudes change when:

1. A person receives new information from others or media- Cognitive change.
2. Through direct experience with the attitude object – Affective change.
3. Force a person to behave in a way different than normal – Behavioural change.

Attitude can be changed by a number of sources including other people, family, media, church, or the object itself. "In analyzing the attitude-change process, you must consider the effect of who says what, how, to whom, with what effect".

The Power of Positive Attitude

The way we think, affects all aspects of your life. Learning to listen to your "internal dialogue" will help us recognize our thought patterns and how they may be affecting the way we handle the stressful situations of daily living.

Positive attitude helps to cope more easily with the daily affairs of life. It brings optimism into our life, and makes it easier to avoid worry and negative thinking. If we adopt it as a way of life, it will bring constructive changes into our life, and make us happier, brighter and more successful. With a positive attitude we see the bright side of life, become optimistic and expect the best to happen. It is certainly a state of mind that is well worth developing and strengthening.

Developing Positive Attitude

Here are some ways to help one develop a more positive attitude:

- *Listen to internal dialogue*: Divide one or more sheets of paper into two columns and, for a few days, jot down in the left column all the negative thoughts that come into head. Rewrite each thought in a positive way in the second column. Practice doing this in mind until it becomes a habit.

- ***Learn to communicate***: Not saying the things we feel can lead to a sense of frustration, hurt, anger or anxiety. If we find communicating difficult, or are afraid a arguments, take a course in communicating effectively.
- ***Get back to basics***: Reconnect with old friends, take the dog for a walk, visit an art gallery or listen to favorite music. Enjoy a long, relaxing bath, read a book, tell child a story, or ask an older relative to tell a one! It is the simplest thing in life that gives us the most pleasure.
- ***Help someone out***: The simple act of helping others helps us to feel good. Picking up groceries for an aging neighbour, volunteering at local hospital or reading to someone with failing eyesight.
- ***Finding Spirituality***: Research has shown that those who have developed their spirituality through associating with other spiritual individuals or having their own personal and unique beliefs live longer, more satisfying lives. The secret is practising those beliefs, live longer, more satisfying lives. The secret is practising those beliefs, either through organized spiritual associations, or simple meditation in a quiet place.
- ***Allowing oneself to be loved***: The ability to love and be loved is the most basic need. Finding ways to reconnect with others is extremely helpful in developing a positive attitude.
- ***Allow to laugh and find humour in the simplest of things***: Laughter is a powerful mood elevator. If one is feeling down, reading some jokes, watching a funny movie or just act ''silly'' once in a while. At times, it is a good thing to let oneself see the world through a child's eye.
- ***Participate in new physical and mental activities***: To improve confidence levels and coping mechanisms. It could be as easy as learning the meaning of new words or learning about new topics to build confidence.
- ***Follow the principles of holistic health***: Better nutrition and exercise help improve mood and attitude.

Obstacles in developing positive attitude: The common obstacles to developing a positive attitude are the types of negative thinking that distort the evaluation of situations.

Exaggerating Overestimating problems and under estimating abilities.

Over generalizing Taking an isolated event and assuming it always happens.

Personalizing Thinking everything revolves around oneself.

Thinking Seeing things as mutually exclusive, even when they're not.

Jumping to conclusions Drawing conclusions from limited information.

Ignoring the positive Focusing on one negative and forgetting about all the positives.

Setting yourself up for success

Nothing enhances a positive attitude more than success, so regard success as the normal state of affairs – and the lack of success as an exception.

Avoid perfectionism

Very few things are perfect in this world, so recognizing that falling short of perfection is not failure.

Helpful approach

Remember that there is no such things as failure – only outcomes. If one efforts produce an outcome that is less successful than we had hoped, it is not a failure. Instead, it, will change what we did wrong and next time we will do better.

Focus on the future

We can't change the past, but if we can decide where we want to go in the future, we will give oneself the best chance of getting there. Always aim high, and we too will make it a winning life.

Negative attitude

Negative attitudes are customary ways of thinking that eventually cause a misperception of a situation or event. Constant frustration can prompt negative attitudes. When frustrated, thinking can become increasingly inflexible and may ultimately become distorted. Negative attitudes could account for 90% of the times we experience frustrations.

Five types of negative thinking:

- **Magnifying:** Magnifying turns the consequences of an event into a catastrophe such as, ''I'm going to be fired.''
- **Destructive labelling:** This is an extreme form of over-generalization, making someone or a particular situation totally negative.
- **Imperative thinking:** Think of this as a list of flexible rules about how one should act-usually based on negative past experiences.
- **Mind reading:** This attributes to motives that explain other people's actions towards a person or event.
- **Divide and conquer:** This happens because of over-magnification and wanting others to support a particular position.

***Overcoming negative attitude*:** Working on personal strengths: **(F.R.E.S.H)F:** Finances; **R:** Relationships; **E:** Environment; **S:** Spirituality and **H:** Health try to write down one task for each of the Five letters (F.R.E.S.H) that can make an immediate improvements in.

- Stop listening to that little negative voice. Counter-punch the negative self talk with a substantial dose of positiveness.
- Look for alternative explanations. In many cases there could be hundreds of them. Unfortunately when we are frustrated we are unable to see the forest for the trees.
- Moods are contagious, both positive and negative. Do something that will place ourself in a positive mood, such as a daily delicious habit or speaking with anyone who is known for having a positive attitude or mood.
- Control the frustration. Speak slowly, breath deeply into your heart, choose to remain calm or excuse oneself from the discussion-take a time out. Stop drinking coffee/tea.
- Stop looking at the rear view mirror. The vision of each event or situation is totally up to us. We can either choose to be negative or positive it is our choice. Or, we can either focus on the negative past or look forward into a positive future. Remember this quote: "Past failures don't equal current successes."

Thus, positive attitude can be turned a powerful skill forrendering better performances at work places in general and specifically in the extension front.

TEAM WORK AND TEAM BUILDING

Organizations are essentially about people working together; yet so often they fail to capitalize upon the full potential of this. A team can accomplish much more than the sum of its individual members.

A team is a group of people working for a common purpose and goal. When groups perform operating tasks, they act as a team and seek to develop a cooperative state called teamwork. In other words, a team is "a small number of people with complementary skills, who are committed to a common purpose, goals and approach for which they are mutually accountable.

Teamwork is about individuals working together to accomplish more than what they can do individually. The success of the team would depend not only upon individual skills, but on the way those individuals support and work with each other.

Over the past few years we have seen many approaches aimed at increasing organizational effectiveness. Organizations today pay increasing attention to the training and development of its employees, particularly those who hold executive positions. Most of that activity is centered on improving individual skills, knowledge and experience. But many organizations are finding that this is not enough, that the real key to success is the way in which groups of people relate to and work with each other. Teamwork improves these things.

Competition is the life blood of many organizations. But there is just rewards of their deserved success and others can accept that the best man, system or policy succeeded. A meeting is another key indicator of good teamwork in an organization. The main reason for a meeting is to utilize the collective skills of a group of people whilst working on common problems or opportunities. The effective team uses external help constructively by recognizing the unique contribution and viewpoint that it can bring, but will always maintain ownership of its own problems and its own destiny.

Characteristics of Effective Teamwork

- People can and do express themselves honestly and openly. Conversation about work is the same both inside and outside the organization. Mistakes are faced openly and used as vehicles for learning and difficult situations are confronted.
- Helpful competition and conflict of ideas are used constructively and team members take pride in the success of their team. Unhelpful competition and conflict have been eliminated.
- There is a good relationship with other teams and departments. Each values and respects the other and their respective leaders themselves comprise an effective team.
- Personal relationships are characterized by support and trust, with people helping each other whenever possible.
- Meetings are productive and stimulating, with all participating and having ownership of the actions that result from the decisions made, new ideas abound and their use enables the team to stay ahead.
- Boss subordinate relationships are sound, each helping the other to perform his role better, and the team feels that it is led in an appropriate way.
- Personal and individual development is highly rated and opportunities are constantly sought for making development happen.
- There is clear agreement about and understanding of objectives and of the roles that the team and its individual members will play in achieving them.
- External help is welcomed and used where appropriate
- Finally, the team regularly reviews where it is going, why it needs to go there, and how it is getting there. If necessary, it alters its practices in the light of that review.

All of these mean that the office is a happy place to be in. people enjoy themselves wherever possible but this enjoyment is conducive to achievement, not a barrier to it. They get satisfaction from their working lives and work is one of the places where their needs and aspirations are met.

Team Building

Commitment : Commitment to the purpose and values of an organization provides a clear sense of direction. Team members understand how their work fits into organizational objectives and agree that their team's goals are achievable and aligned with the organizational mission and values. Commitment is the foundation for synergy in groups. Individuals are willing to put aside personal needs for the benefit of the work. To enhance team commitment, leaders might consider inviting each work team to develop team mission, vision and values statements that are in alignment with those of the corporation but reflect the individuality or each team. These statements should be visible and 'walked' every day. Once a shared purpose is agreed upon, each team can develop goals and measures, focus on continuous improvement, and celebrate team success at important milestones. The time spent upfront getting all team members on the same track will greatly reduce the number f derailments or emergency rerouting later.

Contribution : The power of an effective team is in direct proportion to the skills members possess and the initiative of its members. Work teams need people who have storing technical and interpersonal skills and are willing to learn. Teams also need self-driven leaders who take responsibility for getting things done. To create confidence in a team work leaders can highlight the talent, experience and accomplishments of team members and emphasizes on past team successes. The confidence of team members can be bolstered by providing feedback, coaching, assessment and professional development opportunities.

Communication : Communication is equally important to a team's success. To assess work performance, members must provide honest feedback, accept constructive criticism and address issues head-on. To do so requires a trust level supported by direct, honest communications. Positive communication increases the energy of a work team. Members can then talk about what they like, need or want to enhance team communication, leaders can provide skill training in listening, responding, and the use of language as well as in meeting management, feedback and consensus – building.

Cooperation : Success depends upon recognizing the degree of interdependence within the team. Leaders can facilitate cooperation by highlighting the impact of individual members on team productivity and clarifying valued team member behavior.

Conflict Management : It is inevitable that in teams with bright, diverse thinkers, there will be conflict among team members from time to time. If leaders help teams manage conflict effectively, the team will be able to maintain trust and tap the collective power of the team. Teams manage conflict better when members learn to change their paradigms (mindsets) about conflict in general, about other parties involved, and about their own ability to manage conflict.

Change Management : Teams must not only respond to change, but actually initiate it. To assist teams in the management of change, leaders should acknowledge any perceived danger in the change and then help teams to see any inherent opportunities. They can provide the security necessary for teams to take risks and the tools for them to innovative; they can also reduce resistance to change by providing vision and information, and by modeling a positive attitude themselves.

Connections : Team develops strong connections among its members, peer support manifests itself in many ways. Colleagues volunteer to help without being asked, cover for each other in a pinch, congratulate each other publicly, share resources, offer suggestions for improvement, and find ways to celebrate together. A few ideas for developing and maintaining such connections are : allow time before and after meetings for brief socializing, schedule team lunches, create occasional team projects outside of work. Circulate member profiles, take training together, and provide feedback to one another on development.

Techniques in Team Building

Role Analysis Technique (RAT) intervention is designed to clarify role expectations and obligations of team members to improve team effectiveness. In a structures series of steps, role incumbents, in conjunction with team members, define and delineate role requirements.

Interdependency Exercises is done if the team members express a desire to improve coordination among themselves and among their units.

Role Negotiation Technique is used when the causes of team ineffectiveness are based on people's behavior that they are unwilling to change because it would mean a loss of power or influence. The change effort is directed at the work relationships of the members. It avoids probing into the likes and dislikes of the members for one another and their personal feelings for one another.

Appreciation and Concerns Techniques are use if the interview data suggests that one of the deficiencies in the interactions between members of the group is the lack of appreciation, and that another deficiency is avoiding confronting concerns and irritations.

Responsibility Charting helps to clarify who is responsible for what on various decisions and actions.

Visioning is a term used for an intervention in which group members in one or more organizational groups develop and / or describe their vision of what they want the organization to be like in the future. The time frame may be anywhere from six months to five years in the future.

Force-Field Analysis is a device for understanding a problematic situation and planning corrective actions.

Team building and development is the process of getting the best out of individual members of the organization and therefore has to be intelligently done. There are certain prerequisites for the team building and development that needs to be not only carefully learnt, but intensely practiced. The knowledge and skills of the team builder are important, apart from those of the team members. Another important step in team development is overcoming obstacles to effective teamwork.

References

1. **K.Vijaya Raghavan & Y.P.Singh (1988),** "*Managing human resources within extension (Chapter 14); Improving agricultural extension:* A reference manual by FAO and International programme for agricultural knowledge system, US".
2. **Indu Grover, Lali Yadav & Deepak Grover (2002),** "*Extension Management*" published by Agrotech Publishing Academy, Udiapur.
3. **PC Tripati & P.N. Reddy (2004)** "*Principles of Management*" published by Tata McGrawhill Publishing Company Ltd., New Delhi.
4. **Dr.B. Rathan Reddy (2005),** "*Team development and Leadership*" published by Jaico publishing house, Mumbai.
5. **Ali Hasan Obaid Khalil et al (2008),** "*Extension worker as a leader to farmers : Influence of extension leadership competencies and organizational commitment on extension workers performance in Yemem,* The journal of International Social Research, Vol.14.
6. **C.S.G. Krishnamacharyulu & Lalitha Ramakrishnan (2010),** "*Personality development interpersonal skills & career management*" published by Himalaya Publishing Home, Hyderabad.
7. **C.S.G. Krishnamacharyulu & Lalitha Ramakrishnan (2010),** "*Personality development, interpersonal skills and career management*" published by Himalaya Publishing House, Mumbai.
8. **Uday Kumar Halder (2011),** "*Leadership and team building*" published by Oxford University Press, New Delhi.
9. **Dr. K. Alex (2011),** "*Soft skills*" published by S.Chand & Company Ltd, Ramnagar, New Delhi
10. **S.P. Sharma (2012),** "*A youngsters guide to personality development*" published by V&S Publishers, New Delhi.
11. **M.S. Neelam S Bhargava & Mr. Gaurave Bhargava (2012),** "*Team building & leadership*" published by Himalaya Publishing Home, Hyderabad.
12. **Ishita Bhawn (2014)** "*Improve your presentation skills*", published by V& S publishers, Hyderabad.
13. **Sani Yakube Gombe Turiman Suandi et al (2015),** "*Extensin worker competencies needed for effective management of self help groups in Gombe state;* International Journal of Education and Training (INJET) 1 (1) June, 2015.

18 Capacity Building of Extension Service Providers for Professional Excellence

- *Prof. S. V. Reddy*

Preamble

India`s agricultural sector is faced with severe challenges. A better performance in Agriculture and alied subjects namely; Animal Husbandry, Horticulture, Fisheries, Sericulture etc., (at least 4 % growth) is considered as one of the key instruments to achieve more inclusive and sustainable growth. Better performance in agriculture depends on many factors and importantly knowledge and skills, extension and advisory support place a lead role. Hence capacity building of extension service providers assumes a critical input in Indian agriculture.

Indian Extension System has considerably weekend in terms of Human Resource and Capacity over the last two decades. To meet the demand of Futuristic agricultural Extension, there is a need to operationalise the vision and build the needed capacities.

The efforts in building capacities in extension are adhoc, disjointed and reaching only a fraction of actual clients. Hence there is need to expand the coverage to unreached and also to maintain the quality input during the process.

The Capacity Building - Concept

Capacity Building of local, regional and National level has become central to the goals of development organizations to reduce poverty and improve livelihoods.

Capacity is the power or ability of something, a system, association, groups and individual to conduct and produce results appropriately (UNDP 2002).

Capacity building is an ongoing process through which individual, group's, organization and societies enhance their ability to identify and meet development changes and challenges.

Capacity building in general relates to enhance or strengthen a persons or organization capacity to achieve their goals. The capacity building also increase the abilities and resources of persons, communities and organizations to manage change.

Building Capacity of Extension service providers namely managers of agricultural extension programmes, extension personnel, NGOs / CBOs / civil society organizations, input dealers and entrepreneurs, farmers and farm women at National, State, District, Block, Village and Community Level is central to this process. Many of the extension service providers need capacity on emerging subject matter areas and lack extension management as well as communication skills to address Futuristic Extension Challenges.

Many Recommendations and Few Actions

Indian agricultural in general and capacity building programmes in particular had the benefit of valuable recommendation of so many expert committees and commissions. The study of all these reports would reveal that all things possible to improve the state of capacity building programmes and training institutions have been recommended repeatedly. However, many recommendations were ignored and not implemented due to various considerations.

The capacity building programme of Extension Services Providers at all levels are essentially to enhance sustainable livelihood of farmers, to ensure farming an attractive profession and to make the Nation primarily self reliant and also an important exporter of value added farm produces. For about 80 million poor families of India, the cost – risk – return structure of farming is becoming adverse. There fragile poverty economy further trumbles beyond immediate recovery from the signal shock of either a crop failure or uneconomic market prices in one season. While capacity building in terms of knowledge and skills are crucial to improve the farm economy and the livelihood of the farm production dependent families, this alone with out complementary policy support and services is unlikely to impact Indian Agriculture.

Future Scenario

Capacity building programmes are inter related with **knowledge generation, Management and transfer.** The futuristic extension demands empowering farmers on the advance knowledge and skills. The future scenario of agriculture with developed pockets and contract farmers will be ranging from place of seed by aero tractors, monitoring of irrigation and plant health by computers, milking of cows by robots and GPS receivers to lazar leveling for precession, farming. Biotechnology, eco –technology and information technology will play a decisive role. Farmers tend to be more efficient and analytical with labour time and money. The voice of young men and women will be heard demanding for quality services. Therefore, the transition from a resource based to a technology based system of agriculture place a greater responsibility on the capacity building of extension service providers with the latest knowhow and do how. This again needs an input from a vibrant research and extension systems.

Revitalizing and strengthening research system

The knowledge in agriculture is generated by the formal and informal systems. The ICAR, SAUs and private sectors places a major role in knowledge generation. However, the formal systems are constrained by bureaucratic rigidity, Duplication in research and inadequate resources including man power. If a quality and capacity building programmes have to be designed and implemented there is need for promoting relevant knowledge generation by blending modern and traditional knowledge's. Shift in research priorities and approaches focusing more on farming systems research, evolving Extension Methodologies / Methods and their combination for transfer of technologies. Institution of effective mechanism to minimize duplication in research, strengthening basic and strategic research is important. Introduction of private public partnership arrangements in knowledge generation and providing adequate funding infrastructure and human resource is vital to strengthen research system to generate technologies for addressing the needs of futuristic extension.

Reinventing Extension

A new direction for agricultural extension is needed that capture the institutional reforms towards market oriented, privatizing innovations and non market decentralizing reforms. The new direction must also accommodate many paradigm shifts taking place in the world; the green revolution has shifted to ever green revolution; the commodity approach has moved to an integrated systems approach; mono-disciplinary to multi and interdisciplinary approach, Thrusts in technology research have shifted to eco-technology, biotechnology and information technology which are congruent with productivity, profitability, equity and sustainability. The supply-driven approach has become demand-driven. Farm employment has sifted to off-farm and non-farm employment. The definition of food security is evolving; there is increasing, importance on economic access, on self –reliance rather than self-sufficiency, Extension role has attained empowerment rather than confining to adoption and on skills, self-learning, information technology, nutrition, health awareness and environment protection.

In current times, the main global forces of change that are affecting or likely to affect the existing structure, mandate and practices of national agricultural extension systems in developing countries are globalization and market liberalization; privatization; commercialization and agribusiness; democratization and participation; environment concerns; disasters and emergencies; information technology breakthroughs; rural poverty, hunger and vulnerability; the HIV / ads epidemic; sustainable development; echo technology / biotechnology, biodiversity and genetic engineering; market led extension, Integrated farming systems diversification and value addition and integrated, multidisciplinary and holistic development. The capacity building programmes have to address the above emerging challenges.

Therefore, in order to build the Capacity of Extension Service Providers to meet the demands of Futuristic Extension there is needed for revitalizing research and reinventing extension for providing needed knowledge and skills to empower the farming communities.

Capacity Building Methods and Approaches

Many view capacity building has synonyms to training. To me Capacity building includes a wide range of methods and approaches such as Training Programes, Workshops, Seminars, Staff Meeting, Mentoring, field trips, Farmer Field Schools (FFS) / Farmer Life Schools (FLS), demonstrations Exposure visits, use of documents, library, on the job attachment village learning activities, ICT tools and mass media demonstration.

Capacity building methods, experimental learning / attachments need to be carefully combined to take advantage of their relative strengths and weaknesses to empower Extension Service Providers with quality input.

Training

As described in previous pages, the challenges of extension in the twenty first century are plenty. Training as a key input for human resource development can contribute a lot to face the challenges by all concerned. Although there are significant effort made during the past decade to improve the quality and relevance of training, still there is a greater scope for further improvement since the training needs are dynamic.

(a) Institutional Training

The institutional training is being imparted at National, Regional, State and District Level to top level, middle level, district level, community level Extension Service Providers. While National and Regional Level Institutes such as NAARM, MANAGE, EEIs, NIRD, SAUs, ICAR institutions on different crops and enterprises cater to Training of Trainers; the State and District Level institution mostly KVKs , FTCs and SAMETIs shoulder responsibility of training the District, community level Extension service providers including farmers / CBOs / NGOs. Thus, there is need for clear-cut roles and clientele demarcated to avoid duplication. The MANAGE, NAARM and EEIs have to play a crucial role with emerging needs of Futuristic Extension as Master Trainers.

The are major training institutions and training providers at National, Regional, State and District Levels which can able to caterer to address training needs of futuristic agricultural extension are given in (Annex 1).

In this direction, MANAGE, NARAM, NIRD, ICRA institutions, NIPHM, Extension Education Institutes, SAUs, SAMETIs, KVKs, ATMAs, NGOs/CBOs, Private Sector including input dealers and Agriprenuers play a major role in discharging the training functions. Oflate, farmer's organizations and panchayats are coming up as training providers. Farmer to Farmer Extension plays a critical role in this process.

The National Institute of Extension Management (MANAGE) at National level Extension Education Institutes at regional level, SAMETIs` at State Level and Krishi Vignan Kendras (KVKs) at District Level have been playing an important role in training Extension Service Providers in subject matter and extension management. The manage programmes on ACABC and Desai are commendable. The Extension Education Institute since 50 years has rendered great service in this direction. The EEIs have emerged as leader, in providing quality training to extension personnel and as a result of which extension workers were able to understand the dynamics of extension planning, management communication and monitoring. Besides, the trained Extension personnel have in turn emerged as a good communicators and also motivators of farmers towards change. EEIs were also able to provide consultancy to various organizations thus, a new direction to training has emerged.

(b) Field Based Training

There are several methods / approaches for field training. The following are some of the important methods.

1. Farmer Field Schools (FFS) is an approach through which farmers undergo a field oriented, discovery based training that enable them to grow a healthy crop / efficiently mange a farming system and promote better quality of life in a healthy environment. FFS is recognized as an effective extension tool, which can be used for empowering the farming community, developing self confidence of people and improvement in human, economic and social capital. Farmers learn by doing. Agri Eco System Analysis (AESA) for taking rational decision on resource Management, Group Dynamics and Life Skills in addition to Farming Skills forms a Central subject for training. The multiplier effect is realized through Farmers to Farmer Extension.
2. **Farmer Life School (FLS)** is an extension of FFS where in farmers undergoing training through FFS are imbibed knowledge and skills on resources confronting social / economic concerns at personal and community level. This will be a life long problem solving exercise with participatory approach mainly to instill confidence among them to deal with socio-economic problems. It is a good tool for changing attitudes towards farming and life as a whole.
3. **Demonstrations:** It is a age old method used for imparting knowledge and skills on farming related aspects. Mostly the production potentiality of different crops / livestock / farming systems well as critical skills are demonstrated as a transfer of technology tool.

(c) ICT / Mass Media Enabled Services

The ICT and mass media enabled services plays a crucial role in capacity building of extension services providers. Infact this initiative has to take a primary role in imparting knowledge and skills in future.

- Internet access to all extension service providers
- Video conferences facilities in apex training institutions
- Strengthening community radio and farm TV programmes
- Kisan Call Centres / Kisan Portals
- Distance Learning Programmes
- Establishing Village Knowledge Centres and State / National Knowledge Centers
- SMS based ICTs
- E-Sagu concept
- Use of house hold devices for spot data entry and subsequent updating through voice recognition etc

(d) Exposure Visits: Field Visits and Exposure Visits are used in Combination of Training demonstrations, FFS etc to primarily expose the participants to learn new knowledge and skills. It also provides motivation for them to adopt the technologies.

(e) Other Methods for Capacity Building includes;

- Workshops / Seminars
- Meetings
- Exchange visits of staff and farmers
- Attachments to leading farm science spots
- Melas / Exhibitions
- Publications / library

The above are only some of the methods / approaches for capacity building. No single method can give desirable results. A combination of different methods can able to yield better results for which Extension Research plays an important role.

The role of NGOs / CBOs / Panchayat / Farmers organizations / Farmers have to be well recognized in the capacity building effort and the training Institutions need to build partnership for effective extension specially to meet the demands of Futuristic Extension.

Training Areas for Futuristic Extension

Based on the overall priorities, and taking due cognisance of the emerging challenges, of Futuristic Extension, the following priority areas are proposed for HRD and training of farmers and extension personnel during XII plan[1].

- Climate change implications and coping mechanisms
- Scaling up of water productivity in agriculture
- Scaling up resource use efficiency in agriculture and allied sectors
- Scaling up energy use efficiency in agriculture
- Selective mechanization in agriculture
- Secondary agriculture (post harvest handling, storage and processing)
- Innovative extension models and approaches
- ICTs for knowledge and enterprise management
- Group based approaches for production, processing and marketing
- Micro-level implication of IPR, GATT , etc.,
- Bio-diversity conservation and management for sustainable use and benefit sharing
- Integrated farming systems
- Dryland agriculture
- High value Horticulture crops
- Conservation agriculture
- Organic farming
- Modern Animal Husbandry Practices
- Dairy Models
- Supply Chain Management
- High value farming
- Aquaculture
- Marketing - domestic and export, agricultural finance and insurance planning
- Management of development (CDAP, BAP, SAP) project formulation, implementation, monitoring and evaluation.
- Agri clinics and Agri business
- Entrepreneurial Skill Development Programmes
- Negotiation skills for farmers and other stakeholders

[1]Most of the training observation and suggestions are taken from recommendation Subgroup III on HRD, Training and Accreditation of the working group on Agricultural Extension in Agriculture and Alied Sectors for formulation of XII plan

- Public awareness on development programmes for politician and media personnel
- Participatory Approaches for Resource Management
- Institutional Development in Agriculture
- Result Oriented Management

A quick analysis of details of existing institutional framework for training in agriculture reveals the following:

Observations, Critical Gaps and Issues of Concern

- The training efforts of various agencies / institutions reach only a fraction of the actual client groups. Hence, there is a need to extend both the reach and coverage of training to the unreached by evolving multiplier mechanism like farm schools, open and distance learning, information and communication technologies (ICTs) augmented training (web-based, on-line) as well as through conventional Radio, television, etc)
- Training infrastructure, by and large, is very poor in terms of facilities like hostels, classrooms, laboratories, audio-visuals, etc. There is need for immediate attention for this to improve training effectiveness.
- Training process also needs a re-look in most of the training institutions. While training needs assessment is done mundanely, training design and delivery are left much to be desire, while the post training evaluation and follow-up are a rarity. Training is done as a ritual more for fulfilling the `target numbers` rather than as a powerful HRD tool. Creativity and innovation in training needs identification, curriculum development, training design, training methods, implementation, evolution and follow-up are the need of the hour. It is observed that limited human, financial, infrastructural and institutional capacities have resulted in poor state of visibility of most of the training institutions.
- Accreditation of training institutions and trainers is non-existent even in its most rudimentary format. This, need to be addressed by evolving appropriate institutional, structural and operational framework for a rigorous accreditation mechanism for institutions and trainers alike.
- Absence of a well-stated HRD policy at all levels in agriculture and allied sectors
- Institutional base for training is
  - Adequate in terms of number in agriculture
  - Evolving in fisheries and dairying
  - Inadequate in veterinary and animal husbandry including poultry, piggery, goatery, horticulture.

- Number and quality of trainers is inadequate to support increasing training demands.
- Lack of training data base at all levels.
- Training in business as usual approach to meet target numbers
- Training components of the central schemes, centrally sponsored schemes and state schemes overlap and are diffused.

Teaching

There is need for revitalizing and reorienting the teaching curriculum to meet the challenges of futuristic agricultural development. This calls for substantial changes in the curriculum of undergraduates and postgraduates programmes. While there is need to make modifications in the existing courses, it is also suggested the new courses could be introduced wherever necessary based on situational analysis and based on the subject areas suggested above.

The Dean`s committee on agricultural education has recommended the following for strengthening agricultural education which will further support the futuristic agricultural extension needs and priorities.

- Experiential learning programmes have been recommended in all disciplines. For Home Science, a tow year professional programme has been suggested with one year of industrial / agribusiness attachment.
- Flexibility in the course curriculum has been recommended.
- Courses on entrepreneurship development, communication skills, computer knowledge, agribusiness, environment science, biotechnology, etc., to e included in all disciplines,
- Practical content to be increased substantially to 50-60% of the total course load.
- Major strengthening of Food Technology programme recommended by HRD to meet emerging needs.
- Initiation of M.Tech in Food Technology in all Colleges recommended.
- Recommendations have been made for organizing non-formal education programmes in all SAUs
- Full support for the creation of one new development in any one merging areas in each faculty has been recommended.

Recommendations/Suggestions For Revitalizing Training Programmes

The following strategies need to be initiated for improving the quality and relevance of the training programmes.

- In order to meet the needs of training there is also a need for;
- Capacity Building of all staff of the Training Institutes periodically through different mechanisms.
- Building Staff specialization in some important areas.
- Providing opportunities for staff to undertake action research programmes which will enable them to be in touch with Live Extension and emerging needs.
- Documentation of success stories in the field for use in training as case studies.
- Promotion of Consultancy Services which can help them for field level experience and also to provide supplementary funds for Institutes Promotional activities.
- Exchange visits of staff from one institute to other for a limited period to expose to different learning situations.
- Providing equal emphasis for follow-up of training programmes in terms of resources and time.
- Encouraging outsourcing and / or organizing programmes in public – private partnership arrangements by identifying partner institutions.
- Provision of incentives to trainers like special pay for attracting best talent to training institutions.
- Filling up all vacancies of staff in training institutions and upgrading the training infrastructure.
- Testing and certifying staff and farmers competence for empowerment and upgrading their skills in recognized institutions.
- Developing Human Resource Development and Training Policy for Extension System in Agriculture and alied Sectors.
- Extension Education Institutes have done great service so far and they have major role in future for undertaking Distance Education, Enterprenial skill Development Programmes and also to organize certificate / PG diploma courses in a selected extension fields of specialization.
- MANAGE AND NAARM to act as centres of Excellence for Extension and Research Management respectively providing policy and management support to extension training providers.

- Undertaking Capacity Building programmes to politicians and media personnel on Futuristic Extension Schemes, Models and Approaches is by MANAGE, NAARM and EEIs.
- Revamping KVKs and SAMETIs in terms of staff strength and capacity building for Futuristic Extension needs.
- Development of quality resource materials printed and electronic version for capacity building activities.
- Optimizing training process for effectiveness (Planning, monitoring and evaluation) MANAGE as National hub and EEIs as regional hubs and SAMITEs as District levels need to be developed.
- Strengthening human resource for Extension in allied sectors such as veterinary, animal husbandry and fisheries.
- Develop a capacity assessment of institutions and service providers once in five years and then send them for training in deficit areas or recruit consultants to fill up the gap.
- Initiating a comprehensive study on training need assessment and manpower development in order to formulate training programme at various levels based on the study findings.
- Introduction of innovative training programmes methods and strategies to make the extension programmes more useful and interesting.
- Rationalizing and harmonizing the training programme among the training providers at various levels.
- Provision of adequate resources for research and development in several areas of extension training.

Conclusion

In the present context, India`s agricultural sector is faced with severe challenges. These include arresting decline in productivity, producing quality products with less costs for highly competitive external markets. The nation is marching towards sustainable agriculture development. India requires agricultural growth of 4 percent GDP within an expected economic growth of 7 to 8 percent. Poverty reduction will be possible only when smell and marginal farmers and farmers from rain fed areas participate fully in economic growth. Agricultural extension has to play a pivotal role in meeting these challenges. We shall be able to transform subsistence farming to modernized action of agriculture and eradicate poverty. This calls for restructuring, retooling and revitalizing the teaching, training, research in agricultural extension as well as retraining the trainers and extension service providers. There is greater chance now for extension to bridge the gap between technical know-how through intensive capacity building mechanisms and farmers do-how. There is a need for instituting rewards and awards for the Capacity Building trainers extension service providers moving on the path of sustainable agricultural production and those who reach the unreached. There is lot of duplication of efforts with multiplicity of agents doing training and extension work without convergence. Similar, is the case with Capacity Building Issues. The motivated trainers extension workers and the result- oriented managers are bound to meet the futuristic challenges.

We are now sitting in a time bomb. Unless we act in right earnest, it may explore on our children faces. I thought it is important to quote the Father of Nation Mahatma Gandhi words;

"Civilization, the real sense of the term consists not in the multiplication, but in the deliberate and voluntary reduction of wants. This alone promotes real happiness and contentment and increases the capacity for service"

- Mahatma Gandhi

19 Research in Extension – Current Scenario, Problems and Challenges

- Dr. R.M. Prasad

Introduction

In research, knowledge is acquired using scientific method. Scientific method is a body of techniques for investigating a phenomenon and acquiring new knowledge, as well as for verifying, correcting and integrating previous knowledge. It is based on gathering observable, empirical and measurable evidence, subject to the principles of reasoning. Extension research proposes specific hypotheses as explanations of social phenomena and design studies that test these predictions of accuracy. The research process has to be objective so that the scientists do not bias the interpretation of the results or change the results outright. How far extension scientists are objective and sincere in doing research and guiding research? This is a serious and resounding question that has to be clearly answered. This is really a challenge faced by our extension scientists.

Extension scientists have to move beyond exploratory research, which answers "What is where?" The results of exploratory research are not usually useful for decision making. Descriptive research answers the question "What is what?". Extension research employs three main types of descriptive methods- observational methods, case study methods and survey methods. These are the common methods used by the extension scientists in their research methodology. Here, it is to be borne in mind that descriptive methods can only describe a set of observations or the data collected, but cannot draw conclusions from that data about which way the relationship exists. Explanatory research has to be employed to know the cause and effect relationship. Though extension scientists use different statistical tools to explain the cause-effect interaction, in many cases, the relationships are not properly inferred and explained, which presents only loose inferences, which may not be valid. While research tools and techniques in core disciplines from which extension borrowed its research methods have evolved significantly, extension research still depends heavily on many of the outdated tools (Sivakumar, 2013).

Researchers can employ experimental research to present strong evidence for causal interpretation. One important feature that differentiates experimental research from explanatory research is that instead of simply measuring two variables, the researcher can manipulate one of them in the case of experimental research. Extraneous variables can also be controlled in experimental research.

Present Scenario of Extension Research

Extension education originally was conceived as a service 'to extend' research-based knowledge to the rural sector to improve the lives of farmers. It thus included the concept of technology transfer. Over the years, the field of extension has gone beyond technology transfer to facilitation, beyond training to learning and also includes forming farmer groups of farmers dealing with marketing issues, and partnering with a wide range of service providers and other agencies. Here, it encompasses the concept of agricultural advisory service. One of the important challenges facing extension research is that while the concept of extension as practiced in the field had gone beyond technology transfer, extension research is still embedded in the concept of technology transfer, to a considerable extent. There are exceptions, but are negligible in number.

As everyone knows, the growth of any discipline is directly proportional to the 'creation' of knowledge in that discipline. Research is the means for 'creation' of knowledge. However, we as extension practitioners, extension administrators and extension teachers, had not paid serious attention to extension research. This is quite evident from the way we look at and practice extension research in our country. For instance, MANAGE (national level institution for giving policy direction to extension functionaries) has not seriously considered research as one of its key mandate and activity, as evident from its website and reports. Though some efforts were initiated by Suresh Kumar Committee as part of 12th Plan exercise, no tangible results could be achieved. Likewise, ICAR, an institution which spearheads and co-ordinates the teaching, research and extension functions in the country has not initiated any co-ordinated research project on extension research till now. (There are many All India Co-ordinated Research Projects of ICAR on various themes/subjects operating in the country). Extension research undertaken by the State Agricultural Universities in the country is largely confined to research by the PG (M.Sc and PhD) students. In this context, it is to be mentioned that under NATP and NAIP scheme of ICAR, there were some innovative and field oriented research projects in extension undertaken by our extension scientists. Compared to research output of other disciplines, extension research often cuts a sorry figure. This also reflects the limited growth of the discipline of Extension education.

Lack of self identity of the discipline of Extension is pointed out as a constraint for Extension research, as Extension as a discipline has drawn its contents from various other disciplines, which is a weakness, as well its strength. In research, it can be considered strength in the sense that there is much scope and space for inter-disciplinary research to be carried out. Limited scope for experimental research is also pointed out as another constraint. Experimental research is not the only answer for quality research output. This can be well demonstrated from the works of Amartya Sen and Elinor Ostrom.

Amartya Sen won Nobel prize in Economics in 1998 for his contributions to welfare economics and social choice theory. Social choice theory is a theoretical framework for measuring individual interests, values or welfare as an aggregate towards collective decisions. It is methodologically individualistic, in that it aggregates preferences and behaviours of individual members of society. Sen proposes interpersonal comparisons based on a wide range of data. His theory is concerned with access to advantage, viewed as an individual's access to goods that satisfy basic needs, freedom and capabilities. Welfare economics as propounded by Sen is more related to Extension than any other discipline.

Elinor Ostrom won Nobel prize in Economics in 2009 for her analysis of economic governance, especially the commons. Ostrom's work emphasised how humans interact with ecosystems to maintain long term sustainable resource yields. Ostrom and her many co-researchers have developed a comprehensive 'Social Ecological System (SES)' framework, within which much of the still evolving theory of common pool resources and collective self governance is now located. There are basic resources like water and land which assume greater global attention because of their decreasing availability and gradually low carrying capacity respectively. These will have significant adverse effect on future agricultural production and food security for the growing human population. In this sense, SES framework of Ostrom has much relevance and application for extension research related to NRM.

It is satisfying to note that of late, there is a shift in the outlook and approach of our extension scientists, who earlier viewed the practice of Extension as 'knowledge applied' or 'knowledge extended'. Recently, the extension scientists have begun to recognise Extension's role in 'creating knowledge', which is a welcome step (Warner *et al*,1998). Extension scientists have an important role to play, especially in the case of Participatory Action Research (PAR). PAR not only incorporates the collective knowledge of the community, it increases the likelihood that research results will be applied by giving the community ownership over the research process and its results. In this sense, PAR is not just a method or toolkit, but an entire approach to research, a cyclical process, which is empowering, researching with and for, rather than, on people. It builds participant's capacities and brings to centre stage the margins of society,economy and culture (Predota, 2009).

What is Wrong with Current Extension Research?

1. The concept and practice of present research designs in most academic research settings needs to be redefined, to make research relevant to the present context. Currently, the research paradigm in most academic institutions tends to be linear in design, as below (Smith and Helfenbein,2009)

Researcher generates questions	Researcher collects data and measures using structured designs and tools ('drop in' fashion)	Researcher publishes results in peer-reviewed journals, read primarily by other researchers

2. In most cases, extension research looks at farmers or extension personnel as subjects and crop/farming system as settings of the study.
3. The dissemination of research findings into the public sphere is very often limited to sharing of results with other scientists or students.
4. Of late, though farmer participatory research has gained popularity among researchers, it is observed that in farmer participatory research, research or extension are too dominant, farmers too deferent, to arrive at genuinely joint decisions on research topics, designs, analysis and dissemination of results (Katz *et al*, 2007)

Issues and Concerns

(a) In many cases, the problems selected for research in the field of Extension are based on convenience, easiness in conducting research and replica of studies already conducted elsewhere. Original, field-oriented and need based research for addressing the problems and the results of which could give directions for policy, etc are either lacking or are too little. There is narrow focus on the research problem taken up for research, while most of the extension problems are multi-dimensional in nature.

(b) In the case of PG research, Extension students often select research problems that have been studied earlier by other researchers, and pursue their research by changing the crop or locale or sampling unit of the previously studied problem. In many cases, even the results of the study and discussion are merely copied (cut and paste) as such from the original.

(c) Ideally, review of literature establishes context for the interest of the researcher in the problem and introduces and provides insights into the range of techniques and tools that can be employed in analysing the research idea. However, in practice, it is seen that most of the researchers do not undertake proper review of literature. It is considered as a 'ritual'. The literature review in many cases lack rigour and consistency, context and breadth, clarity and brevity and effective analysis and synthesis. As a result, the researcher will not be in a position to develop 'theoretical orientation' for the study, which affects the results.

(d) Selection of a pilot site and conducting a pilot test using each data gathering method will help in obtaining better results. However, it is observed that in many cases, conduct of pilot study at the field level is not done, but is shown in the report as being done.

(e) The extension research is considered as largely "ex-post facto". Experimental or hypothesis testing approaches are often viewed as luxury by academicians and scientists. Choosing a "right" research method for a specific research problem is a concern in the extension science.

(f) The results of the study are analysed without considering the suitability of the test and its relevance. It is observed that many of the extension researchers are interested in getting the results analysed using many tests and preparing large number of tables. However, in many cases, the results are not properly explained or interpreted. This often happens because the researcher feels that his/her quality of research can be enhanced only by using more number of statistical tests and presenting the results using many tables. This mindset of researchers should definitely change in favour of selecting and applying only the most relevant and appropriate test for the study.

(g) I have observed that discussion on the results of the study by the researchers in large number of cases is quite shallow, superficial and not supported by theories or relevant concepts. The readability of the discussion chapter in many theses/ research reports is very poor and is just repetition of the findings in many cases. In fact, this chapter can contribute much to the knowledge building process. Researchers need to employ 'hermeneutic principles'. Hermeneutics refer to the theory and practice of interpretation. Many of our extension scientists lack this skill.

(h) The research reports are prepared by the researchers in a routine and mechanical way, and many a times, it is observed that sincere attempt is not made to make the reports meaningful to the target users. The target users could be sponsors of research, administrators, policy makers or academic community. The form, content and style of the research report should be chosen to suit the level of understanding, experience and interest of the targeted users as well as to make the readers to apply the findings in their respective areas. However, this is not seen in many of the research reports prepared by the extension scientists.

Though extension scientists in the NARS can potentially engage in useful research that can influence the technology development process, they are mostly engaged in organizing training programmes, conducting events, dealing with visitors to the institute and handling documentation responsibilities (Gowda et al, 2014). Lack of clarity on the role of extension scientists (extension work vs. extension research) has also contributed to this situation. Though participation in extension related activities provide lot of scope for conducting research and publishing them, academic publishing is a low priority in extension.

Quality of Publications

A very serious problem related to extension research is the absence of standard research articles published in reputed national and international journals. Publications primarily help researchers working in similar areas to gather information on recent developments in their field of study. It also helps the authors to get feedback on their work and get motivated to pursue their research in new directions. Academic publishing in extension has suffered mainly due to these limitations in quantity and quality of extension research. Selection of appropriate journals is always a challenge for researchers. In the case of extension, though the number of extension journals is few, there are many social science and general agriculture related journals which also publish extension articles, which many researchers are not much aware.

There is scope and space for publication of extension research in reputed international journals. I may substantiate this with samples of research articles published in three journals, viz;, Agricultural Systems, Rural Sociology and Outlook on Agriculture.

The following two articles published in Agricultural Systems may be taken as examples of how a social process could be scientifically studied, analysed and presented to the readers for wider application.

Cardenas and Ostrom (2004) in their article relating to experiments in the field about co-operation in the Commons had indicated that experimental research has enhanced the knowledge acquired from theoretical and field sources of when and how groups can solve the problem of collective action through self-governing mechanisms. Widespread agreement exists that co-operation can happen, but little agreement as to how. The article attempts to answer this question.

Sultana and Thompson (2004) in their article on methods of consensus building for community based fisheries management portrays a methodology (Participatory Action Plan Development- PADP) which recognises diversity in livelihoods and works through a structured learning and planning process that focuses on common interests. It works with each category of stakeholder separately to providing the natural resources problems that their livelihoods are largely dependent on, they then share and agree common priorities.

Likewise, the following research articles published in Rural Sociology may be taken as good cases as to how the current and emerging areas of concern and significance to the rural community could be properly analysed and presented.

Glenna et al (2007) in their article portrays the industry perceptions of University-Industry relationships related to Agricultural Biotechnology research. Following a rise in University-Industry relationships (UIRs), scholars began questioning the efficacy of those relationships, as well as whether industry and university research interests and integrity are being compromised. While many of the studies focus on the University, there are only few studies that have examined the perceptions of industry participants. Industry representatives listed many advantages of UIRs, but some also expressed an interest in drafting policies to preserve a public interest emphasis for University research.

Holleman (2012) in his article gives an overview of bio-fuels and their role in ecological change. While a growing body of research indicates the severe ecological and social costs of bio-fuel production worldwide, the US Government continues to promote the expansion of this fuel sector. The article is a sociological investigation into the practices of key political actors and their roles in promoting and discouraging ecologically oriented change. This research is intended to contribute to alternative energy debates, and add to our understanding of the role of various actors in energy policy formulation and further theoretical developments within Sociology.

Dunlap (2010) in an interesting article on Climate change and Rural Sociology discuss how Rural Sociology should be able to make broader contributions to understanding the human dimension of climate change. The article suggests a broader research agenda for rural sociologists and other social scientists interested in climate change and refer to a wide range of literature that shall provide guideposts for pursuing research related to climate change.

If we critically analyse the contents of the articles mentioned above, we can see that there is flavour of extension research in these articles. The sad part is that though extension scientists may also be conducting quality research, output of the research in terms of standard publications is not forthcoming. However, there are exceptions to such generalisations, which are very less. Two interesting research articles on methodology evolved by a group of extension scientists are presented as good cases.

An article published in Outlook on Agriculture by Bonny et al (2005) describes in detail a model for participatory technology evolution in farming. The model named as Participatory Learning, Experimentation, Action and Dissemination (PLEAD) advocates that integration of local knowledge, experience of farmers and quality assessment of evolutional strategies help in developing technology that promote the long term sustainability of the farming system. The model was developed through field experimentation among the rice farmers of Kerala using 18 Farmer Research Groups (FRGs) and tested through field interventions. This research work was undertaken under NATP.

Similarly, there is a very good attempt for developing multidimensional scale for Effective Measurement of Rural Leadership (Mohanty et al, 2009). It is a practice that many social researchers measure multidimensional variables, such as rural leadership, by aggregating the scores of individual dimensions / sub-dimensions. These dimensions / sub-dimensions are highly correlated and have inherent problem of multi-collinearity which causes attenuation in measurement. In order to overcome this problem, the M-K-J-B-D method was developed by a group of researchers, which has been used to construct multidimensional scale of rural leadership. The advantage of multidimensional rural leadership scale is demonstrated in the article by comparing it with uni-dimensional scale in further statistical analysis.

Other related problems related to Extension research

1. There is no appropriate and desired link and connectivity between field extension and extension research (Lack of functional integration). The information derived from research needs to be "in context" to be used by the extension system, which is totally absent
2. Perils of deficient theorizing- as a result, extension research cannot make analytical contributions to the subject
3. Lack of will and dedication to practice values and ethics in the research process.
4. Lack of high profile interdisciplinary research projects linking technology, society, environment, etc – Scientists are interested only in protecting disciplinary boundaries.
5. We are still happy and feel comfortable with the old and outdated methods, scales and tests and mentally obsessed to develop and try new tools. Extension researchers still use age, education, etc as independent variables. The scales developed by Trivedi (1963), Supe (1969), Moulik, Ambastha, etc still are used as such in many cases, without any modification.
6. In the case of our extension scientists, various ideological and conceptual blockages and lack of proper understanding of the concepts prevent deeper analysis of social phenomena and problems.

Research on Impact of Extension

Evenson (1997) based on economic studies of extension services in a number of countries have given some insight on researches related to impact of extension. There are two conceptual themes taken as relevant for analysing extension impact. The first is the awareness-knowledge-adoption-productivity (AKAP) sequence. The second is the 'growth gap' interrelationship between extension, schooling and research.

Studies of extension impacts have measured farmer awareness (and sources of awareness), knowledge (and testing of practices), adoption and productivity. Not all studies have examined all parts of the sequence. Most studies have shown a statistical relationship between the quantity of extension services made available to the farmers and increases in awareness, knowledge, adoption and productivity.

In the case of productivity gaps, four yield levels are depicted for each type, as below:

A - Actual yield
BP - Best practice yield
BPBI - Best practice, Best infrastructure yield
BPBIRP - Best practice, Best infrastructure, Research potential yield

The above yield levels in turn define three gaps, as indicated:

G (P) – A practices gap between the best practice (BP) yield and actual (A) farmers' yield

G (I) – An infrastructure-institution gap between the best practice, best infrastructure (BPBI) yield and best practice (BP) yield

G (R) – A research gap between the research potential yield (BPBIRP) and the best practice, best infrastructure (BPBI) yield

The depiction of the above gaps provides a way to classify the contribution of extension activities and to show how research and extension are linked.

Though there are means to measure the impact of extension, estimation of extension impact is subject to a number of problems which are also faced in the evaluation of other public sector investments. The approach commonly used is statistical analysis relying on data measuring extension activities at the farm level. Alternatively, statistical analysis can be used where observations refer to aggregate extension services offered to a given region in a specified time period. Studies assessing extension impact at the individual farm level that use farm level measures of extension may be affected by two basic estimation problems- the problem of statistical 'endogeneity' in extension farmer interactions and the problems of indirect or secondary information flows.

Evenson (1997) had also indicated that all the studies reviewed by him were conducted by economists in academic programmes. He also pointed out that it is important to extension policy makers that the competence of their monitoring and evaluation programmes be upgraded to enable more effective and meaningful evaluation. It is a reflection of lack of attention paid by the extension scientists and their poor involvement in such type of research work.

What is Needed?

1. The concept of 'research' in extension needs to be broadened, recognising that beyond the public research and extension organisations a range of actors have important and vital roles in the generation and dissemination of agricultural innovation.
2. The understanding of innovation needs to change, as it is increasingly recognised that non-technological innovations such as ways to access to more profitable markets, value chain development or organisation of producers are equally, if not more important than technological innovations.
3. The agricultural innovation system (AIS) landscape has a wide range of actors going well beyond formal research and extension institutions, but the research in extension is still stuck with the typical actors and has not moved beyond R-E-F linkages. Of late, marketing is also added. There is diversification of actors and without a functional interface between the various actors, neither research will be

able to make innovations that benefit farmers, nor can extension offer services that resolve all the problems of farmers.

4. The practice of Extension has been described as 'knowledge applied' or 'knowledge extended'. What about 'knowledge created'? Of late, the research community has begun to recognise Extension's role in 'creating knowledge' which is a welcome step. But this has to yield results.
5. Research institutions need to provide researchers with the right incentives to engage effectively, enable them to contribute to policy and political processes and develop realistic expectations as to what they can collectively achieve.
6. Researchers need to alter their own mindsets. This may mean working in inter-multi-and/or trans disciplinary research teams, admitting to being part of a value based system. The research agenda is usually decided by the researcher, which needs a paradigm change. Defining the research agenda is about defining the problem with research users, who they are- not just farmers, but scientists, entrepreneurs, environmentalists, policy makers, journalists, etc whoever is part of the 'innovation system' that affects research uptake and use.
7. Knowledge brokering is absent in the current research system. This is a central component of knowledge transfer that involves bringing people together, helping to build links, identify gaps and needs, and sharing ideas. It encourages the use of research in planning and implementation and uses evaluation activities to identify successes or improvements. Thus it helps to bridge the gap between research and policy development.
8. Translational research has to be promoted. It is defined as engineering research that aims to make findings from basic science useful for practical applications that enhance human health and well being. It focuses on multi disciplinary collaboration.

Way Forward

(a) The problems selected for conducting extension research should have relevance and applicability in the field situation. There should be data bank of researchable problems in Extension available for our extension scientists. Some of the concepts can be studied across the nation, while some concepts may be confined to selected geographic locations. Some of the suggested current areas for research by our extension scientists include social capital, cash transfer, micro finance, convergence as a social process, management of CPRs, climate change adaptation, public private partnerships, agricultural innovation systems, etc.

There are five thematic areas for agricultural research within an Agricultural Innovation System (AIS), as suggested by World Bank (2012). They are:

- Designing Agricultural Research Linkages within an AIS Framework

- Building and strengthening public-private partners in Agricultural research
- Regional research in an AIS framework- Bringing order to complexity
- Co-designing innovations: How can research engage with multiple stakeholders?
- Organisational change for Learning and Innovation.

It is necessary that extension scientists may give proper attention to the above areas in formulating research problems.

(b) Extension research should cover all the following types of research.

- Basic Research- inquiry focused on basic concepts and theories with a view to revisiting the existing concepts/theories and developing new theories. For instance, relook on innovation decision process, communication behaviour, etc.
- Developmental Research- in terms of contributing to the development of the discipline by way of developing innovative methodologies, good practices, effective tools of measurement, etc.
- Adaptive Research- studying the applicability and usefulness of the new practices, tools developed, etc and testing their effectiveness.
- Academic Research- focussing on the process and methods of developing tests, scales and new approaches in the field of extension and getting trained as a professional.
- Applied Research- focussing on the problems of conducting research in terms of data collection tools, measurement, experimentation, etc.

(c) Lack of network projects and co-ordinated projects in Extension is a serious handicap. To overcome this, ICAR under the Division of Agricultural Extension should have a separate wing for Extension research and should organise All India Co-ordinated Research Projects (AICRP) on Extension. This should be initiated within the 12th Plan period itself.

(d) MANAGE at the national level and SAMETIs at the state level should earmark at least 10 per cent of the funds for extension research. EEIs should also earmark a fixed percentage of funds exclusively for extension research. MANAGE and EEIs may award research projects based on the call for proposals from qualified extension researchers. An advantage is that they can use the research results from such awarded research projects as input for their training programmes.

(e) There should be steps taken to enhance the quality of extension research in the country. Conduct of refresher courses on research methodology for teachers and scientists, short courses on research methodology for Doctoral students in similar lines with ICSSR courses, research seminars by leading extension researchers in the field, etc could be organised.

(f) The quality of research publications has to be drastically improved. Some of the research articles published in extension journals lack readability, focus, methodological rigour and clarity, and in many cases, carry repetitive nature of findings and outdated concepts. There should be a rigorous system of screening the articles, peer review, conforming to minimum standards, etc and our extension scientists should give serious attention to this aspect.

References

1. Bonny, P B., Prasad, R M., Narayan, S S and Varughese, M., 2005. Participatory Learning, Experimentation, Action and Dissemination (PLEAD): A model for farmer participatory technology evolution in agriculture, Outlook on Agriculture, 34 (2): 111-115.
2. Cardenas, J C and Ostrom, E., 2004, What do people bring into the game? Experiments in the field about co-operationn in the Commons, Agricultural Systems, 82: 307-326.
3. Dunlap, E R 2010 Climate change and Rural Sociology: Broadening the Research Agenda, Rural Sociology, 75 (1): 17-27.
4. Evenson, R., 1997, The Economic contributions of Agricultural Extension to agricultural and rural development. In Swanson, E B et al (1997) Improving Agricultural Extension- A Reference Manual, F A O, Rome.
5. Glenna, L L ., Welsh, R., Lacy B W and Biscotti, D 2007 Industry perceptions of University-Industry relationships related to Agricultural Biotechnology Research, Rural Sociology, 72 (4): 608-631.
6. Gowda C M.J., Dixit S, Burman R and Ananth P N. 2014, Extension Research and Technology Development Blog 29, Agricultural Extension in South Asia. (available at http://aesa-gfras.net/Resources/file/FINAL-M_J_Chandre%20Gowda-13-FEB.pdf).
7. Holleman, H 2012, Energy Policy and Environmental Possibilities: Biofuels and Key Protagonists of Ecological Change, Rural Sociology, 77 (2) : 280-307
8. Katz, Elisabeth, Lydia Pluss, Urs Scheidegger, 2007, Capitalizing experiences on the Research-Extension Interface- Forging Strategic Alliances for Innovation, Linden and Zollikofen.
9. May, Tim., 2001, Social Research: Issues, Methods and Process, Open University Press, USA.
10. Mohanty, AK., Kumar, G A K., Singh, B B and Meera, S N ., 2009, Developing multi-dimensional scale for effective measurement of rural leadership. Indian Res. J.Ext.Edu. 9(2); 57-63.
11. Prasad, R M, 2014, Extension Research: Random Thoughts from a Well Wisher, September 2014. (available at http://aesa-gfras.net/Resources/file/Prasad%20Sir-%20Blog%2039-FINAL.pdf).
12. Predota, E., 2009, Spotlight on Participatory Action Research Approaches and Methods, Geography, 94 (3): 223-225.

13. Smith, S Joshua and Helfenbein, J Robert 2009, Translational Research on Education (Chapter 5), education.iupui.edu/CUMER/pdf/Collaborativeturn: 2009 pdf.
14. Sivakumar, Sethuraman 2013, Research in Extension: New tools to reinvent its future. Blog 4, Agricultural Extension in South Asia (available at http://aesagfras.net/Resources/file/Blog%204%20Enhancing%20the%20potential%20of%20quality%20videos%20for%20farmers.pdf).
15. Warner, E M., Hinrichs, c., Schneyer, J and Joyce, L., 1998, From Knowledge Extended to Knowledge Created: Challenges for a New Extension Paradigm, Journal of Extension, 36 (4):
16. World Bank, 2012, Agricultural Innovation Systems- An investment sourcebook, The World Bank, Washington DC.

20 Participatory Monitoring and Evaluation

- Prof. M. Surya Mani

Monitoring and evaluation are critical for building a strong, global evidence for assessing the wide, diverse range of interventions being implemented. At the global level, it is a tool for identifying and documenting successful programmes and approaches and tracking progress toward common indicators across related projects. Monitoring and evaluation forms the basis of strengthening understanding around the many multi-layered factors.

At the programme level, the purpose of monitoring and evaluation is to track implementation and outputs systematically, and measure the effectiveness of programmes. It helps determine exactly when a programme is on track and when changes may be needed. Monitoring and evaluation forms the basis for modification of interventions and assessing the quality of activities being conducted.

Monitoring and evaluation can be used to demonstrate that programme efforts have had a measurable impact on expected outcomes and have been implemented effectively. It is essential in helping managers, planners, implementers, policy makers and donors acquire the information and understanding they need to make informed decisions about programme operations.

Monitoring and evaluation helps with identifying the most valuable and efficient use of resources. It is critical for developing objective conclusions regarding the extent to which programmes can be judged a "success". Monitoring and evaluation together provide the necessary data to guide strategic planning, to design and implement programmes and projects, and to allocate, and re-allocate resources in better ways.

Input: Goods, funds, services, manpower, technology and other resources provided with the expectation of outputs and results expressed in threebroad categories.

Output: Specific products or services which an activity is expected to produce from its inputs in order to achieve the set objectives.(increased irrigation, fertilizer use, health facility etc.)

Effects: Outcomes of the use of Project outs – (incremental yield, income, etc.)

Impact: Outcome of Project Effects (broad long term objectives: Std. of living and reducing poverty both at individual and community level)

Monitoring: A Continuous / periodic review and surveillance by management **at every level** of the implementation of an activity to ensure that input deliveries, work schedules, targeted outputs and other required actions are proceeding **according to plan.**

Evaluation: Is an assessment of end result or impact of a project with reference to the objectives set in the project.

Evaluation provides answers to 'why' programme is success / failure and helps us to identity the strengths and weakness / merits and demerits.

1. Basic features of Evaluation

- It is always with reference to a stated criteria
- It always with reference to a point of time
- It starts where progress reporting / monitoring / estimation surveys end.
- It establishes relationship between policies / methods and results
- It investigates and find out factors for success / failures and suggest remedies
- More qualitative in approach.
- More purposive and less aggregative
- Focus on policy issues, organizational and administration practices, technical content and people cooperation, attitude and impact.

2. Why Evaluation

- To record achievement / failures
- To measure progress in terms of project and results
- Improving monitoring
- Identify strength and weaknesses
- Cost benefits
- Collecting information
- Sharing experience
- Improving effectiveness
- Allowing for better planning

Types of Evaluation

Formative Evaluation done during the programme development Stages (mid term appraisal) on going/concurrent.

Summative Evaluation taken up once the programme achieves a stable state of operation (Ex-post evaluation) Terminal.

Participatory Monitoring and Evaluation

As institutions become more inclusive in the 'front-end' of project development – that is, in promoting participation in appraisal and implementation-then questions of who measures results and who defines success become critical. Monitoring and evaluation is vital if governments and aid organizations are to judge whether development efforts have succeeded or failed. Conventionally, it has involved outside experts coming into measure performance against pre-set indicators, using standardized procedures and tools.

Participatory monitoring and evaluation (PM & E) has emerged because of recognition of the limitations of this conventional approach. It is attracting interest from many quarters since it offers new ways of assessing and learning from change that are more inclusive, and more in tune with the views and aspirations of those most directly affected. PM and E provides an opportunity for development organizations to focus better on their ultimate goal of improving poor people's lives. By broadening involvement in identifying and analyzing change, a clearer picture can be gained of what is really happening on the ground. It allows people to celebrate successes, and learn from failures. For those involved, it can also be a very empowering process, since it puts them in charge; helps develop skills, and shows that their views count.

Different PM & E Approaches

- Participatory evaluation (PE)
- Participatory monitoring (PM)
- Participatory assessment, monitoring and evaluation (PAME)
- Participatory impact monitoring (PIM)
- Process monitoring (PM)
- Community monitoring (CM)
- Stake holder based evaluation

Principles

- **Participation** – which means opening up the design of the process to include those most directly affected, agreeing to analyses data together;
- **The inclusiveness of PM&E** requires negotiation on to reach agreement about what will be monitored or evaluated, how and when data will be collected and analyzed, what the data actually means, and how findings will be shared, and action taken;

- This leads to learning which becomes the basis for subsequent improvement and corrective action;
- Since the number, role and skills of stakeholders, the external environment, and other factors change over time, flexibility is essential.

Participatory Monitoring

Participatory Monitoring is the systematic recording and periodic analysis of information that has been chosen and recorded by insiders with the help of outsiders. The main purpose Participatory Monitoring is to provide information during the life of the project, so that adjustments and modifications can be made if necessary.

Participatory Monitoring measures progress: Take the example of a bus trip from one community to the other. When passengers can see out of the windows, they can monitor progress by observing the passing landscape, reading the road signs and watching the movement of the sun across the sky. Monitoring these kinds of information on a bus trip lets them know whether they are heading in the right director. Participatory monitoring is like having all passengers on the bus know their destination and decide how they will measure their progress.

Information is periodically analyzed: Participatory Monitoring is not only keeping records. It is also stopping at set times to analyze (add up, discuss, integrate) information. The time to stop and analyze will vary according to the nature and / or seasonality of activities.

Agreement on the objectives and activities is necessary: Before Participatory Monitoring begins, the community must understand why they are monitoring. Information should keep everyone informed of progress (or lack of progress) toward planned objectives and activities.

Insiders choose the terms of measurement: When the terms of measurement, (kilos, grams, guntas, sacks, cans, pounds, bundles, etc.) are chosen by insiders, the information is better understood. The chances of the monitoring continuing in the future are more likely.

Broadly examines progress towards objectives and activities: Insiders, given the opportunity, have the ability to combine qualitative (descriptive) information with quantitative (numbers) information, providing a more complete analysis.

Steps in Participatory Monitoring

Take time to prepare and plan monitoring. It helps everyone why they are monitoring, and how it will be done. 1) can include all those directly involved in the activities as well as other interested groups. But it will be concentrated on those directly involved or those selected by the groups who will be responsible for monitoring. Planning for monitoring can use a framework much like those used for Participatory Baselines and Participatory Evaluation. This framework is explained in the following steps.

1. **Discuss reasons for monitoring:** Review the benefits and purpose of monitoring, so that insiders can decide for themselves whether monitoring will help them.
2. **Review objectives and activities:** If PAMA has been continually used, the insider objectives and activities will have been established during the Participatory Assessment. If insiders have not previously been involved, the objectives and activities as established by outsiders can be reviewed and discussed by insiders. A Participatory Assessment may be necessary if insider and outsider objectives are very different.
3. **Develop monitoring questions:** After objectivities and activities are reviewed, discuss the information needed to help know if activities are going well. Focus on the questions "What do we want to know?" and "What do we monitor that will tell us this?"
4. **Establish direct and indirect indicators:** For each monitoring question, determine direct and / or indirect indicators that will answer the monitoring question.
5. **Decide which information gathering tools are needed:** For each indicator or monitoring question, the most appropriate information gathering tool must be chosen. Remember one tool can gather information that answers many monitoring questions. Some of the information gathering tools are:
    - Community Environment Assessment
    - Survival Surveys
    - Farmer's Own Records
    - Nursery Record Books
    - Community Financial Accounts
6. **Decide who will do the monitoring:** Monitoring may require people with specific skills such as book keeping or mathematics. It will also require a certain amount of labour (time) from people. Those with the skills and the time can be identified. They may have to be compensated for the task of monitoring. It might be a part of the job of a paid nursery person, or the community members responsible for monitoring might receive free seedlings from the nursery.
7. **Analyze and present results:** It is important that information monitored be analyzed at specific times throughout the activities. The analysis can be discussed at community meetings, posted or put in community newsletters. The community will then know whether or not activities are progressing as planned or if changes or modifications are required.

Benefits of Participatory Monitoring

- Provides an ongoing picture
- Problems are identified and solutions sought early
- Good standards are maintained
- Resources are used effectively
- Complete picture of product is produced
- Information base for future evaluations

Participatory Evaluation

A Participatory Evaluation is an opportunity for both outsiders and insiders to stop and reflect on the past in order to make decisions about the future. Insiders are encouraged and supported by outsiders to take responsibility and control of:

- planning what is to be evaluated
- how the evaluation will be done
- carrying out the evaluation
- analyzing information and presenting evaluation results

Insiders already intuitively and informally evaluate in the light of their own individual and / or group objectives. This is because:

- Development activities often require involvement and inputs from insiders
- It is ultimately insiders who reap the benefits and bear many of the costs of the project
- Insiders choose whether to continue or discontinue activities when the outsiders leave

Thus, it makes sense for outsiders to help insiders conduct an effective evaluation. With the results of evaluation, insiders may choose to continue activities, modify all or some, change the strategy, change the objectives, or discontinue activities.

Information to guide management decisions: A Participatory Evaluation should not be thought of as a final judgement on whether activities are successful or unsuccessful. The information should encourage changes and adjustments either during the life span of the activities, for future phases of the activities, or for future new activities.

Both objectives and activities are considered: The overall and immediate objectives, their continued relevance, and the effectiveness of the activities are all taken into account.

Other methods contribute to participatory evaluation: Much of the information from Participatory Assessments, Participatory Baselines, and Participatory Monitoring can be used in Participatory Evaluation. Information from Participatory Monitoring will give progressive trends and total amounts.

Why Participatory Evaluation?

Participatory Evaluation may be done: Because

- They have been planned
- They is a crisis
- A problem has become apparent

Steps to Participatory Evaluation

The time that is taken to carefully prepare and plan a Participatory Evaluation is time well spent. It helps everyone know why they are evaluating and how they are going to do it. The first meeting to prepare and plan the evaluation should be open to all interested groups. This meeting could include beneficiaries, others in the community, as well as groups from outside the community who have an interest in the project.

1. **Review objectives and activities:** The larger group, as described above, decides why an evaluation is necessary.The community's long-term and immediate objectives and the activities they have chosen to meet these objectives can be reviewed at this meeting. If PAME has been used, the objectives and activities established during the Participatory Assessment can be reviewed. If the activities have not been participatory, the objectives, as established by outsiders, can be reviewed.
2. **Review reasons for evaluation:** After objectives and activities are reviewed, discussion can focus on the questions "Why are we doing an evaluation?" & "What do we want to know?"
3. **Develop evaluation questions:** The facilitator can write (or draw), the evaluation questions on large sheets of paper or a black board. The group should agree on each question. If many questions are generated around each objective and activity, they can be ranked in order of importance.
4. **Decide who will do the evaluation:** In the larger group meeting, decide who will do the evaluation, and who will want to know the results. It may be decided to include the whole community, or only the beneficiaries, or delegate the responsibility for the evaluation to an evaluation team.
5. **Identify direct and indirect indicators:** Taking the evaluation questions that were generated in the first meeting, direct and indirect indicators are chosen for evaluation questions.

6. **Identify the information sources for evaluation questions:** For each evaluation question and indicator that is chosen, the evaluation team identifies where the information is available, or if it is not available, how it will be obtained. Some information may be available in an unanalyzed form, and require some effort to analyze. Other information may not be readily available, and will have to be gathered. The information that is required may be available from either Participatory Assessment, Participatory Baselines, and / or Participatory Monitoring.

   If information is not readily available, it must be decided which information gathering tool will be used to obtain information. Some of the information gathering tools useful in Participatory Evaluation are:

   - Community Case Studies
   - Semi-structured Interviews
   - Ranking, Rating and Sorting
   - Community Environmental Assessment
   - Survival Surveys
   - Farmer's Own Record
   - Nursery Record Books
   - Community Financial Accounts
   - S.W.O.L. Analysis

7. **Determine the skills and labour that are required to obtain information:** The assistance of people with specific skills, such as interviewing, mathematics, and other art and / or drama, as well as a certain amount of labour (time) will be required. The Participatory Evaluation team must decide which skills and resources are available to them by asking the questions:

   What resources do we need?

   What resources do we have, or can we develop?

   What resources do we need to get?

8. **Determine when information gathering and analysis can be done:** It is important to assure that information will be gathered and analyzed within the time frame that is given to the evaluation team, so that the results can reach decision makers on time. The timing of the evaluations must take into account factors such as: seasonal constraints; religious holidays; field staff availability; and community labour demands.

9. **Determine who will gather information:** When the specific dates, the required time and skills are known, then the tasks can be delegated to individuals or small working groups.

10. **Analyze and present results:** When all the tasks have been completed, it will be necessary to analyze and synthesize information for presentation. Some of the information may have already been analyzed. It will simply have to be put in its place in the presentation.

Benefits of Participatory Evaluation

Better decision-making by insiders: Participatory Evaluations, by examining the activities individually and relative to objectives, give insiders relevant and useful information. Helping them decide whether the objectives and / or activities should stay the same or change.

Insiders develop evaluation skills: Participatory Evaluation reveals community skills that were undervalued, and / or develops analytical skills needed to make good decisions. It helps insiders better organize and express their concerns and interests in ways outsiders can understand. This strengthens two-way communication.

Outsiders have better understanding of insiders: Outsiders benefit from Participatory Evaluation as it complements and enriches their own evaluations. This is especially so when the outsiders' objectives are self-help and sustainability. A Participatory Evaluation will let them know whether or not the community is likely to continue the activities when outsiders are left.

Insider to insider communication is strengthened: Participatory Evaluation can be used for local extension, with results presented to other communities who are expecting the same kinds of problems. In this way insiders learn from insiders.

Information is useful for ongoing management of project: Information from Participatory Evaluations can be used by insiders and outsiders to identify strong and weak points of activities. If activities are to be continued, or phased over to insiders, information can be used to modify activities, and make them more effectively meet the objective, and better respond to real community needs and priorities.

Entry point for the participatory approach: In a community where participation has not been a feature, Participatory Evaluation may be the beginning of a participatory approach. It may be that including school children in the process not only helps the community gather information, but also helps the children develop analytical skills and experience.

PM and E has created new ways of measuring change, while helping build the monitoring and evaluation capacity of the people involved. Nevertheless, problems have been encountered. Common mistakes are:

- Assuming that all stakeholders will be interested in taking part;
- Imposing inappropriate indicators and methods in effort to standardize and save time

- Being unclear about how information will be used, and by whom;
- Collecting unnecessary information;
- Starting too big, too soon.

Social Audit

A Social audit is

- Process by which the people, the final beneficiaries of any scheme, programme policy or law, are empowered to audit such schemes, programmes, policies and law.
- An ongoing process by which the potential beneficiaries and other stakeholders of an activity or project are involved from the planning to M & E of the activity or project.

Principles

Transparency: Complete transparency in the process of administration and decision-making with an obligation on the government to suo moto give the people full access to all relevant information.

Participation: A right based entitlement for all the affected persons (and not just their representatives) to participate in the process of decision making and validation.

Representative Participation: In those rare cases where options are pre-determined out of necessity, the right of the affected persons to give informed consent, as a group or as individuals, as appropriate.

Accountability: Immediate and public answerability of elected representatives and government functionaries, to all the concerned and affected people, on relevant actions or inactions.

How is a Social Audit different to other Types of Audits?

As government or institutional audits do not significantly involve the affected persons, or the intended beneficiaries; they end up at best assessing outputs rather than outcomes and are also not able to assess whether the decision making processes had the inputs and support of all the critical stakeholders. Such systems are also very corruptible, as those involved in the audit do not have a real stake in the outcome of the process that they are auditing.

Public audits do not have these problems, because the affected persons and or the intended beneficiaries usually conduct them. However, the findings of a public audit might not be easily acceptable to the government and other implementing institutions as they are not intrinsically involved in the process of audit and, despite best efforts, might not participate. Besides, without participation of the implementing departments / institutions their side of the story does not get told.

On the other hand, social audits do not have a uniform approach and methodology and many local factors affect their efficacy. To conduct social audits a huge amount of public mobilization is necessary and, in the absence of that, social audit might not be effective.

When should Social Audit be Conducted?

Social audit can be done at any point of time during the planning and implementation of a scheme / programme. For instance in the case of National Rural Employment Guarantee Scheme (NREGS), social audit can be taken up.

- **Planning Stage:** to ensure that the Gram panchayat Plan is need based covering productive / investments and drawn up in consultation with community serving the poor and disadvantaged.
- **Preparation Stage:** to ensure that estimates are proper and are in tune with the approved quantum of work.
- **Implementation Stage:** to ensure that wages are paid rightly, properly and to right people.
- **After the completion of work:** to ensure that quality of work is in tune with quantity and estimated cost.

Who will carry out Social Audit?

It is carried out by the community of stake holders. This will include beneficiaries / participants, implementing agency, Gram Panchayat representatives, etc,. The entire Gram Sabha is expected to participate in Social Audit. Since this may not be always possible, a group can be formed voluntarily (with encouragement by panchayats and officials) with representatives from beneficiaries, SHGs, village level organizations (VEC, FPC, etc) respected local / community leaders, youth clubs, marginalized sections (SC / ST / Women) etc. This group along with Gram Panchayat representatives and officials can carry out social audit and present their findings in the Gram Sabha.

Benefits of Social Audit

Social Audit helps in

- Awareness generation
- Monitoring the implementation
- Impact in processes
- Grievances Redressal and follow up corrective actions

e-Extension – ICTs & Mass Media

21 21st Century Agricultural Extension:

ICT Imperatives for Success

- *Dr. Richard W. Oliver*

As India and other developing nations experience the projected acceleration of urbanization, agricultural production must increase at least in proportion to that growth. While traditional educational and communication tools used in agricultural extension services have proven very successful, agricultural productivity must increase exponentially in this century. Agricultural Extension must re-think its role and practice to meet this need. Biological and information technologies offer the single best opportunities to meet this critical demand for massive agricultural productivity increases. Importantly, the use of such technologies will require a rethinking of the traditional approach of agricultural extension from a limited model (i.e., communicator/applied practice facilitator between basic researchers and farmer) to an extended model (applied research communicator to producer, distributor, logistics and support functions and consumer) model that encompasses the entire agricultural "value system.

Social Media and Information Technology: Catalyst for Developing Country Growth

The formal creation of agricultural extension services is generally acknowledged to have occurred that the time of the potato famine in Ireland in the mid-1800s. Since then, it has become a critical part of the agricultural development of nearly every country, particularly the Western countries, all with well-devolved agricultural research, technologically sophisticated corporate and private farming and agri-businesses. In the West today, extension services plays less of a role than it did historically, as large corporate businesses, and well-developed research and communications networks have supplanted the need for such services. In less developed countries, agricultural extension still plays a vital role in the development of sustainable agriculture, particularly within countries with rapid urbanization.

By the mid-2000s, the United Nations believed that more than half the world's population lived in urban areas, and by mid-century, nearly 90 percent of the developed world and 65 percent of the developing world will live in cities (some 3 billion people). The rapid urbanization of India, starting after independence, reached over 30 percent of the population by the turn of the century, and should reach over 40 percent by 2030, according to UN estimates. While not minimizing the immense problems associated with the speed and size of urbanization, the ability of agriculture to feed and clothe urban populations in the developing world will become acute.

There has been significant improvement in the productivity of agriculture in the developing world, and agricultural extension services can be justifiably proud of their role. While many statistics can be cited to underline that growth, wheat yields serves as just one example. The productivity growth of wheat yields in the developing world has been nothing short of phenomenal, as shown in the following chart.

There are many reasons for the overwhelming difference in agricultural productivity in the U.S. (principally public policy, mechanization, chemical technology, and consumer influences, etc.), but agricultural extension services have often been cited as a major contributor. By the turn of the century, less than 2 percent of the U.S. workforce was employed in agriculture (down from more than 40 percent in 1900).See: The 20th Century Transformation of U.S. Agriculture and Farm Policy. http://www.ers.usda.gov/media/259572/eib3_1_.pdf)

Today, U.S. agriculture continues to improve, for many of the same reasons cited above, but two new technological advances have the potential to drive even more dramatic change: information technologies and biotech. These monikers were chosen to illustrate the dynamic nature of technological change on agriculture in the developed countries. It is beyond the scope of this paper to discuss all the implications of those descriptions, but they highlight the tremendous potential for these technologies to dramatically improve agricultural productivity in countries like India. Importantly, creating this change must engage the entire agricultural "value system," stretching "from field to table, wardrobe and factory," but my basic thesis here is that agricultural extension is the central actor in bringing about that change.

Agricultural Extension and Education Technologies

Today, two broad categories of education are generally acknowledged: formal, degree/diploma (primarily from public and private universities, often basic-research driven) education; and, non-degree learning (from either traditional or non-traditional educators). Both have a critical role in the agricultural improvements in developing countries. However, dramatically increasing non-degree learning, such as agricultural extension services, may be the key to agricultural productivity growth in the developing world. Agricultural extension is essentially an educational activity; it is and will increasingly be, driven by the dramatic changes in education technology.

Education in the U.S. one of the fully online University in the world went through a decade of significant change beginning just before 2000. That change palled in comparison with the changes of the next five years, from 2010 to 2015. They will be even greater in the next five, to 2020. The educators who form the nucleus of the agricultural extension service in the developing world are not immune to these changes. The compelling cost and effectiveness advantages of various forms of digital education technologies will ultimately override any consumer, government or industry attitudinal or structural resistance to change.

Of course, there are countervailing trends which function to slow adoption, including policy, regulations and social/cultural constraints (including lack of digital fluency). In the context of agricultural extension however, it is important to acknowledge significant barriers to adoption: policy; regulatory; institutional; leadership; and funding (for services and devices); to name just the most obvious. It must be pointed out however, that those same barriers have existed and in some measure continue to exist in education in the West. The dramatic advantages of speed, quality, service (flexibility), variety (choice) and price, are sweeping away barriers daily. Of all the various information technologies, social media appear to have the greatest opportunity for agricultural extension.

Social Networking: Huge Opportunity for Change

There is no universally accepted definition of social networks or social media. To me, they can be defined as:

Social networks or media are internet-based, primarily mobile services that use dedicated websites and applications to provide self-selected groups of a small or large number of users with the ability to virtually interact and exchange information, entertainment and education, in text or visual formats, instantaneously.

Social media platforms are growing exponentially and are among the fastest adopted technologies of all time, with Facebook now exceeding some one billion users and Twitter more than half that number. The chart below demonstrates its explosive growth.

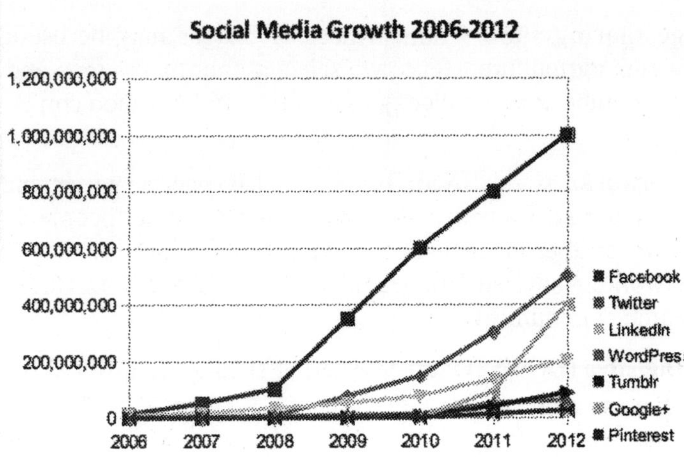

Kayvan Kousha and Mahshid Abbdoli, in Using Social Networking Sites, Blogs, and Online News Services in the Agriculture Research: ACitation Analysis. (See: http://conference.ifla.org/ifla78), investigated the role social media plays in agricultural research and publication. They found that

A stable upward trend in citing different types social networking sites within academic publications in agriculture and biological sciences. Nevertheless, the absolute numbers of citations to the selected websites is relatively low... we found that Wikipedia (45%), online news contents (25%) and blog posts (19%) are more commonly cited in agriculture publications than general social networking sites (less than 1%) which mostly connects people (e.g., Facebook, twitter and Myspace) and document (7%) or video sharing sites (3%). One explanation is that more scientific or scholarly-related contents are available through Wikipedia, news agencies (e.g., science news in BBC, CNN and Reuters) and blogs (e.g., agriculture blogs).

In their paper, they also provide a useful overview of the social networking services commonly used in non-academic journal publication of agricultural and biological research:

"As the first study of its kind, we selected a range of different social networking sites and online news contents:

Wikipedia: Wikipedia is a free online collaborative encyclopedia which contains over 21 million articles in various subject areas (Wikipedia, 2012). These articles can potentially be used for research communication. However, not much is known about the extent of citations to Wikipedia articles from agriculture research.

Blogs: blogs or weblogs are the fastest and the easiest method for online publishing, where people (or bloggers) can write about different topics or leave messages and comments. We selected 63 major blogs or blogs software from different lists (e.g., list of social networking websites, 2012) and assessed how these blogs may be cited or used by agriculture research.

Video and image sharing sites: Online videos or images may be useful for scholarly communication within agriculture. Hence we selected four main video and image sharing sites including youtube.com, video.google.com, video.yahoo.com, ted.com and flickr.com.

General social networks sites (SNS): Facebook, Myspace, Twitter and LinkedIn are among the most popular social networking sites which connect people with friends and others people. However, there increasing number of institutions creating their own pages in these social networking sites such as USDA (www.facebook.com/USDA) which might be useful for research communication.

Document sharing sites (DSS): Document sharing sites specifically allow users to share different types of documents such as PDF, DOC, PPT or PS. We selected six major document sharing sites including Slideshare.net, Mendeley.com, Scribd.com,

Dropbox.com, DocStoc.com and Delicious.com to examine how documents uploaded in these sites were formally cited by agriculture publications.

Online news contents: Every day millions of people around the world listen to or watch the news and feature stories broadcasting by the major news agencies such as BBC, CNN and Reuters. These major news agencies not only disseminate daily political and economic news, but also are reliable and up-to-date sources for tracking science, technology, health, entertainment features and analysis. We selected four major online news agencies including BBC, CNN, Reuters and The Associated Press. Major document sharing sites including Slideshare.net, Mendeley.com, Scribd.com, Dropbox.com, DocStoc.com and Delicious.com to examine how documents uploaded in these sites were formally cited by agriculture publications."

According to the 2015 Horizon Report (See: http://www.nmc.org/publication/nmc-horizon-report-2015-higher-education-edition/), which annually reports on current and emerging technologies "enabling" education development. They list the following technologies and trends, virtually all of which the author believes have exceptional potential for use in agricultural extension.

"Enabling technologies are those technologies that, like location awareness, have the potential to transform what we expect of our devices and tools. The link to learning in this category is less easy to make, but this group of technologies is where substantive technological innovation begins to be visible. Enabling technologies expand the reach of our tools, make them more capable and useful, and often easier to use as well."

Consumer Technologies

3DVideo

Electronic Publishing

Mobile Apps

Quantified Self

Tablet Computing

Tele presence

Wearable Technology

Digital Strategies

Bring Your Own Device (BYOD)

Flipped Classroom

Games and Gamification

Location Intelligence

Makerspaces

Preservation/Conservation Technologies

Internet Technologies

Cloud Computing

The Internet of Things

Real-Time Translation

Semantic Applications

Single Sign-On

Syndication Tools

Learning Technologies

Badges/Microcredit

Learning Analytics

Massive Open Online Courses

Mobile Learning

Online Learning

Open Content

Open Licensing

Personal Learning Environments

Virtual and Remote Laboratories

Key Emerging Technologies

Social Media Technologies

Collaborative Environments

Collective Intelligence

Crowd funding

Crowd sourcing

Digital Identity

Social Networks
Tacit Intelligence

Visualization Technologies

3D Printing/Rapid Prototyping
Augmented Reality
Information Visualization
Visual Data Analysis
Volumetric and Holographic Displays

Enabling Technologies

Affective Computing
Cellular Networks
Electrovibration
Flexible Displays
Geolocation [Geographic Information Systems]
Location-Based Services
Machine Learning
Mobile Broadband
Natural User Interfaces
Near Field Communication
Next-Generation Batteries
Open Hardware
Speech-to-Speech Translation
Statistical Machine Translation
Virtual Assistants
Wireless [electrical] Power

Extending the Traditional Value System

Often referred to as a "value chain" by writers, the system of delivery of value from the first level producer to the ultimate consumer, is more rightly called a "value system." For clarity, a value chain is the description of value creating activities within an organization, while a value system is, more properly, the entire group of organizations that create value for a consumer.

Social media in other areas is intensifying the desire of all members of a value system to be involved in information generated anywhere and everywhere in the system. Social media makes it relatively easy, efficient and cost effective to do so. In the case of agriculture, food and fuels derived from agricultural products, some consumers are demonstrating a marked desire to know information from start to finish in the system. As such, agricultural extension services would do well to begin to operate as "captains" of the information flow within the system. This would expand their role and create additional political capital. They are ideally suited for this role and are likely to be the most "trusted" suppliers of information. They are also best situated to manage the two-way flow of information, if they manage the social media system. The following chart helps illustrate.

Few Case Studies across the Globe

Across the globe

An exhaustive search of the Internet yielded comparatively little (see For More Information at the end of this chapter) in the way of research or other information about the use of social media in agriculture in developing country situations. The following, however, are illustrative of the limited types of activities underway in several locales, including the U.S. Another example (Tesco/Buzzfeed) is included to help illustrate the need to include all payers in the value system as an important part of the future for agricultural extension. Two specialized global social media services are also briefly described.

Caribbean

By 2012, despite significant use of social media in the same countries in the Caribbean by individuals (e.g., Trinidad and Tobago had nearly half a million Facebook users; Jamaica with over 688,000 users) and business interests (notably energy, tourism, commerce and fast food) there was comparatively little use by agro-institutions/stakeholder sector. Some countries such as Guyana, are making use of other communications technologies (e.g., cell phone networks). Other countries, such as Jamaica (Rural Agricultural Development Agency), use a Facebook page with some 300 users, as well as other IT services such as podcasts and weblogs.

There has been some success using social media among the young, for example in the West Indies, where youth groups have been sharing information via Twitter and YouTube. Among the most active are country chapters (most active include Dominica, St. Lucia and St. Kitts and Nevis) of Caribbean Agriculture Forum for Youth. None of these are strictly defined as agricultural extension activities.

The University of the West Indies, however, has established the Virtual Outreach. This private service provides farmers answers to agricultural questions, field diagnostics and recommendations, and an opportunity to collaborate with experts. (See: Keron Bascombe. Social Media and Agricultural Extension in the Caribbean

http://blog.gfar.net/2012/10/24/social-media-and-agricultural-extension-in-the-caribbean/)

Zimbabwe

Nearly 75 percent of the population of Zimbabwe have mobile phones and Web 2.0 services are becoming more prevalent. The University of Zimbabwean's mobile service (see: http://m.uz.ac.zw/library/opening_hours_main.html) is assisting students and researchers involved in agriculture as well as other pursuits. New services research indicates that mobile technologies are now supplementing traditional media such as radio and message boards. Other developments include the wider use of "telecenters," which facilitate access to social media. These telecenters, the Southern African Network (SATNET) help promote regional interaction. Given the traditional cultural desire for community social interaction, social media is expected to be adopted initially by young farmers. (See: Collence T. Chisita. Knotting and networking agricultural information services through Web 2.0 to create an informed farming community: a case of Zimbabwe. http://conference.ifla.org/ifla78)

China

China already has huge numbers of mobile phone users and social media has been widely adopted. There has also been extensive development of web services aimed at agriculture. One of the most important is the China Agriculture Information Network that provides various types of information and business services of value for production and business activities. Research has indicated that younger users (18-30 years old) account for 70 percent of the total users up to 2010. Services such as weather information, sales information, bank and tax information as well as statistical data is provided to individual famers, agricultural commodities brokers, agricultural insurance brokers, farm machinery salesmen and agricultural product dealers. Much of the information is provided by various mobile phone and PDA services. (See: Xiaowen Ding and Li Qing. The vision of Agriculture network community information integration: A case study of China agriculture sustainable development information service mobile interconnection .http://conference.ifla.org/ifla78)

India

Researchers have noted the growing use of social networking, particularly among younger age groups, among staff, faculty and research scholars in Indian national agricultural research institutes. Social media is increasingly preferred over traditional media such as print for basic information, and for academic discussions rather than in classrooms. The researchers however found little use of social media among agriculturalists in Indian villages due to the lack of infrastructure, technical support and devices. (See: Ajay Pratap Singh and Mayank Yuvaraj. Creating global village of agricultural practices through Social networking: opportunities and threats. http://conference.ifla.org/ifla78

Nigeria

Agriculture is critical to Nigeria, as it employs some 70 percent of its workforce and accounts for 40 percent of its GDP. Research among six agricultural research institutes and a university found that the most used social media site was Facebook, followed by LinkedIn, Google, Academia and Twitter. Users cited connections with professional colleagues as the major reason for usage and noted that the benefits of using the site were exposure to the latest skills and knowledge in their profession, and professional problem solving.

Abiola Abosede Sokoya, Fehintola Nike Onifade, and Adefunke Olanike Alabi. Establishing Connections and Networking: The Role of Social Media in Agricultural Research in Nigeria. http://conference.ifla.org/ifla78

Jordan

Queen Rania Al Abdulla of Jordan has spearheaded a move to bring MOOCs to Jordan and the Arab world to "democratize access" to education. "Engaging, fresh, relevant – and, most importantly, in Arabic – MOOCs on Edraak will open up a world of possibility for intellectually hungry Arab youth," Queen Rania has said. Her foundation has formed Edraak in partnership with EdX: "We want to make sure the Arab world is caught up." Although important for everyone in the region, this new service will be particularly valuable for providing up-to-date education for under-represented populations, such as women. According to the CEO of the Queen's foundation, "we do not want to leave anybody behind. Our hope for this is that people in poor Arab countries will have similar access to people in wealthier Arab nations."(See: http://venturebeat.com/2013/11/18/how-jordans-queen-plans-to-democratize-access-to-education)

Tesco/Buzzfeed

Buzzfeed (an Internet news organization) achieved 15.2 million fans just four months after launching, Tasty, its video food site, on July 31, 2015. It is based on the experience of Facebook with "native video." Tesco, a British food store, is following their lead, capitalizing on the growing trend in the U.K. and other places to tie local foods and ingredients producers directly with consumers who increasing want to be part of the sustainable "food chain." Importantly in this regard is the overwhelming consumer preference for short video over photos. (See: www.Socialbakers.com)

Auburn University, U.S.A.

At Auburn University in the U.S., Dr. John Fulton, a precision agriculture extension specialist sees social media as a means of enriching his efforts, not as a hindrance. Dr. Fulton says: "If I restrict dialogue only to a one-on-one conversation, then only that person can take advantage of it." The sites include: http://www.youtube.com/PlantNutritionInst; Twitter @Plant Nutrition #aghttp://info.ipni.net/Y53U6.

The various services allow Auburn "immediate dissemination of important emerging issues and the sharing of positive information among producers and consumers of agricultural products … providing science-based plant nutrition and fertilizer use information to industry, farmers, agricultural and environmental leaders, scientists, and public policy makers." (See: https://www.ipni.net/ipniweb/pnt.nsf/5a4b8be72a35cd46852568d9001a18da/a6f93b4c674d8128852577ec0074c03d!OpenDocument

Digital Green : International

According to their website, Digital Green is a not-for-profit international development organization that uses an innovative digital platform for community engagement to improve lives of rural communities across South Asia and Sub-Saharan Africa. According to their website, "Since 2008 and as of April 2015, we have reached over 660,646 individuals across 7,645 villages through 3,782 videos, which showcase and demonstrate best practices. These videos were produced by 522 community members trained by us and screened in an interactive format by 6,403 trained community members. As many as 343,609 of the viewers have adopted one or more of the best practices promoted through these videos."

International (Penn State, U.S.A.): Plant Villages

According to their website, Plant Villages is an open access database of 50,000+ images of healthy and diseased crops. According to their website, "We are launching and growing this database in order to enable the development of open access machine learning algorithms that can accurately classify crop diseases on a smart phone ... it is a user moderated Q and A forum dedicated to the goal of helping people grow their own food. It is an open freely available resource that helps … solve plant related questions." (See: https://www.**plantvillage**.org/)

Conclusion

Agricultural extension services have declined in importance in developed countries, but they are growing in importance in the developing world as those countries rapidly urbanize. While agriculture productivity in the developing world has grown dramatically, the rate of improvement must accelerate in proportion to the growth of urbanization. Agricultural extension services are uniquely positioned to be the catalyst to achieve the needed growth. The availability of new educational media, particularly social media, offers the single best opportunity to educate the entire agricultural sector value system. There are two key imperatives for the 21st century agricultural extension services: widespread adoption of social media; and the expansion of their focus to include all segments of value added providers.

These imperatives would involve a significant shift in government policies and funding (e.g., people, personnel, infrastructure and smart mobile devices) in regards to the agricultural value delivery process in developing economies. In addition, it implies a significant shift in cost and value added expectations for various segments. It also presupposes a significant increase in the number of agricultural extension service personnel, and, in particular, the skills needed (i.e., social media and information technology) for effective execution of the function.

22 Tele Agri-Advisory Services to Farmers:

A Case of Kisan Call Centre in Telangana and Andhra Pradesh

- *Dr. V.P. Sharma*

Introduction

Access to information and improved communication is a crucial requirement for sustainable agricultural development. Modern communication technologies when applied to conditions at grass root level can help improve communication, increase participation, and disseminate information and share knowledge and skills. However it is observed that the rural population still have difficulty in accessing crucial information. The challenge is not only to improve the accessibility of information and communication technologies but also to make available the repository of information / knowledge at once place to serve the rural and farming community.

Agriculture is and will continue to be one of the key concerns of Indian Economy with Agriculture for foreseeable future. Agriculture Sector contributes about 14% to Gross Domestic Product (GDP) and supports over 50% of the population for their livelihood. The farming communities need to be supported on agriculture technologies on regular basis to sustain self-sufficiency in food, and achieve the same Oilseeds and Pulses.

To achieve this, there is a need for effective dissemination of information among the key stakeholders. All the information relevant to Farmer should be made available to them at a single source. The number of Farmers (127 million as per 2011 Census) in India is too large to be reached by face-to-face Extension. The Mass Media particularly the Radio played a key role in enhancing the outreach of Extension system during the Green Revolution. The Television has also added to the technology dissemination with a number of Farmer-centric programmes on Doordarshan and a few private channels like e-T.V. Both these channels (Radio and T.V.) are mainly one-way communication, with a few interactive programmes (phone-in). With launch of DD-Kisan Channel –a 24 x 7 dedicated Channel to address Indian Farmers, on 26th May 2015, the Mass-Media support to Agricultural Extension has got a huge boost.

Kisan Call Centres (KCCs): The Mass-media reaches all the corners of the country, but it still remains primarily a Öne-Way Communication". Broadcast or Telecasting by its very nature has to be generic and hence the specific queries of the Farmers need Face-to-Face communication or Tele-communication or Video Communication. The Face-to-Face communication is the best process, but very costly for the system. As per recent AESA (Agricultural Extension for South Asia) study the ratio of Farmer to Extension-worker in the country is around 2400:1. The video interaction between the Farmer and the Agricultural Scientists/ Extension Workers is the next best option, but this is and will remain a far-cry as not even 20% of the Farmers have access to smart-phones. The most optimal and workable option remains the Tele-Advisory Services or Farmers Help-Lines. Most of the State Agriculture Universities in the country have started Farmers' Help Lines during last 10 years. To harness the vast potential of rapidly improving Telephone Infrastructure in the Country, Ministry of Agriculture, Government of India, took a new initiative by launching the scheme "Kisan Call Centres (KCCs)" on January 21, 2004 aimed at answering farmers queries on a telephone call in farmers own dialect. These Call Centres are now working in 14 different locations covering all the States and UTs (Table 22.1). A countrywide common eleven digit Toll-free number 1800-180-1551 has been allotted for Kisan Call Centre. The number is accessible through all mobile phones and landlines of all telecom networks including private service providers. Replies to the farmers' queries are given in 22 local languages. Calls are attended from 6.00 am to 10.00 pm on all seven days of the week at each KCC location. KCC Farm-Tele-Advisors (FTAs) known as Level 1 FTAs respond to the farmers queries instantly. who are graduates or above (most of them being PG or Doctorate) in Agriculture or allied (Horticulture, Animal Husbandry, Fisheries, Poultry, Bee-keeping, Sericulture, Aquaculture, Agricultural Engineering, Agricultural Marketing, Bio-technology, Home Science etc. offered by Agricultural/ Horticultural/ Veterinary Universities) disciplines and excellent communication skills in respective local language. Queries which cannot be answered by Level 1 FTAs are transferred to higher level experts in a call conferencing mode. These experts are subject matter specialists of State Development Departments, ICAR and State Agricultural Universities.

KCC Call Handling Mechanism: The KCC operates at three levels – viz., Level I, Level II and Level III (Fig. 22.1):

- **Level - I:** This is a basic Call Center interface, consisting of high quality telecom infrastructure, adequate data processing equipment and local language proficient Agriculture Graduates designated as KCC Agents (now called Farm-Tele-Advisors or FTAs). The FTAs receive call from the farmer after a short welcome message. The basic details regarding information about the farmer and the query regarding the problems that he / she is facing with crop or animal or agri-related governmental schemes, are fed into the computer in order to document all the received calls. FTAs at level-I answer farmers queries, with their agriculture knowledge based on educational background, referring to the general Books on

Agriculture, literatures and brochures on agriculture or the Agricultural Yearbook published by concerned Agricultural University (e.g. VYAVASAYA PANCHANGAM: solutions to various crops pests / diseases including package of practices- published yearly by Acharya NG Ranga Agricultural University of Andhra Pradesh).

- **Level – II:** If the call cannot be answered by FTAs at Level I, then it is escalated to Level II where Subject Matter Specialists, in the areas of crop technology (agricultural university / ICAR) and about programmes / schemes (technical officials of departments of agriculture, animal husbandry, horticulture, fisheries, marketing etc.) respond to the farmer. There are 4 to7 level-II experts identified for every state for answering KCC Calls.
- **Level – III:** If the call cannot be answered even at Level II, it is escalated to Level III Nodal Institution / Office, where an Institute or agency which looks after the working of KCC in the concerned state responds to the query. The response is sent promptly by post or telephone within 72 hours of receipt of the question, by Level III.

During the working hours there would be immediate response. Beyond working hours or on holidays, the call would be recorded and the query will be answered later by post or telephone.

The KCC call escalation process has been restructured during April 2011 with an emphasis to involve (i) State Agricultural Department right from Block to state level,(ii) State Agricultural Universities and KVKs as well in facilitating FTAs to answer farmer's queries by way of call conferencing with the experts from these organizations in the event of the FTAs not being able to answer the farmers' queries. Active involvement of Common Service Centres and other Stakeholders has also been envisaged as detailed at Fig. 22.2. The States have been requested to take up following steps to implement the revised escalation matrix:

- To proactively involve in KCCs for supervising the quality of extension services provided by the FTAs and ensuring the revised escalation matrix under KKMS is put in place and higher level officers keep a track of the answers given at lower level.
- To improve functioning of KCCs, a major ad-campaign is being undertaken on electronic media.
- To coordinate with the State IT Department to get the scheme rolled out through the CSCs and appointment of a Nodal Officer for KCCs.
- Online monitoring by using yahoo messenger/skype id.
- Keep the FTAs apprised about new scheme/programmes/contingency plans taken by GOI and State Governments.

- Provide the FTAs with latest versions of guide books and booklets brought out by the State Government and the local Agricultural Universities.
- Organize monthly video conference on pre-announced dates in the university campus or through State Information Centre for interaction of FTAs with the Divisional/Zonal level officers of the State Agriculture and allied departments.
- The KCCs to give a weekly feedback to the State Department of Agriculture and allied departments regarding the nature of call including area specific prevalence of crop diseases, pest infestation etc.
- To create login IDs to various officers of the State Governments, KVKs, and SAUs from block level upwards.

KKMS: A Kisan Knowledge Management System (KKMS) to provide correct, consistent and quick replies to the queries of farmers have been developed by putting therein validated information on Agriculture and allied sectors of all States. Kisan Knowledge Management System (KKMS) has its independent web site, www.dackkms.gov.in The web site contains knowledge database on Package of Practices on Agriculture, Horticulture and Animal Husbandry of all the States. The Kisan Call Centre (KCC) FTAs working at various KCC locations throughout the country have access to this web site through their specific ID's & Pass-Word provided to them.

Whenever a farmer calls at KCC location by dialling the toll free number 1800-180-1551 raising his query for answer to be given by the Kisan Call Centre Agent, the KCC agent, apart from using his own knowledge/ experience, printed literature from concerned State Department of Agriculture/ State Agriculture University or Govt. of India guidelines, may access to the KKMS web-site search for information asked by the farmer in the knowledge database and pass on correct answer to him instantly. In this way the farmers get benefited with the KKMS.

Besides the KCC FTAs, farmers and other stake holders can also have direct access to the knowledge database search on package of practices in the KKMS web site by simply clicking on the 'Kisan Login' displayed on the front page of web- site to search for desired information available on the web-site for their own use.

KCC – Number of Calls Received: In order to create awareness amongst farmers about KCC programme, audio/video spots on KCC are being broadcast through All India Radio/ Doordarshan and private TV Channels. With consequent increased awareness among farmers, calls at the KCC have increased over three fold during the last three years.

Recently KCCs have been further revamped and restructured. The restructured KCCs are now more professional with the following technological innovations:

(a) Voice/Media Gateways (IPPBX based decentralized system).

(b) Dedicated MPLS leased line network with dedicated bandwidth.

(c) Call barging.

(d) SMS to caller farmers providing a gist of advisories given to them on phone.

(e) Voice mail system for recording farmer's queries during idle time of KCC or during call lines busy, with provision for call back to the caller.

(f) Soft phones in every personal computer with caller ID facility.

(g) Up scaling the knowledge of Farm-Tele-Advisors (FTAs) by way of providing latest versions of guide books and booklets issued by the State Agricultural Department or the Agricultural Universities. Facility of video conferencing of each KCC for interaction of FTAs with the Divisional/Zonal Level Officers of the State Agriculture and allied departments as well as on line monitoring for the working of KCCs.

(h) Provision for registering the farmers for receiving SMS messages on agri-advisories and mandi prices of different commodities as per their priority.

The new arrangements have been operational from May 1, 2012.

Future Plan (upto March 2017): The Ministry of Agriculture and Farmers Welfare aims to have an efficient, effective and a Kisan Call Centre Service based on a dynamic database and regularly updated knowledge (through experts in research and extension system) for each National Agriculture Research Project (NARP) Zone to rapidly enhance successful call inflow by the end of 12th Five Year Plan to such an extent that at least one third of the cultivators call KCCs once in a year on an average. This will necessitate about substantive increase in seats (up to 2000 from current around 500 seats) in the KCCs and corresponding augmentation of IT infrastructure of KCCs.

Table 22.1 Kisan Call Centre Locations and Languages

Sl. No	Location	States/ Uts Covered	Language
1.	Hyderabad	Andhra Pradesh, Telangana	Telugu
2.	Patna	Bihar	Hindi
		Jharkhand	Hindi
3.	Jaipur	Delhi	Hindi
		Rajasthan	Hindi
4.	Ahmadabad/Anand	Gujarat	Gujarati
		Dadra & Nagar Haveli	Gujarati
		Daman & Diu	Goan
5.	Chandigarh	Haryana	Hindi/Haryanvi
		Punjab	Punjabi
		Chandigarh	Punjabi
		Himachal Pradesh	Hindi
6.	Jammu	Jammu & Kashmir	Dogri, Kashmiri, Ladakhi
7.	Bangalore	Karnataka	Kannada
		Kerala	Malayalam
		Lakshadweep	Malayalam
8.	Jabalpur	Madhya Pradesh	Hindi
		Chhattisgarh	Hindi
9.	Nagpur/Pune	Maharashtra	Marathi
		Goa	Konkani, Marathi
10.	Coimbatore	Tamil Nadu	Tamil
		Puducherry	Tamil
11.	Kanpur	Uttar Pradesh	Hindi
		Uttarakhand	Hindi
12.	Kolkata	West Bengal,	Bengali
		Andaman & Nicobar	Tamil
13.	Bhubaneswar	Orissa	Oriya
14.	Guwahati	Arunachal Pradesh	Adi
		Assam	Assamese
		Manipur	Manipuri
		Meghalaya	Khasi
		Mizoram	Mizo
		Nagaland	Nagamese
		Sikkim	Sikkimese
		Tripura	Bengali

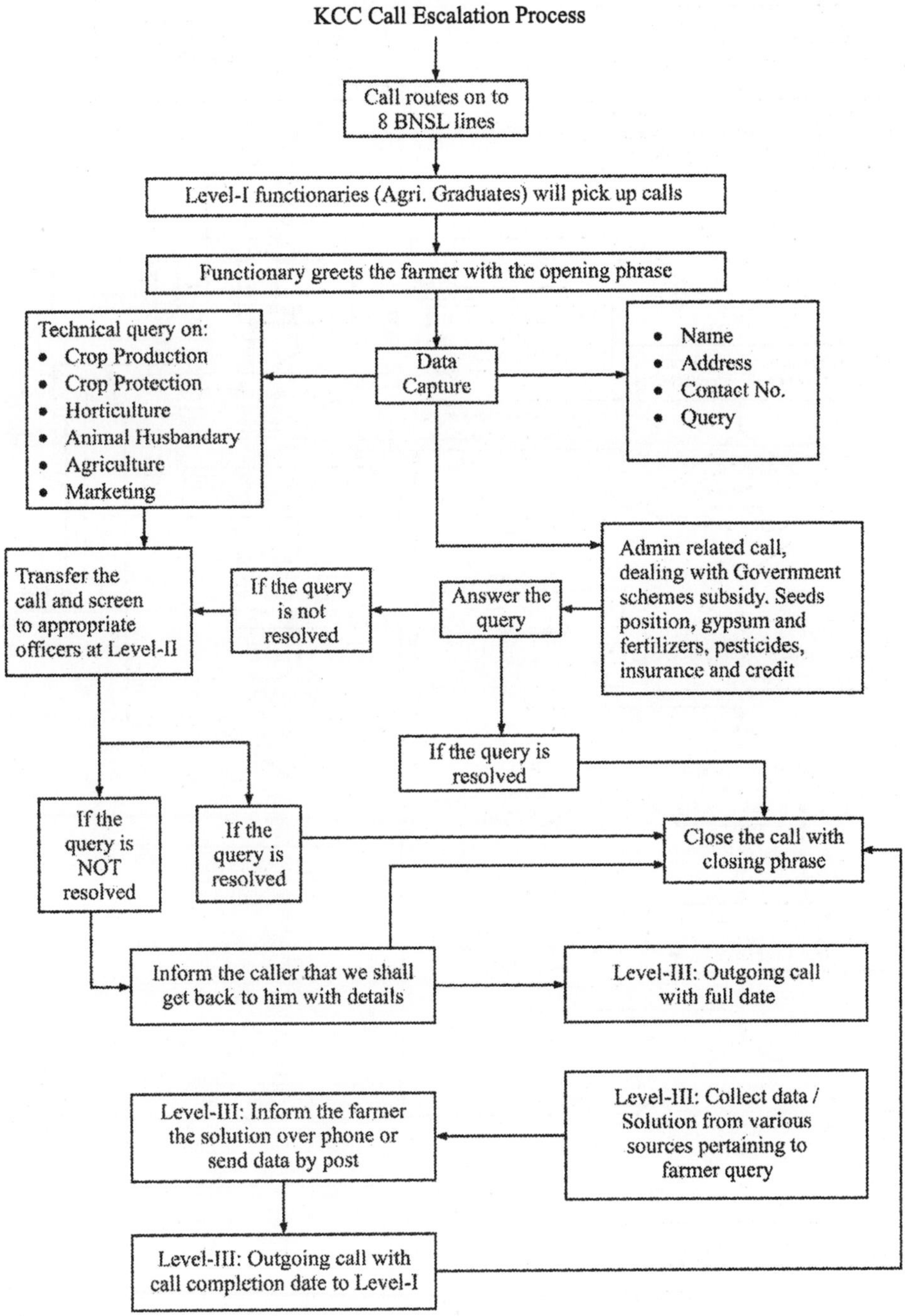

Fig. 22.1 KCC Call Escalation Process (since January 21, 2004)

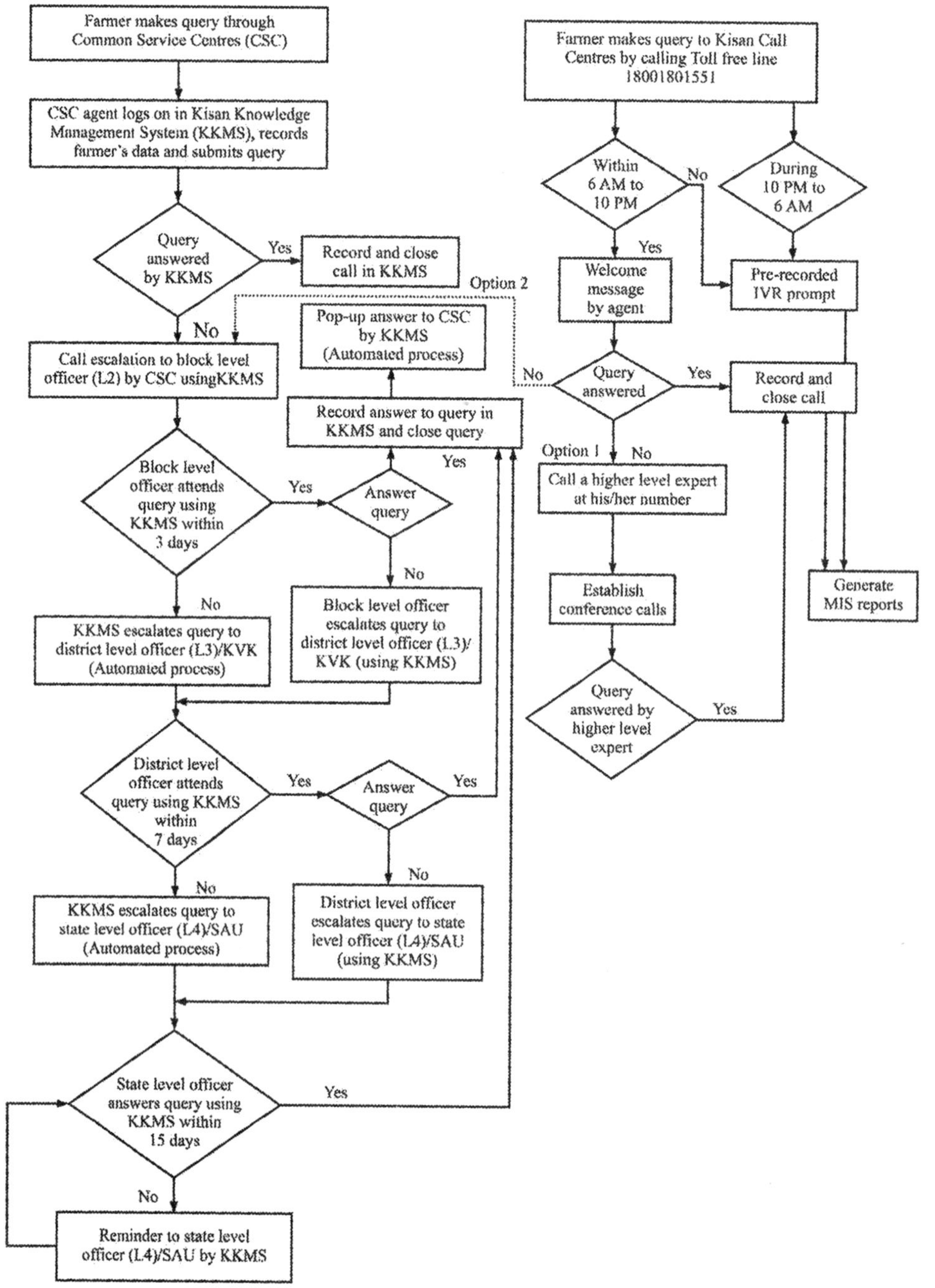

Fig. 22.2 Call Receiving and Escalation Matrix in Kisan Call Centre (KCC)

23 Impact Acceleration with Digital Technologies: Strategic Ideas for Extension Systems

- Dr. Shaik N.Meera

The success of Green Revolution in Asia indicates that giving rural communities access to knowledge, technology and services will contribute to expanding and energizing agriculture and certainly extension systems played an important role in proving that point.

In the present context, when governments across the globe are redefining the extension advisory services, new digital technologies provide a tremendous opportunity to transform linear extension approach (that excludes other stakeholders in the agricultural sector) into a vibrant impact acceleration approach.

Basically in the extension perspective, the impact acceleration is possible with better, faster and cheaper delivery of knowledge, technologies, inputs, market and other mission critical services.

Need

Need for improved agricultural extension and rural advisory services throughout the world has never been greater. Agricultural and rural development and hence, rural extension continue to be in transition in the developing world because of the vulnerability of farming due to climate change, changes in natural resources quality, lack of coping strategies at micro and macro levels of decision making, coupled with globalization, emerging market forces like commodity markets, sustainability constraints etc. The challenges can only be met from knowledge intensive and dynamic service coordination across the value chains and in the extension systems (Meera, Shaik N.et.al, 2010). This necessitates extension systems embrace digital opportunities available with us today. It is by now established that digital technologies can be harnessed effectively for improving the livelihoods and accelerating the impacts in agriculture and rural development.

Although there are evidences that digital technologies can revitalize research-extension interactions in ways that respond to farmers' demands, we need to understand that use of digital technologies is merely one element in the wider transformation of a traditional, top-down, technology-driven extension system into one that is more pluralistic, decentralized, farmer led, and market driven (and thus more effective within the innovation system). Part of the role of digital technologies is to contribute to the many reforms that are urgently needed to empower and support small-scale farmers as developing countries seek to respond successfully to food security, market development, and climate change challenges (Christoplos 2010).

The idea of digital agriculture essentially provides linkages, enhance market access, improve business process, increase product diversity and reduce development cycle time in agriculture. While the idea of Digital agriculture is very exciting concept, for extension systems there is a real challenge in integrating 'knowledge' with 'time critical services' across the agricultural value chain. We should have evidences of use, pattern, purpose, users etc., for digital technologies.

Digital Tools for Extension: A Quick Review

India has rich experience over last two decades, with large number of ICT projects (digital extension projects) providing a range of services in the villages. In spite of large number of ICT projects, there is vast untapped potential for its use in agricultural development by reducing the time and space barriers. The gap between what is possible with digital technologies and what has been achieved in extension systems is perhaps due to our inability to integrate Digital technologies into structural and functional components of extension systems.

A range of ICT pilot projects in the country were conceived and developed by public, private, non-government organizations, co-operatives and individuals. A variety of technologies were/are deployed, tested and modified for providing access to knowledge and services. Initial rural ICT projects were focused on building infrastructure including power supply. Almost all the projects provided the last mile connection using the telecom infrastructure for the sole reason of its affordability. Now the situation changed with the advent of mobile telephony, cheaper video options, common service centres, open source software, social media and smart phones.

As Digital Technologies have become more pervasive, they have become more relevant in agricultural extension advisory systems. The most pertinent developments for research, extension, markets, credit and e-learning are reviewed by many researchers. On a macro level, R&D organizations draw on increasingly connected and extensive digital infrastructure to facilitate collaboration and knowledge exchange nationally, regionally, and globally. On a micro level, extension professionals and farmers - powered by increasingly affordable mobile digital devices such as phones, laptops - connect with sources of information and services. In both cases, Digital Technologies empower individuals and institutions to create, access, and use knowledge and to communicate in unprecedented ways.

The ultimate choice of the ICT enabled extension depends on (1) the ICT policy environment, (2) the capacity of ICT service providers, (3) the type of stakeholders and the ICT approach adopted, and (4) the nature of the local communities, including their ability to access and apply the knowledge and various e-readiness parameters. Much literature is available online about these individual components.

All the digital initiatives undertaken during 1990 - 2015 can be classified into following tools and process. The level of integration of digital media into the governance process in extension at large will determine the fate of Indian agriculture in years to come.

Digital Technologies in Extension - Tools	Earliest / Popular efforts
	Computer Networks
	Audio- Video Conferencing
	Community Radio
	Mobile Telephony
	Landline Phones
	Automated Tools
	Video Dissemination tools
Digital Technologies in Extension - Processes	Hub & Spokes Model
	Kiosk Model, CSC Models
	Knowledge Management Process
	Open and Distance Learning (ODL) process, Massive Open Online Courses (MOOCs)
	Blended
Digital Technologies in Extension - Integration	Integrating Digital technologies into existing extension systems
	Optimizing with minimum change in existing systems
	Maximizing with maximum change in the systems

Digital Tools Scope of Extension Services

So what digital technologies will have bearing on future extension systems? Let us discuss them in brief here. The agriculture can take maximum benefit from few technologies. But how far they would be integrated into extension organizations is something that is seriously thought about by all. We may like to use mobile telephony, community radio, cable tv, video projections, computer networks etc.

ICT Tools	Functions	Content	Delivery Format
TV *Krishi Darshan, Annadata, Jaikisan, Krishi Deepam*	• Information dissemination • Distance learning	• Information on technology; critical farming practices relevant to the season, schemes of the government; advertisements on new products; commodity prices in different markets; talks with experts on new technology	• News announcements; talks by experts; question and answer sessions as part of phone-in-programmes (recorded and live); demonstrations; success stories • Programmes of a 30-60 minute duration every morning and evening. There are also re-runs of some programmes.
Internet-enabled computer centres (Information kiosks/ knowledge centres/ common service centres/telecentres) Ex: Village Knowledge Centres of MSSRF; ITC's e-choupal; The government of India's Common Service Centres	• Dissemination of information • Training in computer skills • Forum for interactive learning when centres are owned/managed by rural development NGOs • Distance learning	• Wide variation in content, depending on the objective of the centre, ownership, governance, revenue model, etc. • Information on government services, market prices, technology, weather, availability of inputs • In a few cases, locally-relevant content in a local language	• Information is generally intermediated through an employee of the computer centre • The information is also accessible through other means/media (notice boards, public address system, mobile phones, etc.) • Information accessed all through the day during a centre's working hours (around 10 hours a day)
Portals Rice Knowledge Management Portal (RKMP) www.rkmp.co.in Farmers Portal Agropedia agriwatch; Agmarknet, aAqua TNAU Agri Portal	• Dissemination of information • e-commerce • Distance learning	• Information on crop production, management and protection • Agricultural statistics, news, information on inputs (sources) • Dissemination of price information in various markets across the country • e-commerce (linking producers to traders/consumers) • a question-and- answer forum	• Portals vary widely in their content, regular updates, user friendliness, use of visuals, etc. • Most provide generic information. Increasingly portals provide dynamic information (e.g., current prices and weather updates)

Contd....

Call Centres Kisan Call Centre-1800-180-1551	• Dissemination of information and interaction with experts, especially advisory communication (e.g., specific problems answered by experts)	• Information on technologies, crop protection, sources of information, etc.	• Answers on specific queries from experts located at call centres and at other locations • Available during fixed hours during the day and in some cases 24-hour service
Mobile Phones Reuters Market Light (RML); IFFCO-IKSL; Tata m-Krishi	• Information dissemination • Different kinds of information provided by the service provider	• Information mostly on weather; prices of commodities in different markets; crop and animal husbandry advisory services; information on government schemes;	• Mostly paid services available to subscribers • Mostly as text messages or voice-mail
Community Radio Sangham Radio, Kongu FM radio, Mandakini ka awaaz, Krishi Community Radio	• Information dissemination • Raising awareness • Advocacy communication	• Wide range of information on rural life, agriculture, forests, health, handicrafts, etc. • Greater the ownership by the community, greater the involvement of the community in content generation and the content becomes more locally-relevant	• Format varies; Broadcast in most cases, but some communities have gone for narrow casting and cable casting until they collect enough resources to establish full-fledged stations • Content broadcast in local dialects for a few hours in the morning and evening every day. In a few cases broadcast throughout the day
Video Digital Green, Video SEWA	• Information dissemination • Advocacy communication • Training & Capacity Building • Mobilisation	• In agriculture, used to promote new technologies and good practices in farming • Used to raise awareness on women's issues; to mobilise communities around issues of common concern and also used as a training tool	• Screening instructional videos prepared locally to promote specific technologies with the support of a trained facilitator • Screening videos in community meetings, training programmes and workshops with policy-makers

Contd….

Interactive CD-ROMs/ Touch Screen Technologies Touch screen kiosks	• Information Dissemination	• Mainly related to production of different crops or enterprises	• Interactive multimedia CD-ROMS distributed to agricultural extension agencies for use by farmers in cyber units or communication centres of the Department of Agriculture/ veterinary hospitals/ clinics/ knowledge centres
Digital Photography India: e-Seva and e-Velanmai	• Providing information, mainly on plant management	• Advice based on digital photos depicting the growth of crops and symptoms of pest and disease attack	• Digital pictures mailed to experts and advice received through e-mail, on mobile phones or as printouts • Turnover time 1-3 days
Video and Teleconferencing India: Virtual academy for semi-arid tropics	• Information Dissemination • Knowledge exchange	• Depending on the nature of the problem being presented and availability of the expert (advisory on cropping, water and soil management) • Quality of interaction and exchange based on the facilitation and intermediary skills of the rural infomediary as well as his/her understanding of practical agriculture	• Interactive discussions, question and answer sessions with experts and feedback on problems and technologies facilitated by an intermediary organisation in the field • Usually once a week or fortnight at a designated time when the satellite bandwidth is specifically allotted for this activity

Adapted from Rasheed sulaiman et.al., (2011)

Mobile Telephones

Mobile phones can be defined as Basic (with talk and text), Featured (with talk, text and some other applications) or Smart - As above plus they can access the Internet. Mobile phones have emerged as a widely available option; providing multiple opportunities for improving communication.

Mobile phones can be used for

1. Voice to voice - call and ask specific questions (e.g., market prices, input availability, talk back radio, call centers)
2. Voice to machine - call and get an automated response
3. Text to pre-packaged response
4. Text to call center for individualized text or recorded voice response
5. Video screens

6. Accessing the Internet
7. Mobile apps (e.g., Diagnostic Tools, Nutrient Managers, Rice Crop FAQs, RKMP Apps)

There are several examples from India and elsewhere as how mobile phones could be used for credit, service provision, market and technical knowhow. Some of them include

- Banking: m-pesa ; IFFCO Kisan Sanchar Limited (IKSL)
- Market information: Esoko ;
- Technical information; e-Krishok ; iCow ; mKisan

Community Radio

Radio is a powerful means of (primarily) one-way communication . But the advent of Community Radio (FM radio owned & operated by community) has brought lot of relevance in current Indian agriculture. Prime Minister Modi has called for establishing one community radio in each agricultural college that will transform the way local agricultural knowledge and experiences are shared. There are several examples from India and elsewhere as how community radio could be deployed in agricultural development. Some of them include Deccan Development Society, Farm Radio International ; FarmerVoice Radio; Mali Shambani (Kenya).

In most cases, community radio can provide agricultural information that is topical, relevant and of interest to farmers' problems. Further follow-up , interviews with farmers who have successfully improved their farms etc., will speed up modern farming.

Social Media

Social Media can be a powerful communication tool to open lines of communication, engage a large number of people quickly, and reach a broader and (likely) younger farmers and stakeholders. Facebook and Twitter are two well known social media sites whose usage in agriculture field will witness radical changes in near future. On the other hand, crowd sourcing refers to the practice of obtaining needed services, ideas, or content by soliciting contributions from a large group of people. It often uses an online community, rather than from traditional employees or suppliers. Think about impacts of having crowd sourcing in the whole agricultural value chain. Similarly *whatsapp* can transform the problem solving strategies in agriculture. Kenya Seed Company Facebook and Twitter, Google Baraza Agriculture, Ushahidi, different facebook fora in India are few examples.

Smart Devices

Smart phones and other smart devices have the advantages of mobile phones but with greater interactivity and wider access to other media (e.g., Internet). Tablets (e.g., ipad,) or "Audio computers" can also be used in agriculture. Nutrient Manager of International Rice Research Institute, Philippines is an example for this.

Video and TV

The power of pictures is very important in effective communication. Video and TV are a powerful way to raise awareness, provide technical information and help in training. Some of the examples include Access Agriculture and Digital Green. Articulating a simple clear practical message with 5-6 or fewer key points and effective Storyboard will definitely make a difference in modernising the Indian farming.

Web - Internet Tools

The web offers a powerful way to reach the "web-connected" fraction of millions of farmers. Given issues of access, Extension and Development workers will be the primary target group - who then pass on the information they get to the rural communities. Most comprehensive examples being Rice Knowledge Management Portal (RKMP), TNAU Agritech portal, Agropedia and AgrisNet etc.

Big Data Tools

Africa Trial Sites (http://africats.org/) is a portal that enables national and international research organizations to electronically pool their extensive information on trial sites and provides numerous tools (based on ICT advances in bioinformatics, GIS, and data management) that help farmers, plant breeders, and agronomists to evaluate new varieties more efficiently in the field and gain more useful data from field trials. Users can search the website for trial sites and data by country, design trials to evaluate cultivars, obtain tools to manage trials (from developing a budget to estimating water stress during the growing season), analyze trial data, view results of spatial analyses, examine data on an interactive Google map, and report results online. The combination of African trial site data and interactive data analysis tools has made valuable information much more widely available and useful for the agricultural research, development, and extension community.

Rice Knowledge Management Portal (www.rkmp.co.in) has similar data repository comprising the data related to multi location performance of rice cultivars over last 49 years. (www.rkmp.co.in/data-repository). Frontline Demonstrations results are also made available at www.rkmp.co.in/FLDs. We have another example of mapping the submergence prone areas for better targeting of swarna sub-1 variety developed by International Rice Research Institute, Philippines.

For extension as a system, there is a need to build the huge repositories using GIS and state of art data management systems. The initiative for this to come from frontline extension efforts such as Krishi Vigyan Kendras (KVKs), Frontline Demonstrations, Private input databases, weather based agro-advisories, markets and post harvest options.

All the integrated data platforms should have inter-operability and have capability to get accessed with mobile platforms.

Functional Components of Digital Technologies

Farmers require relevant knowledge and information, including technical, scientific, economic, social, and cultural information. To be useful, that information must be available to users in appropriate languages and formats. At the same time, it must be current and communicated through appropriate channels. In the context of rural advisory services that accelerate impact, digital technologies have four broad functions.

1. To deliver or provide access to localized and customized information—adapted to rural users in a comprehensible format and appropriate language—to give small-scale producers as well as providers of advisory services adequate, timely access to technical and marketing information, as well as information or support on new technologies and good farming practices. It is not just a matter of getting information out. A key aim is to give rural people the facilities and skills to find the information and answers they need.
2. To organize the knowledge base by documenting and storing knowledge (experiences) for future use. In many cases, information and knowledge on technologies and good practices is available only in hard copy or in people's heads, and data are incomplete, scarce, or inaccurate. The challenge is evident from the scattered nature of the information, its multiple "formats," and the general lack of attention to documentation and learning in this area. While researchers are rewarded for publishing, extension workers, advisors, and farmers are motivated to deliver "practical" results; documentation is only a potential by-product. But as extension system as a whole the 'knowledge management' across the agricultural value chain is yet to be harnessed effectively. ICT mediated Knowledge Management has been in vogue in the conferences, but it is time for extension systems to develop a strategy to this effect.
3. To connect people and to facilitate networking- locally, regionally and globally—thus leading to collaborative and interdisciplinary approaches to problem solving and research based on shared knowledge and collaboration (Nyirenda-Jere 2010). Many NGOs, research organizations, and national ministries have used digital technologies to improve access to technologies and knowledge in their rural advisory services.
4. To empower rural communities with digital technologies should help farming communities "gain a voice" so that they can convey their needs and demands, negotiate better deals with other actors in value chains, and generally get practical benefits from the services intended for them (and otherwise avoid being exploited). A key element is to use Digital technologies to give rural people the skills and tools to tell their own stories, in their own words and languages, in ways that reach and influence others.

Impact acceleration through Digital Technologies: Tools & Cases

Research studies on assessment of information and communication technologies on farming revealed that the farmers not only accessed the information quickly but also enabled them to save cost of cultivation and realizing better market prices to the produce undertaken. The yield gain was to the up to five per cent and cost of cultivation reduced up to 15 per cent in different crops. However, Digital technologies helped farmers save travel time and transaction costs by more than 80-90 per cent. In an ICT study conducted recently in 4 states of India by Rice Portal team, in a conventional crop like rice, productivity increase was realised to the tune of 22% due to ICT interventions. It was also found that, the timeliness of providing information (pest diagnostics, online fertilizer recommendations) led to reduced cost of cultivation to the tune of 18% .

Mobile phones have increased daily wage earning fishermen's incomes from reduction in price dispersion, reaching new markets to avoid crowding at a market and there by elimination of unsold as waste. In a typical market it could increase per capita GDP by two per cent.

For greater impact of digital technologies on poor and marginal farmers, ICAR-IIRR introduced RiceCheck programme (participatory extension methodology) in 4 provinces in India. Digital Interventions act as backstop for farmer to farmer real time experience sharing (for example, Whatsapp is effectively used for extensionist-assisted farmer-group self learning). Various International (based in Australia, Philippines) and national partners (both public and private sector) were involved in this programme.

One of the critical factors contributing to maximizing technology impacts in farmers' fields is bringing in synergy between extension advisories (Knowledge or information) and availability of technologies. An innovative and integrated extension approach involving private sector (such as Savanna Seeds, Bayer Crop Sciences) was carried out at ICAR- Indian Institute of Rice Research, Hyderabad in bringing in the evidences of impact of digital technologies.

Many ICT experiences like this, revealed that farmers get higher returns on the basis of the timely access to market information and also save their time, travel charges and avoid unnecessary administrative hurdles. A number of ICT projects demonstrated their potential for poverty alleviation, health, education, agriculture, marketing and sustainable development. Opportunities of Digital technologies in agriculture and rural development are being viewed as an investment rather than expenditure. Empowering the rural people using Digital technologies is seen as yet one more policy instrument for economic development.

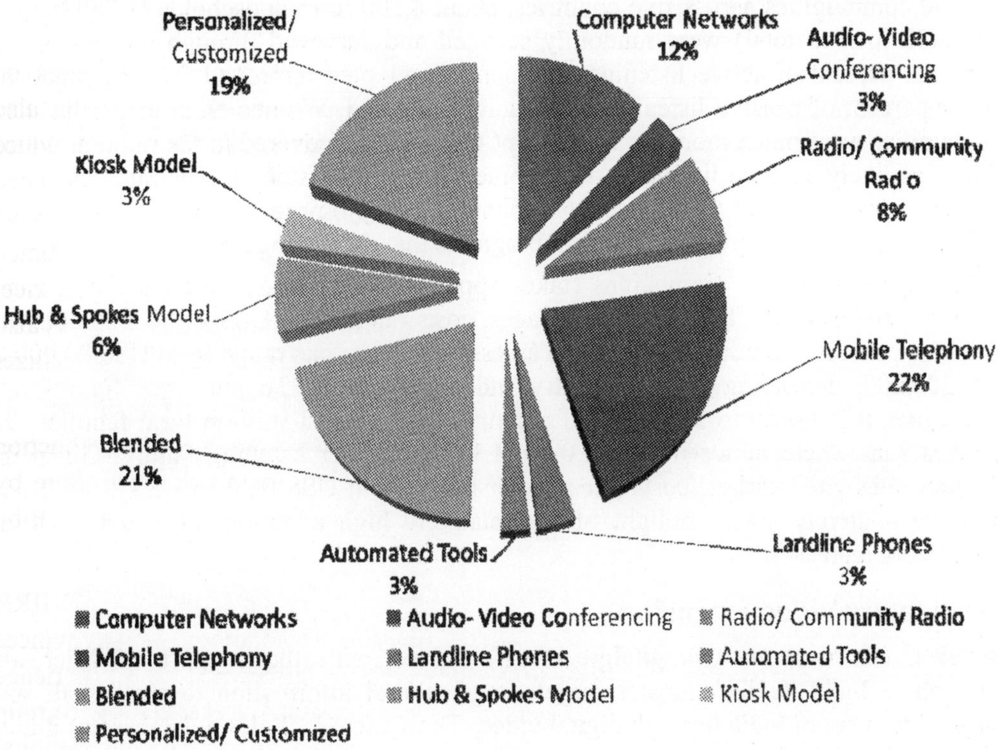

Farm Radio International

Farm Radio International, a Canadian NGO funded by the Bill and Melinda Gates Foundation, has created a new model of radio broadcasting that seeks to overcome some of challenges extension face. Farm Radio International partners with 360 radio stations in 39 African countries and reaches more than 200 million smallholders in more than 100 African languages. It offers a number of services but primarily develops Participatory Radio Campaigns, theme-based radio programs that continue for four to six months. The most innovative aspect of the Participatory Radio Campaigns is the broad base of farmer participation. First, men and women farmers help to develop the scripts, and a number of communities are invited to participate during implementation and evaluation. Second, programs are broadcast on a consistent schedule to keep farmers engaged. Third, Participatory Radio Campaigns feature voice response systems and call-in options that have proven remarkably successful in retaining listeners. The information elicited in this way helps extension staff and local NGOs identify the challenges, understand the perspectives, and gain the knowledge associated with a given community or area. Finally, women farmers are regularly included in the broadcasts and participatory aspects of the programs, improving their visibility and importance in the local

In 90 communities across five countries, about 4,500 farm households (1,988 women and 2,452 men in total) were randomly selected and surveyed through questionnaires. Thirty-six percent of active listening communities adopted improved farming practices, and 21 percent of passive listening communities adopted. Women from active listening communities were much more likely to adopt the practices covered in the radio programs (almost as likely as male listeners) than women in passive listening communities. These adoption rates are higher than those from many other radio programs, demonstrating that participatory radio is more effective than programs that do not engage farmers directly.

Participatory Radio Campaigns take approximately 12–18 months to design, distribute, and evaluate. For themed packages, costs range from US$ 25,000–50,000. For the whole process, including training and assessment, costs can range from US$ 80,000 to US$ 200,000, depending on the country and other factors. To put these figures into perspective, it is useful to know that if a campaign reaches 1 million farm families, the cost lies somewhere between US$ 0.08 and US$ 0.20 per listening family. Given the adoption rates cited earlier, costs per adopter range from US$ 0.20 to US$ 1.00. These costs are relatively small in light of the relatively high adoption rates and resulting productivity increases.

Participatory Video Extension

Digital Green (http://www.digitalgreen.org/) started with the support of Microsoft Research in India. It disseminates targeted agricultural information to small-scale and marginal farmers in India through digital video. The system includes a database of digital videos produced by farmers and experts. The topics vary, and they are sequenced in ways that enable farmers progressively to become better farmers. Unlike some systems that expect ICT alone to deliver useful knowledge to marginal farmers, Digital Green works with existing, people-based extension systems to amplify their effectiveness. The videos provide a point of focus, but it is people and social dynamics that ultimately make Digital Green work. Local social networks are tapped to connect farmers with experts; the thrill of appearing "on TV" motivates farmers. Although Digital Green requires the support of a grassroots-level extension system and other partners, it is effective because its content is relevant and it maintains a local presence. This local presence makes it possible to connect with farmers on a sustained basis.

The Digital Green approach is underpinned by various technological innovations (http://www.digitalgreen.org/tech). For example, its COCO (Connect Online, Connect Offline) software supports data tracking for organizations with sizable field operations, even where Internet service is intermittent and/or poor. COCO, a standalone application in the Internet browser, requires no additional desktop software installation or maintenance. It has an open-source, customizable framework and can be used without support from professional IT or engineering staff. Digital Green's Analytics System provides day-to-day business intelligence on field operations, performance targets, and basic measures of returns on investment relevant to an organization (see http://analytics.digitalgreen.org).

E-Learning in Extension Systems

Learning — formal and informal — is central to all innovation systems, including those for agriculture, and in sustaining the capacity to innovate over the long term.

In theory, e-learning enables governments, agricultural advisory services, NGOs, farmer organizations, private companies—in fact, any actor in the innovation system—to reach large numbers of producers. Content can be updated quickly and accommodate rapidly changing needs. E-learning does not require the complex online workflows associated with standard learning management systems, but a priority in promoting e-learning in extension systems is to build ICT capacity in personnel at all levels of agricultural education, training, and extension. Digital technologies and virtual interactions are not sufficient to form cohesive learning communities. Peer-to-peer contact significantly improves learning, and mobile phones can provide useful support.

A study conducted at Indian Institute of Rice Research focused on E-readiness and Information Literacy among Rice (extension) Workers and tried to assess the feasibility of e-learning strategies for agricultural development in general and rice sector in particular. A total of 18 e-courses were developed on rice production, protection and other technologies and are available online at *www.moodle.learnrice.in*. Both online and offline training is being imparted using this platform. The online version of e-learning platform is converted into offline version (Portable *Moodle*- Poodle).

A massive open online course (MOOC) is an online course aimed at unlimited participation and open access via the web. In addition to traditional course materials such as filmed lectures, readings, and problem sets, many MOOCs provide interactive user forums to support community interactions between students, professors, and teaching assistants (TAs). MOOCs are a recent and widely researched development in distance education which was first introduced in 2008 and emerged as a popular mode of learning in 2012. Extension systems can revolutionize the way adult and continuing education is carried out and also for skill development in agricultural sector.

Integrating Mobile Phone-Based Learning and Credit for Women Livestock Producers

VIDIYAL, an Indian NGO, uses L3F to promote community banking among 5,000 women organized into self-help groups (SHGs). During 2008, nearly 300 women from the SHGs became partners and decided to build their capacity through open and distance learning related to various aspects of sheep and goat production. As poor laborers, most of the women felt that attending classes or watching multimedia materials restricted their ability to work and attend to household chores. They asked VIDIYAL and COL to explore the use of mobile phone as a learning tool, because they would not need to be confined to any particular place or time during the learning process.

Through face-to-face and computer-based learning, COL and VIDIYAL encouraged the women to develop a business proposal for rearing sheep and goats. They developed a business proposal in which each member would obtain credit for buying nine female goats, one buck, and one mobile phone. The local bank agreed to the proposal and sanctioned a loan of US$ 270,000. The credit and the legal ownership of the assets are in the names of the participating women.

The 300 women bought simple mobile phones, and VIDIYAL entered an agreement with IKSL, one of India's major mobile network operators, to send audio messages to the women's phones free of charge and enable free calls among group members. The company felt that this strategy would enhance its mobile service in the long run. VIDIYAL and some of the participating women were trained in developing audio content for mobile phone-based learning. Learning materials are prepared within the broad principles of open and distance learning to meet learners' time and geographical constraints. VIDIYAL developed the materials in consultation with the Tamil Nadu Veterinary and Animal Sciences University and contextualized them to the local culture and dialects. The learning materials convey information in granular fashion—in short, concise messages. Three to five audio messages are sent to participating women every day. Each message runs for 60 seconds (Balasubramanian, Umar, and Kanwar 2010).

Digital Technologies and Local Supply Chain

Digital technologies may reduce the number of steps between producer and consumer (reducing the price spread thereby). It refers to either a short distance or a small number of intermediaries between the production and consumption of the food. A short supply chain covers a broad range of organizational strategies, but some examples are:

Farmers' markets, many of which have started to use social and digital marketing to increase the 'reach' of their message.

Food hubs, many of which are moving to 'virtual' methods of operation, using internet and web technologies to coordinate market exchange between producers and consumers, rather than using farmers' markets as a means to distribute produce.

Micro-enterprises with dynamically organized supply chains, which apply lightweight, often mobile, digital technologies in unique ways in order to coordinate supply and demand and manage business processes.

Maximizing the Impact of Digital Technologies

A project was initiated at IIRR critical factors contributing to maximizing technology impacts in farmers' fields is bringing in synergy between extension advisories (Knowledge or information) and availability of technologies. When a variety/hybrid is adopted along with management practices in toto (mostly knowledge based), the visible impacts can be realized. In this backdrop it is proposed to develop an innovative and integrated extension approach by blending a range of traditional and modern extension methods.

During the years 2012-13 and 2013-14, a small action research and surveys were conducted in two major rice growing states in India (Andhra Pradesh and Karnataka) involving 180 rice farmers and 45 extension professionals. Indicators such as knowledge, technology impacts on productivity were studied.

In both the provinces, impact of knowledge interventions was found to be significant when blended with field demonstrations. Out of 32 non-negotiable adoption points, 15 (46%) Information needs were met from various ICT tools. Changes in practices due to ICT interventions were found to be in the range of 11-12% of total adoption points. Majority of the farmers opined that productivity increase was realised to the tune of 30%. About 20% of the farmers benefitted either by reduced cost of cultivation or other advantages.

Precision to Decision Farming: Integration

To formulate strategic ideas for integrating digital technologies into extension systems is challenging job. Such strategies require to answer few basic questions emanating from ICT interventions.

A practical, yet complicated, solution can be found in integration of all these ICT experiences into structural and functional components of agricultural organisations. Business as usual would no longer help. The critical answers lie in the following points.

1. Extension systems, whether public or private, operate in a context that influences the organization, form, and content of transfer activities. The history and recent developments illustrate that digital "prescriptions" are doomed to fail if they are not based on 'farmers needs'. And that must be driven by learning about what works and what does not and by the nature of local circumstances and context. There is an opportunity for Extension division of ICAR to initiate a large scale project to document and extrapolate evidence based digital impacts, processes and to finalize the immediate integration strategies.
2. Most of the ICT reports end up stating "what Digital technologies can do for Extension". Seldom have they addressed the issue of "how extension organizations can harness Digital technologies in existing contexts". What would be the role of various public and private agricultural organizations in addressing such issues?. Content, trust, infomediaries, sustainability issues are very generic lessons that have been reported for last two decades. Rather we should focus on- most practical ways of content development (who, how, processes, scale and depth), capacity building needs of infomediaries. We need to address relevant issues such as what makes public extension workers to become infomediaries.
3. Digital applications alone will not be readily available, accessible and applicable in farmers conditions. It requires higher commitments from all the agricultural professionals. Further we need to build farmers communities (such as radio rural fora) so that ICT services/ content are applied in the field conditions. On large scale

government should plan campaigns for 'zooming in zooming out' farmers learning/ experiences using Digital technologies (take Digital Green for example).

4. State Department of Agriculture is a major public sector extension body. We need to understand the type of constraints grass root extension worker face in accessing (ICT enabled services/ content). Digital India campaign should focus on providing enabling factors. Digital agriculture strategies may warrant few structural adjustments among Research -Extension organizations onto a common platform. Their job chart needs to be transformed radically with a scope for incentives for efficient performance using digital tools. There has been an ambiguity in terms of what public sector externalists role in mkisan (in the absence of clear cut job chart) is well known among the analysts.
5. The main direction of reform in agricultural extension is towards learning rather than teaching paradigm. This learning approach should incorporate new methodologies and approaches that are demand-driven and increase the real, interactive participation of local people at all levels of decision making in an extension delivery network. The latest digital technologies could be effective answer. The e-learning and Massive Open Source Online Courses for extension systems and famers.
6. While focusing Digital technologies, we tend do completely ignore basic extension work/ methods that an extension worker uses. For example, if field demonstrations are conducted by extension workers - how Digital technologies would contribute to maximizing their impact. Whether 'field days' can be captured in video format and then video shows are arranged in neighbouring villages? There are 'n' number of basic extension tools that could be blended effectively with Digital technologies.
7. "Issue of content/ knowledge" is perhaps the most neglected of all. (Even though we know the importance of it - we seldom know how to do). We are yet to grow from addressing connectivity divide (establishing ‘ICT kiosks’) to knowledge divide (Managing the knowledge). Over the years, ICT experts have over simplified the issue of agricultural content/ knowledge - they report when the entire ICT infrastructure is available - within no time 'knowledge can be generated/ digitized/ uploaded). It’s high time; we focus on this as well. Our recent initiative of Rice Knowledge Management Portal (*www.rkmp.co.in*) tried to build, validate, and contextualise the rice knowledge (running in more than 16,000 pages with 18 platforms) in a short span of 2.5 years. But for Indian agriculture as a whole, the effort required is huge. In the absence of such validated knowledge on the web, the junk is being fed to agricultural stakeholders.

Way Forward

Digital technologies into extension systems should go beyond providing the advisories and technologies to the farmers. Rather extension systems can also transform the financing and service coordination at village level.

There is a need to develop an unique ICAR - Frontline Mobile Extension Program to build sustainable models, where farm and crop management tools and financial services are "bundled" in affordable, unified platforms on mobile phone channels to promote mass uptake commercially. This approach can facilitate development of a business model whereby the bundling process provides an increased value proposition for each partner, such as, increased fee income, greater outreach or reduced risks.

There is a need to develop and deploy eVouchers platforms across the country that enable extension agencies provide specific non-cash services. These include development grants, agricultural subsidies and various incentives to improve farming practices. Such eVouchers are much easier to track than cash vouchers, and they also help to avoid fraud, which is a common problem with paper vouchers. Think about clubbing this with the online soil health cards, nutrient manager app, optimum fertilizer recommendation and hence fertilizer supply using e-vouchers. This will dramatically improve the fertilizer demand and supply dynamics.

One of the areas where extension systems can leverage the benefits from mobile phones is - enabling easy, quick and safe payments in emerging markets. A mobile and web application that integrates SMS and mobile money services, will help savings and credit agencies. A mobile finance platform will help farmers to register with financiers, apply for loans and credit and make repayments – all using their mobile phones. The cloud-based management information system may be designed to improve the efficiency and reduce the operating costs of microfinance institutions and saving and credit cooperatives. To achieve this, governments need to create an enabling environment to encourage financial institutions to work more closely with the agricultural sector.

To enable all this on mobile interfaces, we need to work on developing agricultural semantic web portal (comprising dynamic location specific, local language knowledge), local sharing mechanisms using social media (whatsapp and facebook groups), Skill development platforms (video extension), ready to use scripts and experience sharing (community radio), big data platforms (GIS and Remote Sensing), market and input supply platforms (integration of private and public input supply inventory) etc.

The task is uphill and the pathways are complex. But for extension systems / organizations to survive they have to undertake this challenge. If it is done, the digital technologies will give new impetus to extension discipline and service.

24 Mass Media Consumption in Agricultural Extension

- Dr. P. Venkataramaiah

There is no doubt that agriculture is back on the development agenda. But despite the promises from governments worldwide, investment in agricultural development is still lagging. Communication for agriculture is also not seen as a major priority at either national or international level and the role of mass media as an effective player in agricultural development is undervalued. Mass media has a potentially broader role in raising the profile of agriculture amongst decision-makers as well as the wider public, and in communicating farmers' needs to step in to modern agriculture.

Modern agriculture is characterized among other things by the salient role of communication as factor of change and progress. Electronic media transmit the agriculture innovation to the farming community. Undoubtedly, there has been a rapid quantitative diffusion of mass media. The primary conveyors of development information in agriculture are also the persuasive agents of change in rural areas. Communication of development information; and equally persuasive conveyors of change, is the development workers: extension agriculture personnel in agriculture and influential opinion leaders. It is increasingly become aware of the tremendous role that mass media can play in agriculture growth. Agriculture development is the need of time for a better and prosperous nation.

Agricultural extension, or agricultural advisory services, comprises the entire set of organizations that support people engaged in agricultural production, facilitate their efforts to solve problems; link to markets and other players in the agricultural value chain; and obtain information, skills, and technologies to improve their livelihoods. This definition has evolved since the T&V program, where the focus of extension was transfer of technology to improve productivity, especially for staple food crops. While transfer of technology still has relevance, agricultural extension is now seen as playing a wider role by developing human and social capital, enhancing skills and knowledge for production and processing, facilitating access to markets and trade, organizing farmers and producer groups, and working with farmers toward sustainable natural resource management practices. Within this expanded role, the breadth of information that agricultural extension can support through provision and facilitating access and sharing is much larger. In addition, as the agriculture scenario has become more complex, farmers' access to sources of reliable and relevant information has become increasingly important.

Agricultural extension is essentially a message delivery system organized to convey the latest findings of agricultural research to farmers. Effective communication is therefore, the prime requirement in extension work. Three classes of extension methods, namely individual contact, group contact and mass contact, accomplish the task of extension communication. There are various Mass contact methods used to promote advanced agricultural information and techniques to the farmers, such as, agri. newsletters, grey literature, hand bills and walls newspaper, posters, radio programs, television programmers. Mass media are used for mass contact for impersonal transmission of messages to large audiences. The most generalized and widely accepted classification of mass media which is used in actual practice is print media and electronic media. Print media comprised of those forms of printed material which are distributed on a mass scale. These include newspapers, newsletters, books, grey literature (brochures, bulletins, pamphlets, leaflets, hand bills and posters.). Electronic media include radio and television, which have transformed World into a 'global village'. The electronic gadgetry of information technology like transistors, video tape recorders, mobile cinema vans and other audio-visual equipment like sound slide system, slides and film strips and recent programme like **Mann Ki Baat** - an Indian radio programme hosted by Hon'ble Prime Minister Narendra Modi in which he addresses the people of the nation on radio, DD National and DD News included under electronic media.

The success of agricultural development programmes in developing countries largely depends on the nature and extent of use of mass media in mobilization of people for development. The planners in developing countries realize that the development of agriculture could be hastened with the effective use of mass media.Radio, Television have been acclaimed to be the most effective media for diffusing the scientific knowledge to the masses. In a country like India, where literacy level is low, the choice of communication media is of vital importance. In this regard the television and radio are significant, as they transfer modern agricultural technology to literate and illiterate farmers alike even in interior areas, within short time. In India farm and home broadcast with agricultural thrust were introduced to enlighten farmers on the use of various technologies to boost agricultural development. With the main stream of Indian population engaged actively in agriculture, television could serve as a suitable medium of dissemination of farm information and latest technical know – how. The farmers can easily understand the operations, technology and instruction through television.

Among the several mass media, newspaper and farm magazine are commonly used. They have a vital role to play in the communication of agricultural information among the literate farmers. Increasing rate of literacy in the country offers new promises and prospects for utilizing print medium as a means of mass communication. The print media widened the scope of communication. It is cheap and people can afford to buy and read them at their convenience. It is a permanent medium in that the message are imprinted permanently with high storage value which makes them suitable for reference and research. Agricultural journalism is of recent origin in India. It is now gaining importance particularly after the establishment of agricultural universities in India, technical information needs to be provided to the farmers at the right time and in the right way, so that the productivity can be increased. In the view of increase in literacy level in India, print media has acquired a greater role in dissemination of information on improved agricultural practices to the farming community and also to inform the public in general. India has farm magazines in every state, published mostly in local languages. Agricultural department also encourages the publishing of such farm magazines particularly through farmers association.

The coverage of different subject matter by radio, television, newspaper and farm magazine are almost similar with regard to agriculture, horticulture, animal husbandry, agricultural marketing, agricultural engineering and cooperatives.

In other words, Mass media are those channels of communication which can expose large numbers of people to the same information at the same time. They include media which convey information by sound (radio, audio cassettes); moving pictures (television, film, video); and print (posters, newspapers, leaflets). The attraction of mass media to extension services is the high speed and low cost with which information can be communicated to people over a wide area. Although the cost of producing and transmitting a radio programme may seem high, when that cost is divided between the millions of people who may hear the programme, it is in fact a very cheap way of providing information. The cost of an hour's radio broadcast per farmer who listens can be less than one-hundredth of the cost of an hour's contact with an extension worker.

However, mass media cannot do all the jobs of an extension worker. They cannot offer personal advice and support, teach practical skills, or answer questions immediately. Their low cost suggests that they should be used for the tasks to which they are well suited. These include the following:

- Spreading awareness of new ideas and creating interest in farminginnovations.
- Giving timely warnings about possible pest and disease outbreaks, and urgent advice on what action to take.
- Multiplying the impact of extension activities. A demonstration will only be attended by a small number of farmers, but the results will reach many more if they are reported in newspapers andontheradio.
- Sharing experiences with other individuals and communities. The success of a village in establishing a local tree plantation might stimulate other villages to do the same if it is broadcast over the radio. Farmers are also often interested in hearing about the problems of other farmers and how they have overcome them.

- Answering questions, and advising on problems common to a large number of farmers.
- Reinforcing or repeating information and advice. Information heard at a meeting or passed on by an extension worker can soon be forgotten. It will be remembered more easily if it is reinforced by mass media.
- Using a variety of sources that are credible to farmers. Instead of hearing advice from the extension agent only, through mass media farmers can be brought into contact with successful farmers from other areas, respected political figures and agricultural specialists.

Mass media communication requires specialist professional skills. Few extension workers will ever be required to produce radio programmes or to make films. However, extension workers can contribute to the successful use of mass media by providing material to media producers, in the form of newspaper stories, photographs, recorded interviews with farmers, items of information about extension activities or ideas for new extension films; and by using mass media in their extension work, for example, by distributing posters and leaflets or by encouraging farmers to listen to farm broadcasts. Of-course, certain thumb rules can not be forgotten while using the media.

For extension through mass media to be effective, farmers must:

- have access to the medium;
- be exposed to the message: they may have radios, but do they listen to farm broadcasts?;
- pay attention to the message: information must be attractively presented and relevant to farmers' interests;
- understand the message.

Mass media messages are short-lived and the audience may pay attention for only a short time, particularly where the content is educational or instructional. If too much information is included, much of it will soon be forgotten. This means that information provided through mass media should be:

Simple and short.

Repeated, to increase understanding and help the audience to remember.

Structured, in a way that aids memory.

Coordinated with other media and with advice given by extension workers. It is important that what the farmers hear and see via mass media matches what extension workers tell them.

A poster on a wall containing several complex messages Dialogue is also an important part of communication. With mass media, however, there is little opportunity for a genuine dialogue between farmers and those who produce the material. Consequently, media producers are not in a good position to determine farmers' precise information needs, or to check whether their messages are understood correctly.

One solution to this problem is for the producers to carry out research into farmers' existing knowledge, attitudes, practices, and problems concerning farming topics, and for mass media messages to be pretested. This means that a preliminary version of the message is given to a small number of farmers so that, if they have any difficulties interpreting it, revisions can be made before the final version is prepared.

Extension workers can help media producers by keeping them informed of farmers' concerns and information needs, and by reporting any failure to understand the content of the products of mass media. People who produce radio programmes' posters and films are usually more educated than farmers and are not normally in regular daily contact with rural people. They cannot, therefore, easily anticipate how well farmers will interpret the material they produce.

Radio

Radio is a particularly useful mass medium for extension. Battery-operated radios are now common features in rural communities. Information can reach households directly and instantly throughout a region or country. Urgent news or warnings can be communicated far more quickly than through posters, extension workers or newspapers. Yet, despite radio's mass audience, a good presenter can make programmes seem very informal and personal, giving the impression that an individual listener is being spoken to directly. Radio is one of the best media for spreading awareness of new ideas to large numbers of people and can be used to publicize extension activities. It can also enable one community or group to share its experiences with others.

There are, however, a number of limitations to the use of radio in extension work. Batteries are expensive and often difficult to obtain in rural areas, and there may be few repair facilities for radio sets that break down. From the listener's point of view, radio is an inflexible medium: a programme is transmitted at a specific time of day and if a farmer does not switch on the radio in time, there is no further opportunity to hear it. There is no record of the message. A farmer cannot stop the programme and go back to a point that was not quite understood or heard properly, and after the broadcast there is nothing to remind the farmer of the information heard.

A further limitation is the casual way in which people generally listen to the radio. They often listen while they are doing something else, such as eating, preparing food, or working in the field. For this reason, radio is not a good medium for putting over long, complex items of information. A popular format in many countries, therefore, is for short items of farming news and information to be presented between musical records. Radio drama, in which advice is given indirectly through a story or play, is also popular. This can hold attention and interest for longer than a single voice giving a formal talk. Finally, there is little feedback from the audience, except with a live broadcast where it is possible for listeners to telephone in their questions or points of view directly to the programme presenter.

It may be difficult to design programmes that meet particular local needs. Moreover, it may not be possible to cater for variations in agricultural practices and recommendations in different areas. However, the growth in recent years of regional and local radio stations does make it possible for locally relevant information to be broadcast, and for extension workers to become more closely involved in making radio programmes.

Farm broadcasts will only be attractive to farmers if they are topical and relevant to their farming problems. Extension workers can help to make them attractive by sending information and stories to the producers, and by inviting them to their area to interview farmers who have successfully improved their farms, or to report on demonstrations, shows and other extension activities.

Ways by which extension workers can achieve a more effective use of radio include:

Recording farming broadcasts on a cassette recorder for playing back to farmers later. This could greatly increase the number of farmers who hear the programmes.

Encouraging farmers to listen to broadcasts, either in their own homes or in groups. Radio farm forums have been set up in a number of countries; a group meets regularly, often with an extension worker, to listen to farm broadcasts. After each programme, they discuss the contents, answer each other's queries as best they can, and decide whether any action can be taken in response to the information they have heard.

Stimulating the habit of listening to farming broadcasts, and the expectation of gaining useful information from the radio. This can be done by the extension worker listening to the programmes and talking about the contents in his contacts with farmers.

Many extension workers will at sometime have an opportunity to speak over the radio. They may be asked to interview farmers in their area or perhaps give a short talk themselves. The following guidelines for radio talks and interviews may be useful.

Radio Talks

- Decide on the purpose of the talk; in other words, what you want people to know, learn or feel at the end of it.
- Attract attention in the first few seconds.
- Speak in everyday language, just as you would in a conversation, and not as though you are giving a lecture.
- Repeat the main points carefully to help the listeners to understand and remember.
- Give specific examples to illustrate your main points.
- Limit your talk to three minutes; the listeners will not concentrate on one voice speaking on a single topic for much longer than that.

- Make the talk practical by suggesting action that the listeners might take.
- Include a variety of topics and styles if you are given more than three minutes. A short talk could be followed by an interview or some item of farming news.

Interviews

- Discuss the topic, and the questions you intend to ask, with the interviewee beforehand.
- Relax the interviewee with a chat before beginning to record the interview.
- Avoid introducing questions or points that the interviewee is not expecting.
- Use a conversational style; the interview should sound like an informal discussion.
- Draw out the main points from the interviewee, and avoid speaking at length yourself; listeners are interested in the interviewee rather than you.
- Keep questions short; use questions beginning "Why"?, "What?", "How?" to avoid simple one-word answers, such as "Yes" or "No".

Audio Cassettes

Audio cassettes are more flexible to use than radio, but as a mass medium they have their limitations. Cassette recorders are less common in rural areas than radio and are thus less familiar to villagers as sources of information. The cassette also has to be distributed physically, in contrast to the broadcast signal which makes radio such an instant medium. However, agents involved in many projects have found audio cassettes to be a useful extension tool, particularly where information is too specific to one area for it to be broadcast by radio.

The advantages of cassettes over radio are:

(a) that the tape can be stopped and replayed;

(b) the listeners do not have to listen at a specific time of day; and

(c) the same tape can be used over and over again, with new information being recorded and unwanted information being removed.

Information can be recorded on cassettes in a studio, where many copies can then be made for distribution, or it can be recorded on a blank cassette in the field. The possibility of recording farm radio programmes for playing back later has already been mentioned.

Cassettes can also be used for:

Updating the extension workers' technical information. Pre-recorded cassettes, distributed by the extension organization, are a good way of keeping extension workers in touch with new technical developments in agriculture.

Sharing experiences between farmers' groups and between communities. An extension agent can record interviews and statements in one village and play them back in others.

Providing a commentary to accompany filmstrips and slide sets.

Stimulating discussion in farmers' groups or in training centres by presenting various points of view on a topic, or from a recorded drama.

Cassette recorders are light and fairly robust. However, they should be kept as free from dust as possible and the recording heads kept clean by using a suitable cleaning fluid, such as white spirit.

Film

The main advantage of film as a mass medium for extension is that it is visual; the audience can see as well as hear the information it contains. It is easier to hold an audience's attention when they have something to look at. It also makes it possible to explain things that are difficult to describe in words, for example, the colour and shape of an insect pest or the correct way to transplant seedlings. Moreover, by using close-up shots and slow motion, action can be shown in far greater detail than it is to see possible watching a live demonstration. Scenes from different places and times can be brought together in order to teach processes that cannot normally be seen directly. The causes of erosion, for example, can be demonstrated dramatically by showing how a hilltop stripped of trees no longer prevents rain-water running down the slope, creating gullies and removing topsoil. Similarly, the benefits of regular weeding can be shown by filming crops in two contrasting fields at different stages of growth. Once a film has been made, many copies can be produced with the result that thousands can then watch the film at the same time.

Films come in two formats: 16 mm and 8 mm. Most cinema and educational films are in the larger 16-mm format. Equipment and production costs for 8-mm films are much lower, but because the picture quality is not quite so good and the projected picture size is relatively small, 8 mm has until recently been regarded as suitable for amateur domestic use only. As equipment improves, however, more organizations are producing training and educational films in 8-mm format. An 8-mm film cannot be shown on a projector made for 16-mm films or vice versa. Whichever format of film is to be used, it is necessary to have a projector; a screen or a white wall on which to project the film; a loudspeaker for the film's soundtrack (unless it has no soundtrack, in which case the extension worker may need a microphone, amplifier and loudspeaker so that he can give his own commentary); and a power source, which will either be mains electricity or a generator. If a generator is used, it should be as far away as possible from the projector and the audience so that its noise does not distract them from the film.

A Suitable Arrangement for Showing Films or Slides: The Audience must be able to see and hear Clearly

Film has a number of other limitations as a medium for rural extension. A film may take several months to produce since filming, processing, editing and copying all take time. Films are also expensive to make, and are worth making only if they can be shown many times over a number of years. They are, therefore not a good medium for topical information which soon becomes out of date.

The films seen by rural audiences have often been made in areas that are very different from those where they are shown. It may be difficult for the audience to relate their own farming to the crops, livestock, farm implements, people and housing that they see on the screen. The content may therefore seem of little relevance to them. Furthermore, there is no opportunity for a dialogue between film producer and farmer. Extension agents should, therefore, preview a film wherever possible, be prepared to explain the relevance of the information it contains whenever the details may be unfamiliar to local farmers, and be ready to answer farmers' questions afterwards. Finally, like radio programmes, a film is over very quickly and there is no permanent record of what was seen and heard.

An extension worker should only use a film when it fits in with his extension programme. If farmers are interested in dairy farming, then a film on the topic can give some ideas about the equipment, breeds of cattle and forms of organization they might need. Again, if an extn. Worker wishes to spread awareness of the dangers of soil erosion, a suitable film could explain the causes and effects as well as control measures.

When using film for extension purposes, an extension worker should keep the following points in mind.

Select films which fit in with the extension programme.

Publicize the film, after selecting a suitable date and venue in consultation with local leaders. Films are best shown in the evening; if the weather is suitable, the film can be projected against the outside white wall of a school or other building.

View the film in advance, and decide if the information needs to be adjusted to suit local conditions. This can be done either by speaking to the audience afterwards, or by turning the sound commentary off and giving a verbal explanation while the film is being shown.

Try out the equipment, especially if there is to be no technician present. It is useful to know how to change the bulb in the projector, for example, as these occasionally break.

Follow up the film by discussion and questions to help the audience to understand the content, relate it to their own situation and remember it.

Television and Video

Television, like film, combines vision with sound and like radio, it can also be an instant medium, transmitting information directly to a mass audience. Television signals can be broadcast from a land-based transmitter, by satellite or through cables. However, in many countries, television transmission and sets are still restricted to urban areas, and the potential of television for rural extension will remain low until sets become more widely available. Television sets are much more expensive to buy and repair than radios, and programme production costs are also far higher. Where television has been used for rural extension communication, access and impact have been increased by group viewing followed by discussion.

Video combines most of the advantages of film and of audio cassettes. Using a video camera, picture and sound are recorded on a magnetic tape and are then immediately available for viewing on a monitor or television set. This enables the production team to re-record any material that is not satisfactory. As with audio cassettes, unwanted information can be removed and the tape reused.

As a mass medium, video has more to offer than film, since video programmes can be made far more quickly in multiple copies, and the lightweight video cassettes are relatively easy to distribute. As video equipment - television monitors and video cassette recorders - becomes more robust, it will be possible to use mobile units to show up-to-date programmes, made within the country and even within the area, to large numbers of rural families. The tape can be slowed down, wound back to repeat a particular action, or held on a particular frame while the extension agent explains a point. The same mobile units could carry portable video cameras to collect material for new programmes. The main limitation to viewing is that only 20 to 30 people can satisfactorily watch a video programme on a normal television set, while several hundred can see a film projected on to a large screen.

Printed Media

Printed media can combine words, pictures and diagrams to convey accurate and clear information. Their great advantage is that they can be looked at for as long as the viewer wishes, and can be referred to again and again. This makes them ideal as permanent reminders of extension messages. However, they are only useful in areas where a reasonable proportion of the population can read.

Printed media used in extension include posters, leaflets, circular letters, newspapers and magazines.

Posters are useful for publicizing forthcoming events and for reinforcing messages that farmers receive through other media. They should be displayed in prominent places where a lot of people regularly pass by. The most effective posters carry a simple message, catch people's attention and are easy to interpret.

Leaflets can summarize the main points of a talk or demonstration, or provide detailed information that would not be remembered simply by hearing it, such as fertilizer application rates or names of seed varieties.

Circular letters are used to publicize local extension activities, to give timely information on local farm problems and to summarize results of demonstrations so that the many farmers who cannot attend them may still benefit.

Newspapers are not widely available in rural areas. However, local leaders often read newspapers, and a regular column on agricultural topics is useful to create awareness of new ideas and to inform people of what other groups or communities are doing.

Printed media can be either very sophisticated, with colour photographs and a variety of lettering styles, requiring expensive equipment that is only available in large cities, or produced simply and cheaply using equipment found in many local extension offices, such as a typewriter, stencils, a duplicator and a photocopier. This simpler technology makes it possible for extension workers to produce leaflets and circular letters that are relevant to their area and can be made available quickly to farmers. With the use of two duplicators - one with black and one with red ink- quite attractive leaflets can be produced. Stencil duplicators cannot reproduce photographs, so illustrations must be limited to simple outline drawings and diagrams. Modern photocopiers, however, can produce reasonable copies of black-and-white photographs.

Where the extension agent is using printed material that has been mass produced, he should make sure that it complements his extension activities. Posters may be used, for example, to draw attention to a topic related to a later demonstration, but printed material that the farmer does not see as relevant to what the extension worker does or says will have little impact.

Printed media are of little use if they are not distributed. Expensively produced posters, leaflets and magazines should not be allowed to gather dust on extension office shelves: they should be made widely available and farmers should be encouraged to look at and discuss them. Posters should be replaced regularly with new ones. In addition, where printed material proves to be irrelevant or difficult for farmers to understand, those who produced them ought to be informed so that improvements can be made. Posters and leaflets that seem clear to the extension worker may not be fully understood by farmers. Whenever possible, the extn. worker should help to explain their meaning. In time, farmers will become used to the ways in which pictures and words convey information and will find it increasingly easy to interpret printed media.

When the extn. worker is preparing his own printed media, or material is being produced to his specifications, the following stages offer a very useful guide. They apply equally to posters, leaflets, circular letters and newspaper articles.

Define the context: The extn. worker should be clear about the purpose of the material. Is it intended to create awareness and stimulate people to seek more detailed information?

Or to remind farmers of what they have learned? Or to provide detailed technical information and serve as a reference for future use? The extn. worker also needs to know how the material will be used by the audience. Will it be seen casually as people pass by a notice-board? Will it be studied individually in the home, or discussed at a group meeting?

Know the audience: Before planning the content, the agent needs information about the particular audience: their knowledge and attitudes concerning the subject-matter of the information, and their farming practices.

Decide on content: The information must be relevant to farmers' needs, and the content and amount of information should also suit the context in which the media will be used. A poster, for example, should contain one simple message in large, readable type that can be interpreted quickly by a passerby.

Attract attention: The material must be attractive at first glance. Only if a person's attention is caught by a leaflet or a poster will he spend the necessary time to look at, read and absorb the information it contains. This can be helped by short, boldly printed headings, eye-catching pictures and sufficient empty space to prevent it from looking too dense or cluttered.

Structure the information: The extn. worker can help farmers to understand and remember the information by dividing the contents into sections that lead logically from one to another, and by the use of headings and underlining to bring out the main points.

Pre-test: All locally produced material should be pre-tested before use. It can be shown to a few people from the target group, who should then be asked what information they have learned from it. This gives an opportunity to improve the material, if necessary, before beginning final production.

Exhibits and Displays

Apart from being a useful way of sharing information, an attractive, neat display suggests to people that the extension worker and his organization are efficient and keen to communicate. Displays are suitable for notice boards inside and outside extension offices, at demonstration plots (where the progress of the demonstration can be recorded in pictures), and at agricultural shows. Although a good display can take quite a long time to prepare, it will be seen by many people. With displays on permanent notice-boards, it is important that the material be changed regularly so that people develop the habit of looking there for up-to-date information.

A display should stick to a single theme broken down into a small number of messages. It should include several pictures (preferably photographs) and diagrams which must be clearly labelled. If there is a lot of printed text that is not broken up by pictures, the display will look dull and fail to attract attention.

Campaigns

In an extension campaign, several media are used in a coordinated way and over a limited period of time in order to achieve a particular extension objective. The advantage of campaigns is that the media can support and reinforce one another. The disadvantage is that campaigns can take a lot of time and effort to plan. Often the extension worker will be involved in campaigns planned by staff at national or regional level. His role will be to make local arrangements for meetings, film shows, demonstrations advance publicity, accommodation for visiting staff and distribution of printed material.

An extension worker can also plan his own local campaigns. A campaign can be useful in situations where the farmers of an area face a common problem for which there is a solution which could readily be adopted. Campaigns require careful planning to make the best use of all extension methods and media available.

Traditional Media

Traditional forms of entertainment can also be used as extension media. Songs, dances and plays can convey information in an interesting way. Even when they are prepared in advance, they can be adapted at the last minute to cater to local situations and response from the audience. No modern technology is required and these media are especially useful where literacy levels are low. By involving local people in preparing the plot of a play, extension worker can stimulate the process of problem analysis, which is a fundamental part of the educational aspect of extension. In brief traditional media means the medium(s) through which the cultural traits passed from generation to generation. It is born and expressed in the idiom of peoples culture and has always seemed to entertain, educate and propagate the existing ideas and attitudes. It hardly needs to make a mention over here that Folk media channels prove to be powerful tools of communication in the rural society. (Folk means race of people and folk media means people's media).

Different types of Traditional Folk media:

Drama, Puppetry, Nautanki, Keerthana or Harikatha, Streetplay, Folk Dance, Folk Song. Storytelling, Riddles, Proverbs, Bioscope, Munandi (announcements), Wall paintings.

Statewise Traditional Folk Forms:

AndhraPradesh – VasthiNatakam (VeediBhagavatham), Yakshaganam, BevalNatakam, Kuchipudi.

Karnataka – Yakshsganam (North Karnataka), Sannata (Belguam & Dharwad Dts), Doddata and Bavalata., etc.

If the traditional media is infused with new themes, new purposes with financial and other back-up support, it can go a long way in preserving the invaluable treasures which are a source of pride to the people.

Due to the familiar format, content and the usage of local language, traditional media has clarity in communication. The greatest advantage of the traditional media over the electronic media is their flexibility in accommodating new themes.

Many scholars have suggested the integration of traditional media with electronic media for quicker transmission of information as it can motivate the farmers.

However, under the impact of the more 'glamorous' and more 'powerful' electronic media, the traditional media and folk art forms are being influenced and even transformed. At the same time it is heartening to see how skillfully the electronic media exploits the traditional or folk forms to convey contemporary messages on programmes for farmers. It is this integrated approach which will strengthen the efficiency of both technology- based and folk media. A healthy combination of the modern and the traditional mass media makes for a practical approach for agricultural development.

Following is an illustration about MANN KI BAAT: (powerful electronic mass media).

Mann ki baat is a two pronged Indian radio programme to reach out to the rural as well as the urban population by way of getting answers to citizens' questions directly by our beloved PM Sri Narendra Modi.

NMC (New Media Cell) under I&B ministry will gauge the mood of urban public through social media sites.

Ironically the decision to create NMC was taken by the UPA government in August 2013, but it could hardly derive any benefit from it.

The idea behind Mann ki Baat programme is that the PM wants to have a direct talk with the countrymen on various issues. The I&B ministry has titled the slot as "Mann ki Baat" under AIR's canvas programme.

The New Media Cell will bring out weekly reports. Before going to the PMO, it will be vetted by the Union home ministry. A team of 7-8 boys and girls will continuously upload tweets and monitor sites like Twitter, Facebook and YouTube. The I&B ministry has roped in a Hyderabad-based IT solutions company to monitor Twitter handles and FB accounts. A dedicated team under IT unit analyses social media trends.

Having officially started on and from October 3,2014,theMaankiBaat programme aims to deliver the PM's voice to the general masses of India. The first Maanki Baat programme was broadcast on the occasion of Vijaya Dasami on Oct3, 2014.

PM with farmers: Maanki Baat: An illustration.

Greetings to all my dear Farmer brothers and sisters!

Today, I feel so fortunate to avail this opportunity to speak with my farmer brothers and sisters from different corners of the country. When I speak to farmers, I speak to the village and its residents, and also to the farm labourers. I am also speaking to the mothers and sisters who work on the farm fields. And in this context probably this feeling is a little different from all the interactions I had through Mann ki Baat. ....

Farmers in Mann Ki Baat, I never expected that the farmers from far off villages across the country would be willing to ask numerous questions and would provide such information. I was taken aback to receive your numerous letters and questions in such huge numbers. First of all I would like to express my heartfelt greetings to you all. I have read your letters and understood the agony and struggle in your lives. Despite bearing so much misery, I cannot imagine what all you have gone through in your lives.

I desire your love and blessings at all times. You are the elders of the nation, you can never think wrong for others rather you will suffer loses for other's benefit. This has been your tradition. And these farmers do not suffer, should have to be the prime focus of my government. Today, after hearing this "Mann ki Baat" many thoughts may have arisen in your mind. Please do write into me at the Akashvani address. I will talk again. Based on your letters, I will try to rectify all the mistakes of my government. I will try to speed up the efforts and if somewhere injustice is being done, I will try and provide justice.

Ten (10) highlights from the address of our PM under Mann Ki Baat radio Programme:

1. The address on AIR to will become a more regular affair. The Prime Minister said that he will speak to people once or twice a month on AIR on Sunday at 11 a.m.
2. Referring to Mahatma Gandhi and his teachings, he urged the purchase and use of 'Khadi'. He said that by buying Khadi one can contribute towards bringing prosperity to poor.
3. He reminded people of the Swachh Bharat Abhiyaan that was launched on October 2 and urged people to join the mission of making India a clean nation.
4. He also touched upon the recent success of the Mars Mission and praised Indian scientists.
5. He quoted Swami Vivikanand's story about the lion cub brought up by a sheep who later discovered his strengths when he came in contact with another lion. He said that as a 125 crore strong nation, we all have skills and only need to recognize these skills. He emphasized on taking the initiative, integrating our individual strengths and unite to achieve goals.
6. He talked about the need for taking care of specially abled children and implementing schemes for them. "I remember when we started Khel Mahakumbh for specially-abled athletes and I myself would go and see the sports."
7. He highlighted some of the points that people have mailed him in the recent past like having more skill based learning for school children, installing more dustbins in the country and so on.

8. He encouraged people to keep writing to him and share their thoughts on the website www.mygov.in.
9. He expressed his happiness over how the radio has let him reach_out to the remotest and poorest villages of India.
10. He emphasized that India belongs to each Indian citizen and this sense of belonging should bring people together and become a part of the development. He said ," if everyone takes a step forward then India will take 125 crore steps".

In nutshell, the paper/article covers mass media role in agricultural extension at length as given here under:

Mass media consumption in agricultural extension in terms of traditional and electronic media viz., individual, group & mass contact has a potential broader role in agricultural development, which is the need of time for a better and prosperous nation.

Agricultural Extension is essentially a message delivery system organized to convey the latest findings of agricultural research to farmers.

Various mass media such as Print media – agril. newsletter, literature, leaflets, handbills, newspapers, wall news papers, farm magazines, posters, circular letters, exhibits & displays and electronic media like Radio talks, Interviews, audiocassettes, film(s), Television programmes, transistors, video recorders, slides, film strips, mobile cinema and Prime minister's Mann KI Baat radio programme. Traditional media like Puppetry, Drama, Harikatha, Street Play, Folk Dance, Folk Song, Story telling, Riddles, Proverbs, Bioscope, Announcements, wall Paintings; Yakshaganam, Veedi Bhagavatam, Kuchipudi, Beval Natakam, Sannata, Doddata, Koodiyattam, Mudiattam, Tyattu and Tharayattam also play a crucial role in agricultural development. If the traditional media is infused with new themes, new purposes with financial and other backup support , it can go a long way in preserving the invaluable treasures which are a source of pride to the people.

Mass Media

Mass Media is a relatively new idea in human culture. It incorporates all those mediums through which information is distributed to the masses. Generally categorized in to 7 major branches in order of their introduction are: Print, Recordings, Cinema, Radio, Television, Internet and mobile phones. (Internet and mobile are often called as Digital media and Radio & TV as Broadcast media).

Characteristics

Communication is mostly one way. Audience has great deal of choice. Reach large and vast audience. Aim messages to attract largest audience possible. Influence society and are in turn influenced by society.

Functions

Information, Consensus, Entertainment, Symbolic function, Advertising and Development - Development communication that focuses on the needs of the underprivileged and oppressed: their socio-economic, cultural interests & needs.

Advantages

Reaches many people quickly. Low cost per person reached.

Can be used to : tell people about new ideas and services. Agenda setting & advocacy. Created favorable climate of opinion.

Disadvantages

Difficult to make specific to local community. Fixed message. Can be easily misunderstood. Access often difficult. Lacks feedback.

Less appropriate for : Changing behaviours rooted in culture or reinforced by Social norms. Promoting empowerment. Learning practical skills. Developing skills of informed decision making.

Mass Media in Agriculture

Information on Agriculture, both crop and livestock was communicated among farmers from ancient times. However, with the development in agricultural research, need arises to transfer new information and technologies to the farmers. To fulfill this need, mass media like news papers, magazines, radio, TV, film and Internet play a vital role. What we know about the new information on technologies, public figures and public affairs is largely dependent upon what the mass media told us about it. The major objective of mass media in agriculture development is : to communicate the feasible farm technologies in such a manner to attract the attention of farmers. Help them to understand and remember the message and ultimately facilitate them to take appropriate decision.

Print Media

Print medium was the first to be used as mass media for communicating the information. First newspaper published was Bengal Gazette in 1780. Since then the use of newspapers and magazines kept on increasing in terms of their number, variety, circulation and readership. Quite often new newspapers and magazines are introduced while the older ones change their pattern of presentation. A newspaper is a publication containing news and information and advertising, usually printed on low cost paper called newsprint. It may be general or of the special interest , most often published daily or weekly.

Types of Print media

Newspaper, Magazine (general & public interest), Journals, Books, Others are Leaflets, Circular letters, folders, banners and wall newspapers.

Print media characteristics

Gives extensive coverage to a large number of items of interest. Can be read by literate audience.

Print Media in Agricultural Development

Among the several mass media, newspaper and farm magazine are commonly used. Cheap and affordable; read as per convenience. Permanent medium, permanently Imprinted message with high storage value making them suitable for reference and research also. Rate of literacy in the country offers new promises and prospects for utilising print medium as a means of communication. Plays a vital role in the communication of agricultural information among the literate farmers on improved agricultural practices and also to inform the public in general. Agricultural Journalism is of recent origin in India. It is gaining importance, particularly after the establishment of agricultural universities. Technical information needs to be provided to the farmers at the right time and in the right way, so that the productivity can be increased. India has farm magazines in every state published mostly in local languages. Agriculture department also encourages the publishing of such farm magazines particularly through farmers association . Among the various types of mass media sources, newspapers can support extension by publishing news of various extension activities, guidance and recommendations, achievements , market news, research findings, successful achievements and problems faced by farming community etc. The green and white revolution could not have been come so quickly without the use of media like print media and radio.

Internet (world Wide Web)

Information Super Highway & The Network of Networks helps in getting information, Disseminate information by publishing , extension and teaching .

Uses of Internet

Knowledge sharing between the agriculture Research Institutes. Access to International best practices. Information sharing on public domain. Online Trading and Import and Export. Web based approaches and wide range of Web portals act as information repositories. Envisaged to address the problem of accessibility of useful and timely information by small and marginal farmers. It helps in saving their time as well as cost of traveling and enables them to exchange information on real time basis. Wireless reach initiative is a strategic programme that brings wireless technology to undeserved communities globally.

Mobile Phones

Any time, Anywhere Mobile phones are multi functional devices. Extension can reach more clients through mobile-based learning platforms - textual or richer platforms, such as Video – that provide tips to farmers to improve agri-cultural skills and knowledge. MMS, GPRS, WAP & GPS can also be utilized effectively for extension.

Various Roles for Mobiles in Agriculture

Information provided via mobile phones to farmers and extension workers about good practices, improved crop varieties, and pest or disease management. Commodity Prices in regional markets to inform decision making throughout the entire agricultural process. Applications that collect data from large geographical regions. Send and receive data on outbreak warning and tracking.

Benefits

Getting better prices. Connected to markets and Kisaan call centers. Increasing yield. Getting advices from experts. Improves adoption of technologies at early stage. Improves social and business network and help in faster spread of knowledge and technology. Provide market information through SMS or Voice messages, or Question-and-answer capabilities. Lifelines and e-Sagu also use mobile phones in combination with computing technology to provide expert advice based on farmer queries.

Also recent electronic based programmes like Mann kiBaat play a significant role in disseminating information to the rural as well as the urban population by way of getting answers to citizens' questions directly by our PM. Ten (10) highlights from the address by our PM in Mann Ki Baat radio programme are given below:

1. The address on AIR to will become a more regular affair. The Prime Minister said that he will speak to people once or twice a month on AIR on Sunday at 11 a.m.
2. Refering to Mahatma Gandhi and his teachings, he urged the purchase and use of 'Khadi. He said that by buying Khadi one can contribute towards bringing prosperity to poor.
3. He reminded people of the Swachh Bharat Abhiyaan that was launched on October 2 and urged people to join the mission of making India a clean nation.
4. He also touched upon the recent success of the Mars Mission and praised Indian scientists.
5. He quoted Swami Vivikanand's story about the lion cub brought up by a sheep who later discovered his strenghts when he came in contact with another lion. He said that as a 125 crore strong nation, we all have skills and only need to

recognize these skills. He emphasized on taking the initiative, integrating our individual strengths and unite to achieve goals.

6. He talked about the need for taking care of specially abled children and implementing schemes for them. "I remember when we started Khel Maha kumbh for specially-abled athletes and I myself would go and see the sports."
7. He highlighted some of the points that people have mailed him in the recent past like having more skill based learning for school children, installing more dustbins in the country and so on.
8. He encouraged people to keep writing to him and share their thoughts on the website www.mygov.in.
9. He expressed his happiness over how the radio has let him reach out to the remotest and poorest villages of India.
10. He emphasized that India belongs to each Indian citizen and this sense of belonging should bring people together and become a part of the development. He said ," if everyone takes a step forward then India will take 125 crore steps".

References

1. Coombs, P.H. & Ahmed, M. 1974 Attacking rural poverty. World Bank, Johns Hopkins University Press.
2. Foster, G. 1962 Traditional cultures and the impact of technological change. New York, Harper.
3. Lele, U. 1979 The design of rural development. World Bank, Johns Hopkins University Press.
4. Long, N. 1977 An introduction to the sociology of rural development. London, Tavistock.
5. World Bank. 1975 The assault on world poverty: problems of rural development. Baltimore, 1975 Johns Hopkins University Press.

Resource Books on Extension

1. Crouch, B.R. & Chamala, S. 1981 Extension education and rural development. Chichester, 1981 W. Sussex, Wiley.2 vols.
2. Jones, G.E. & Rolls, M.J. 1981 Progress in rural extension and community development.1981 Chichester, W. Sussex, Wiley.
3. Maunder, A.H. 1972 Agricultural extension: a reference manual. Rome, FAO. 1972
4. Savile, A.H. 1965 Extension in rural communities. Oxford University Press. 1965

Extension Practice

1. Batten T.R. 1967 Non-directive approach to group and community work. Oxford University Press.
2. Benor, D. & Harrison, J.O. 1977 Agricultural extension: the training and visit system. Washington, World Bank.
3. Diaz Bordenave, J.E. 1977 Communication and rural development. Paris, Unesco.
4. Fao. 1979 Small farmers development manual. Rome. 2 vols.
5. Fuglesang, A. 1982 About understanding – ideas and observations on cross-cultural communication. Uppsala, Sweden, Dag Hammarskjold Foundation.
6. Havelock, R.G. 1973 Training for change agents. Ann Arbor, University of Michigan, Institute of Social Research.
7. O'sullivan-Ryan J. & Kaplun, M. (eds). 1980 Communication methods to promote grass roots participation. Paris, UNESCO.
8. Stuart, M. & Dunn, A.M. 1982 Extension methods. In Hawkins, M.S. Agricultural livestock extension. Canberra, Australian Universities International Development Programme, Vol. 2.

Journals

1. Agricultural information development bulletin United Nations Economic and Social Commission for Asia and the Pacific (ESCAP), Bangkok, Thailand.
2. Bulletin Agricultural Extension and Rural Development Centre, School of Education University of Reading, UK.
3. Ceres FAO, Rome.
4. Media in education and development British Council Media Department, London, UK.
5. Rural extension, education and training abstracts Commonwealth Agricultural Bureaux, Farnham Royal, Slough, UK.

25 Development Communication

– *I. Sreenivasa Rao & Neema Praveen*

Communication for development is a broad cognitive field of enormous international, national and regional interest attracting attention as a special field of study by students and researchers across disciplines. New media and communication convergence are reshaping the ways in which communication can be used in development infusing renewed interest in the field as a subject of serious academic study.

Development Communication, simply defined, is the use of communication to promote social development. More specifically, it refers to the practice of systematically applying the processes, strategies, and principles of communication to bring about positive social change.

The term "**Development Communication**" was first coined in 1972 by **Nora C. Quebral,** (is a pioneering figure in the discipline of Development Communication in Asia and is often referred to as the "Mother of Development Communication" giving birth not only to an academic discipline but to a new crop of scholars in the field as well) who defines the field as "the art and science of human communication linked to a society's planned transformation from a state of poverty to one of dynamic socio-economic growth that makes for greater equity and the larger unfolding of individual potential." (Paolo Mefalopulos, 2008)

Erskine Childers defined it as Development support communications is a discipline in development planning and implementation in which more adequate account is taken of human behavioural factors in the design of development projects and their objectives (Ogan, 1982).

According to the World Bank, development communication is the "integration of strategic communication in development projects" based on a clear understanding of indigenous realities.

In addition, the UNICEF views it as: "...a two-way process for sharing ideas and knowledge using a range of communication tools and approaches that empower individuals and communities to take actions to improve their lives."

Bessette (2006) defined development communication as a "planned and systematic application of communication resources, channels, approaches and strategies to support the goals of socio–economic, political and cultural development".

Ten Key Issues about Development Communication (World Bank, 2008).

The 10 points presented in this section address some of the myths and misconceptions about communication, especially when related to the field of development. These misconceptions can often be the cause of misunderstandings and lead to inconsistent and ineffective use of communication concepts and practices. The first two points on this list are about communication in general, while the others refer to development communication in particular.

1. ***"Communications" and "communication" are not the same thing***. The plural form refers mainly to activities and products, including information technologies, media products, and services (the Internet, satellites, broadcasts, and so forth). The singular form, on the other hand, usually refers to the process of communication, emphasizing its dialogical and analytical functions rather than its informative nature and media products. This distinction is significant at the theoretical, methodological, and operational levels.
2. ***There is a sharp difference between everyday communication and professional communication.*** Such a statement might seem obvious, but the two are frequently equated, either overtly or more subtly, as in, "He or she communicates well; hence, he or she is a good communicator." A person who communicates well is not necessarily a person who can make effective and professional use of communication. Each human being is a born communicator, but not everyone can communicate strategically, using the knowledge of principles and experience in practical applications. A professional (development) communication specialist understands relevant theories and practices and is capable of designing effective strategies that draw from the full range of communication approaches and methods to achieve intended objectives.
3. ***There is a significant difference between development communication and other types of communication***. Both theoretically and practically, there are many different types of applications in the communication family. In this publication, we refer to four main types of communication, which are represented significantly in the work of the World Bank: advocacy communication, corporate communication, internal communication, and development communication. Each has a different scope and requires specific knowledge and skills to be performed effectively. Expertise in one area of communication is not sufficient to ensure results if applied in another area.

4. ***The main scope and functions of development communication are not exclusively about communicating information and messages, but they also involve engaging stakeholders and assessing the situation.*** Communication is not only about "selling ideas." Such a conception could have been appropriate in the past, when communication was identified with mass media and the linear Sender-Message-Channel-Receiver model, whose purpose was to inform audiences and persuade them to change. Not surprisingly, the first systematic research on the effects of communication was carried out soon after World War II, when communication activities were mostly associated with a controversial concept-propaganda. Currently, the scope of development communication has broadened to include an analytical aspect as well as a dialogical one — intended to open public spaces where perceptions, opinions, and knowledge of relevant stakeholders can be aired and assessed.
5. ***Development communication initiatives can never be successful unless proper communication research is conducted before deciding on the strategy.*** A communication professional should not design a communication campaign or strategy without having all the relevant data to inform his or her decision. If further research is needed to obtain relevant data, to identify gaps, or to validate the project assumptions, the communication specialist must not hesitate to make such a request to the project management. Even when a communication specialist is called in the middle of a project whose objectives appear straightforward and clearly defined, specific communication research should be carried out if there are gaps in the available data. Assumptions based on the experts' knowledge should always be triangulated with other sources to ensure their overall validity. Given its interdisciplinary and cross-cutting nature, communication research should ideally be carried out at the inception of any development initiative, regardless of the sector or if a communication component would be needed at a later stage.
6. ***To be effective in their work, development communication specialists need to have a specific and in-depth knowledge of the theory and practical applications of the discipline.*** In addition to being familiar with the relevant literature about the various communication theories, models, and applications, development communication specialists should also be educated in the basic principles and practices of other interrelated disciplines, such as anthropology, marketing, sociology, ethnography, psychology, adult education, and social research. In the current development framework, it is particularly important that a specialist be acquainted with participatory research methods and techniques, monitoring and evaluation tools, and basics principles of strategy design. Additionally, a good professional should also have the right attitude toward people, being empathic and willing to listen and to facilitate dialog in order to elicit and incorporate stakeholders' perceptions and opinions. Most of all, a professional development communication specialist needs to be consistently issue-focused, rather than institution-focused.

7. ***Development communication support can only be as effective as the project itself.*** Even the most well-designed communication strategy will fail if the overall objectives of the project are not properly determined, if they do not enjoy a broad consensus from stakeholders, or if the activities are not implemented in a satisfactory manner. Sometimes communication experts are called in and asked to provide solutions to problems that were not clearly investigated and defined, or to support objectives that are disconnected from the political and social reality on the ground. In such cases, the ideal solution is to carry out field research or a communication-based assessment to probe key issues, constraints, and feasible options. Tight deadlines and budget limitations, however, often induce managers to put pressure on communication experts to produce quick fixes, trying to force them to act as short-term damage-control public relations or "spin doctors." In such cases, the basic foundations of development communication are neglected, and the results are usually disappointing, especially over the long term.
8. ***Development communication is not exclusively about behaviour change.*** The areas of intervention and the applications of development communication extend beyond the traditional notion of behavior change to include, among other things, probing socioeconomic and political factors, identifying priorities, assessing risks and opportunities, empowering people, strengthening institutions, and promoting social change within complex cultural and political environments. That development communication is often associated with behavior change could be ascribed to a number of factors, such as its application in health programs or its use in mass media to persuade audiences to adopt certain practices. These kinds of interventions are among the most visible, relying heavily on communication campaigns to change people's behaviors and to eliminate or reduce often fatal risks (for example, AIDS). The reality of development, though, is complex and often requires broader changes than specific individual behaviors.
9. ***Media and information technologies are not the backbone of development communication.*** As a matter of fact, the value-added of development communication occurs before media and information and communication technologies (ICTs) are even considered. Of course, media and information technologies are part of development communication, and they are important and useful means to support development. Their application, however, comes at a later stage, and their impact is greatly affected by the communication work done in the research phase. Project managers should be wary of "one-size-fits-all" solutions that appear to solve all problems by using media products. Past experience indicates that unless such instruments are used in connection with other approaches and based on proper research, they seldom deliver the intended results.

10. ***Participatory approaches and participatory communication approaches are not the same thing and should not be used interchangeably, but they can be used together, as their functions are often complementary, especially during the research phase.*** Even if there are some similarities between the two types of approaches, most renowned participatory approaches, such as participatory rural appraisal (PRA) or participatory action research (PAR), do not usually assess the range and level of people's perceptions and attitudes on key issues, identify communication entry points, and map out the information and communication systems that can be used later to design and implement the communication strategy. Instead, these are all key activities carried out in a participatory communication assessment.

Basic Principles of Development Communication

Dialogic

Dialog is the heart of the new communication paradigm. The professional application of dialog, the two-way model of communication, is widely endorsed by most development institutions and should be the basis of any initiative. Development communication should foster dialog to facilitate mutual understanding, to assess the situation, and to seek wider consensus. Dialogic approaches guarantee that relevant stakeholders have their voices heard and that project priorities are aligned with people priorities. Professionally directed, dialog is an invaluable research tool and is absolutely to build trust, optimize knowledge, minimize risks, and reconcile different positions. To facilitate dialog professionally and effectively, a communication specialist must be conversant with proper communication skills, including principles of active listening.

Inclusive

Inclusion is a first step in any situation analysis, whereby Devv. Comm identifies, defines, hears, and understands relevant stakeholders. In this respect, inclusiveness is one of the basic principles of the Dev.Comm methodological frame work, even if the appropriate strategy might focus only on selected groups of stakeholders. Omitting a group from the assessment on a basis that might not seem relevant can cause problems further along and can increase the risk factors in the successful achievement of the intervention. Two-way communication should always pay special attention to groups that are marginalized or at a disadvantage in society. Gender issues are always a primary concern in this context, as well as issues related to the poor, or any other vulnerable group.

Heuristic

The investigative use of communication to discover or solve problems during the initial phases of a development initiative is essential. Communication is often defined as a way of sharing meanings or "as a process in which two or more people share information and converge toward mutual understanding, mutual agreement, and collective action". This definition denotes the sharing of information and knowledge, which usually generates more knowledge that in turn can lead to effective collective action. The heuristic and explorative scope of development communication, strengthened by its analytical and dialogic features, constitutes its main value-added in addressing and rectifying the past failures in development.

Analytical

Going beyond communicating could be a Dev. Comm motto; a large amount of its work, such as the assessment of political risks and opportunities, is analytical. In this context, the communication function is not about relating messages but about uncovering and generating knowledge to design better projects and programs that lead to sustainable change. The effectiveness of diffusion and dissemination activities depends significantly on how appropriately the analytical work is conducted and how effectively people are empowered to voice their perceptions and opinions.

Participatory

While rarely employed in practice to its ideal and fullest extent, participation is applied in different degrees according to the intervention. Its relevance is echoed in virtually all development organizations and communities, at the national and international levels. Only genuine communication can facilitate effective participation, especially in its most advanced forms. Participation can be applied in different degrees, and there are several classifications describing the different types of participation. The World Bank classification illustrated in table is in line with many others in this context and identifies four levels of participation (Aycrigg 1998):

A. Information sharing: One-way communication—basically, people are included by informing them about what is being done

B. Consultation: Primarily one-way communication with a stronger emphasis on feedback—stakeholders provide their input but do not have a significant say in the decision-making process.

C. Two-way communication supporting open interaction in decision making—input in decision making is balanced.

D. Empowerment: Transfer of control over decisions and resources—two way communication ensures shared decision making.

While the most common mode of operation in development practices can be categorized as "participation by consultation," Development Communication also operates at a higher level, by collaboration. This occurs specifically in the research phase, where dialog with relevant stakeholders is sought and promoted and their input valued, especially in community-driven development projects. This application is adopted according to the circumstances; although it is a main feature in communication based assessments, participation is not always a feature of communication approaches in subsequent activities.

Contextual

There is no precooked universal formula applied a priori in development communication. In its recent adoption of the Comprehensive Development Framework, the World Bank acknowledged "country ownership" as one of the main principles of development and as a basis for all of the Bank's work. This means "encouraging participatory processes" that are necessarily rooted in the cultural context of specific countries and their socioeconomic reality. In employing communication- based assessment around the world, DevComm staff are fully aware of the implications of this principle. While investigating a local context and assessing needs, problems, risks, and opportunities, DevComm specialists tap local resources to obtain a better understanding of the relevant situation and to triangulate their findings. Even if the overall process of a communication intervention (that is, executing communication-based assessment, designing communication strategies, and implementing and evaluating related activities) is consistently similar, the tools, content, and modes of applications vary significantly according to the specific situation.

Interdisciplinary

To be effectively applied, a development communication body of knowledge includes a number of principles borrowed from other disciplines. In addition to specific expertise in the theory and practices of development communication, the specialist in this field is often required to be familiar with other disciplines, such as ethnography, sociology, political economy, adult education, and marketing. The specialist might be asked to assess political risks, conduct negotiations to reduce conflicts, or mediate between opposing views. While sector experts could address each of these areas with a specific and narrower focus, the cross-cutting nature of communication makes it an easier and more effective tool to acquire a comprehensive overview of the situation.

Strategic

The principle of strategy, which contains many of the previous elements, emphasizes the professional and timely application of communication techniques and methods to achieve intended objectives. At the risk of oversimplification, a strategy could be defined as a plan to achieve set objectives with available resources in a given time frame. It is surprising how often the basics of a strategy are overlooked, not only by communication specialists, but by all sorts of decision makers. Often this occurs when practitioners jump into strategy design without making sure that the objectives are technically sound, well understood, and relevant to most stakeholders. If the project objectives do not meet all these criteria, no matter what strategy one adopts, the initiative is bound to fail—like building a house on a faulty foundation. The principles of "strategic" imply that all parts of the process, from setting the objectives to selecting the media, are carefully assessed, triangulated, and, if needed, modified to allow the design and implementation of an effective strategy. The strategic use of development communication should not be confused with "strategic communication," the narrower use of communication to persuade individuals to change behaviors.

Persuasive

At times this term has a negative connotation, mostly due to past uses of persuasion techniques taken to an extreme and often associated with manipulation and propaganda. Persuasion per se should not be thought of in negative terms. The renowned Greek philosopher Aristotle considered it as an effective way to communicate. In development communication, persuasion can be used to induce voluntary changes in individuals. The legitimacy for its use is derived from this rationale and the definition of change. To avoid the manipulation connotations of the past and be ethically appropriate, persuasion should be based on accurate information and within a context of two-way communication. Each party can present its points of view with the intention of achieving the most appropriate change. Healthy two-way persuasive approaches ensure that the best available options among the various parties are considered and agreed upon, leading to sustainable change.

Key elements of the development communication approach

- **It is responsive:** It does not provide 'useless' information - that which people did not want to know, but which central planners deemed as crucial. People understand their own needs better and through this approach communication becomes a tool in the planning and development process, not a mechanism to persuade communities once unpalatable decisions have been made ('in their best interests!')
- **It hinges on feedback:** It is not a one-way process but involves dialogue mechanisms about the information which was transferred. It is also fundamentally about consultative processes being managed at community level.
- **Innovation and creativity:** The message must not be dull and boring but show clearly how the information transmitted will make a difference in the life of the recipient - it must not instill doubt or disbelief, but trust and confidence (look for local adopters). Development Communication workers should, however, balance creativity with an understanding of what communities would be prepared to accept and where consideration has been given to the norms and prevailing values of that community.
- **Independent validation:** It is not about 'government speak'. This approach builds participatory mechanisms and functional networks involving NGOs, CBOs,

Traditional Leadership structures while also encouraging links with networks from across the country and indeed all over the world. These can either prove or disprove the validity of the information transmitted.

- It's about **sustainability and continuity:** It is not about dumping information in a community and never going back for months.
- It's about establishing **common ground with communities** who are to be the recipients of the information/message: it is not about the public servant who swoops in and out of a community in his or her GG like some 'phantom expert' to 'educate and uplift' communities. The standards, norms, values, habits of the community are paramount. (This may mean that those accustomed to a liaison style hinging on comfortable hotels with prepared meals and warm fluffy duvets will need to make some adjustments to their style!)
- It's about **community participation:** Development programmes which plan for communities or supply information which planners feel communities need, fail to be relevant initiatives and more often than not fail to be sustainable. A primary emphasis of this approach is to plan with communities, create structures which offer communities and developers equal power, and use communication methods which are fundamentally participatory in nature. This often requires that government planners, developers or community workers have to listen to the advice of communities and change the views they themselves hold.
- It's about **access and visibility of government** where government is no longer a distant and unknown entity in community development experiences. This approach reverses the practice of communities having to travel long distances and at relatively great cost to access government Services and information. This is made worse when government is not clearly and properly identifiable and access is difficult because of inaccessible buildings, unfriendly and unprofessional staff etc. The development communication approach brings government employees face to face with communities so promoting accountability at local level. This is not possible when civil servants are remote and impersonal.
- It's about the use of **simple and relevant language** where concepts are packaged in the experiences of communities, in their own language and where communities themselves have played a major role in the development of material for development communication programmes.

Characteristics of a new 'village level worker' or development communication practitioner:

- Community needs at heart commitment to let
- Communities lead: "I will follow"
- Responsive: "I want to make a difference"

- Multi-skilled and adaptable
- Knowledgeable on many areas of government, structures, programmes, policies - well read (but not an expert on everything rather a referral specialist)
- Good facilitation skills
- Strong knowledge of the district in which I work - history, people, language, economic base, structures, gate keepers, institutions, Contact Details
- Creative: strong knowledge of the creative methods of development communication
- Knows and accepts own limitations: "I know when I should pass on to the departmental expert so as to offer my main client - the citizen - the best service".

Tasks to be performed by the development communication officer

- Networking
- Facilitating
- Interviewing
- Interpreting information (intermediary)
- Techno-mediary: selling and familiarizing communities with the great value technology offers
- Referring
- Researching: specifically, "How do I do informal research - running group discussions, focus groups, easy questionnaires, community meetings, writing up case studies which describe scenarios, secondary or documentary research; where to get it, what this constitutes, how do I use it?"
- Some training
- A number of roles associated with my status as an employee of the public service: administration, reporting, financial issues etc.

The development communication process consists of the following steps in sequence

Step 1: Proximity to the receiver

Step 2: Establish credibility

Step 3: Consultation

Step 4: Involve receivers in planning (message design or info product)

Step 5: The message is developed and the programme runs

Step 6: Evaluate the message/programme

Step 7: Next phase of planning

The History of Development Communication in India

The history of development communication in India can be traced to rural radio broadcasts in the 1940s in different languages. Have you ever heard a rural programme on radio? If you come from a rural area, you probably would have heard. People who present these programmes speak in a language or dialect that the people in your area speak. The programmes may be about farming and related subjects. The programme may comprise of interviews with experts, officials and farmers, folk songs and information about weather, market rates, availability of improved seeds and implements. There would also be programmes on related fields. During the 1950s, the government started huge developmental programmes throughout the country. In fact, when Doordarshan started on 15th September 1959, it was concentrating only on programmes on agriculture. Many of you might have seen the 'Krishi Darshan' programme on Doordarshan. Later in 1975, when India used satellites for telecasting television programmes in what is known as SITE (Satellite Instructional Television Experiment), the programmes on education and development were made available to 2400 villages in the states of Andhra Pradesh, Bihar, Karnataka, Madhya Pradesh, Orissa and Rajasthan. As far as the print media is concerned, after Independence when the Five Year Plans were initiated by the government for planned development, it was the newspapers which gave great importance to development themes. They wrote on various government development programmes and how the people could make use of them. If the print media have contributed to development communication, the electronic media – radio and television especially All India Radio and Doordarshan have spread messages on development as the main part of their broadcasts. However, amongst all the media that are used for development communication, traditional media are the closest to people who need messages of development like the farmers and workers. Such forms of media are participatory and effective. You may have seen construction workers cooking their meal of dal and rice over open fires in front of their tents set up temporarily on the roadside. They need to be educated about the values of balanced nutrition, cleanliness, hygiene and water and sanitation. Have you wondered how messages on such issues are communicated ?

In various parts of India, groups of volunteers use street theatre as a medium for development communication. This is done through humorous skits and plays through which the importance of literacy, hygiene etc. are enacted. The content for the skits is drawn from the audience's life. For example, they are told about "balanced nutrition" . This means supplementing their staple diet of dal and rice with green leafy vegetables known to cure night blindness, an ailment common among construction workers. Similarly, female construction workers and their children are taught how to read and write. However, problems in communicating a message in an effective way has been a matter of concern to development workers. How can people be taught new skills at a low cost? What would be a good way to deal with sensitive topics such as health issues? How can complicated new research, like that in agriculture for example, be simplified so that ordinary people can benefit? One option has been the use of comics. But, in order to achieve the desired results, these comics should be created locally. But what are 'comics' ? You must have all at some point of time read a comic. Comics involve story telling using visuals which must follow local ideas and culture in order to be understood correctly by people. The important thing about comics is that they are made by people on their own issues in their own language. So, readers find them closer to their day-to-day lives. Programmes are organized in the remote areas of Jharkhand, Rajasthan, Tamil Nadu, and the North East to provide training to rural communicators to enable them to use comics in development communication. Information on sensitive health issues such as HIV/AIDS has been communicated through the medium of comics in several states. However, you must understand that development communication using various media is possible only with the active involvement of the following:

(i) Development agencies like departments of agriculture
(ii) Voluntary organizations
(iii) Concerned
(iv) Non governmental organizations (NGOs).

We speak about development, the contribution of voluntary groups, concerned citizens and nongovernmental organizations cannot be ignored. Actually these groups help the government in implementing development programmes. Of course the government, both central and state have various departments to reach out to people on various issues. The NGOs undertake studies, conduct research and develop appropriate messages for spreading awareness on various issues pertaining to development. The sensitiveness and awareness of Development Communication by the Change agents in Transfer of Agricultural Technology, livelihood development, rural development etc. will greatly help in all round development of the people/ clients in turn leads to Nation development.

References

1. Ifenkwel, G.E and Ikpekagou, F. 2012. Noise mitigation for effective agricultural extension print message delivery and utilization. *Journal of Agricultural Extension and Rural Development.* 4 (2): 51-56.
2. Ogan C.L. 1982. Development journalism/communication: The status of the concept. *International Communication Gazette* 29(3), 1-13.
3. Oladele O. Idowu. 2005. Farmers' perception of agricultural extension agents' characteristics as factors for enhancing adult learning in Mezam division of Northwest Province of Cameroon. *Australian Journal of Adult learning.* 45 (2): 223-237.
4. Onkargouda Kakade. 2013. Credibility of Radio Programmes in the Dissemination of Agricultural Information: A Case Study of Air Dharwad, Karnataka. *Journal of Humanities and Social Science.* 12 (3): 18-22.

5. Paolo Mefalopulos. 2008. *Development Communication Source Book.* World Bank. Washington DC 20433.18-20
6. Proggya Ghatak. 2010. Development of mass media and its extension in agriculture: A feedback review of audience research survey in air, Murshidabad, West Bengal. *Global Media Journal: Indian Editio.*1-20.
7. Ray, G.L. 2011. *Extension Communication and Management-* Eighth Edition. Kalyani Publishers.
8. Surabhi Mittal and Mamtha Meher. 2012. An Assessment of Farmer's Information Networks in India- Role of Modern ICT. www.afita.org
9. http://www.thusong.gov.za/documents/artic_pres/dev_comm.htm
10. http://www.unicef.org/cbsc/
11. http://wecommunication.blogspot.in/2015/02/basic-principles-of-development.html

Lives and Livelihoods

26 Skills for Agriculture and Rural Livelihoods – Challenges for International Cooperation Communication

- Prof. S.V. Reddy

Preamble

In many developing nations, agriculture still holds key to reducing poverty and increasing the security of livelihoods. There are many challenges faced by agriculture sector at different levels. In India Agriculture sector accounts for 14% of GDP and about 11% of total exports. It provides employment to 57% of rural work force 32% households are self employed in agriculture and 25.6% are engaged as labourer. However, Agriculture growth remains short of 4.0 percent plan target. Similarly is the pattern in most developing countries. Since agriculture growth is the major source of rural livelihood, there is need for a faster growth in this sector and such a break through requires improvements in knowledge and skills. Besides, rural livelihoods are also becoming diversified, and an increasing share of rural household's income is coming from non-farm activities. This is an important development challenge.

Education and skills increase the ability to innovate and adopt new technologies in agriculture and enhance farmers performance. Besides, Evidence from Asia suggests that better education and training increases the chances to find high – paying non-farm employment, whereas lack of education ends to limits options to agriculture or low-wage non-farm employment.

While some farmers are engaged in high-return agricultural business (for example, agri-business, value chain activities and export – oriented cultivations), in developing countries many are still engaged in low-productivity subsistence farming.

For most of the rural population, skills – life and livelihood, in the area of agriculture are life sustaining, given the scale of this sector an incremental advance in skill development will have phenomenal effect on productivity and gainful employment.

Thus, skills are central to improve employability and livelihood opportunities, reduce poverty, enhance productivity, and promote environmentally sustainable development.

There is a need for coordinated efforts to develop an integrated approach that improves access to relevant, good quality education and training for rural women and men with a goal of ***SKILLS FOR ALL by 2020***

The Rationale for Rural Skills Development

Rural skills development is a necessary priority for a number of reasons. In developing countries in general and sub-saharan Africa in particular, the food security challenge leads to an urgent need to increase productivity, skills will be important here. Globally, the environmental challenge leads to the need for better skills for protecting the environment. An economic challenge highlight the importance of skills that can contribute to economic growth and the international competitiveness of the agricultural sector growth. There is also a social challenge of skills development for poverty reduction, employment generation and social capital creation.

Investing in Development, the need to build capacity has become a major element of current skills thinking. Such capacity development is seen as being necessary at all levels: for individual farmers and their organizations, for skills providers and other service organizations, for national departments and agencies, and for international organizations. In some of the contributions, this capacity agenda is linked to other nations such as social capital, decentralization and participatory development.

The FAO/UNESCO-IIEP Paper also emphasizes the importance of relating skills for agricultural and rural development to the education for all Initiative, which is under UNESCO`s leadership. The paper notes that the *Dakar Framework* for EFA goes well beyond the MDGs in stressing *appropriate skills for adults and youth*. Thus, skills for agricultural and *rural development can be argued* to be an integral part of the EFA approach.

The Facts and the Need

- Access to training is a major constraint among rural people in developing countries. For instance, nearly 90 percent of agriculture workers in India have no formal training, and a study among small scale entrepreneurs in Kenya indicated that over 85 percent of rural informal sector operators have no business or technical training at all.
- Rural girls and women are often the most disadvantaged. The global secondary school attendance rate for rural girls is 39 percent as opposed to 45 percent for rural boys and 59 percent for urban girls.
- Training outside the formal training system is often the most important source of skills training in developing countries. For example in Benin, Senegal and Cameroon, informal apprenticeships account for almost 90 percent of all trades training.

- Many rural youth face great disadvantages when trying to enter urban labour markets because of their low level of education and lack of relevant skills and work experience.
- In many developing countries, training systems tend to operate in isolation from the labour market and employers needs, so training does not always match skills demand.
- More than 700 million people are estimated to be of working age (24-59 years) in India by 2020. Of these, approximately 500 million workers will require some kind of vocational / skill training. Besides about 70 million youth in unorganized and informal sector require capacity building on the expertise for performing better in their jobs.

The Skills and Skill Gap

There is a growing gap in Technical, Professional, Life Skills to support agriculture and rural livelihoods. As agriculture is becoming increasingly competitive there is a demand for on farm skills such as floriculture, horticulture, improved cultivation of crops from monsoon to market, farm mechanization, etc. Similarly off farm skills such as fruit processing, Live stock based micro enterprises etc are also found to be wanting to meet the demands. Furthermore, many rural youth are demands for non-farm skills such as Fashion designing, computer literacy, motor winding, lift irrigation etc., Therefore, there is need for an integrated approach to be evolved based on skill gap analysis combing the on-farm, off farm and nonfarm skills, indigenous and modern skills which will enable rural people (Agriculturists, Agricultural workers and persons with indigenous technologies) to modernize their thinking for increased efficiency and productivity. Besides they can also supplement incomes through self employment and employment. This will lead to improvements in Rural Livelihoods.

The Programmes to Bridge Skill Gap

There are number of massive programmes and schemes launched by the Governments to address the skills gap in rural and urban areas in Agriculture and Livelihoods. In India the following programmes are worth mentioning.

- National Skill Development Mission (NSDM)
- National Rural Livelihood Mission (NRLM)
- National Horticulture Mission
- National Food Security Mission
- National Extension and Technology Mission
- National Rural Employment Guarantee Mission (MGNREGS)

Similarly many developed and developing countries also had their own programme in this direction. In Germany there are one lakh Vocational Training Centres but population wise it is not a big country. About five lakhs companies were identified for apprentaship. But we do not have such massive partnership. In Europe vocational training Centre run by Public Private Partnership mode.

No one doubts the varsality of over arching national frame work to give a kick start to the entire Skill Development Scenario and to improve Livelihoods. However, what needs to be aggressively pushed, covered and monitored on priority is sustainable livelihood creation at grass root level by providing conducive climate and logistics.

Suggestions

Skills development in rural areas requires various types of skills provision, using innovative methods of delivery, and capitalizing on existing social institutions. The following are some of the suggestions in this direction.

- Expand the outreach of both schools and training institutions including KVKs in underserved rural areas.
- Provide free basic education as it is a stepping stone to further skills training and provide financial incentives and non-financial incentives (e.g. meals at school and take-home rations) to improve attendance. This is addressed in India and some of the developing countries but still lot has to be done.
- Provide affordable technical and vocational training by reducing financial entry barriers, and design interventions to include those most disadvantaged in accessing education and training, such as working children, women in poverty, people with disabilities, ethnic minorities etc.
- Compliment technical and vocational training with basic education (literacy and numeracy) and life skills (e.g. confidence building, Decision make, social awareness).This enables participants to benefit more from the technical and vocational training, and may be particularly relevant for those most marginalized. KVKs can take a lead in agricultural on farm and on farm skill developer.
- Promote a gender-responsive learning environment, for example, consider safe transportation and training facilities, separate sanitation facilities, dormitories, cafeterias, and child care facilities.
- Encourage training women and men in non-gender stereotypical trades, promoting for instance training in mechanics for women and textiles work for men.
- Develop flexible, modular training. This will benefit those who cannot afford taking time off (for example, due to household or seasonal work) or paying for longer term training.

- Facilitate access to training materials, toolkits and modern equipment and technology, and invest in teacher trainers.
- Consider outreach measures such as mobile or distance learning through Information and Communication Technologies (ICTs).
- Provide career guidance and practical labour market information (e.g. in schools training facilities and community associations) to enable rural youth to make informed choices about their education, training and employment in the rural context.
- Compliment entrepreneurship training by facilitating rural entrepreneurs access to micro-credit schemes, business development services and market information. This may require expanding the scope of these services and ensuring that the right legal framework is in place.
- Strengthen coordination and collaboration with the private sector in skills development both to increase the relevance of training and to improve and facilitate its delivery. Involve particularly employers organizations, but also workers organizations, NGOs and community groups, in planning and implementing programmes in Private-Public Partnership mode.
- Environmental degradation and climate change present risks to rural livelihoods that need to be managed and mitigated. This requires developing new and innovative Skill Training Modules.
- Develop a capacity assessment of skill development institutions and service providers once in five years for quality skill development.

The institutional changes needed to make the skill development training institutions and services providers more effective including moving from the current approach of meeting targets to a more flexible and responsive system and harmonize, rationalize and integrate the different programmes schemes to achieve mission of **SKILLS for all by 2020** using people centered, pro-poor and holistic approach.

27 Nurturing Grassroots Innovators in Agriculture for Enhancing Livelihood

- Dr. S. Senthil Vinayagam

Indian agriculture is unique with diverse agro-climatic conditions. It is trying towards inclusive growth by ensuring augmentation in productivity sustainability and profitability by integrating experiences and efforts of the concerned stakeholders. To achieve this, contributions of the agricultural scientists in the evolution of modern technologies for transforming Indian agriculture and achieving food security are laudable. Technology transfer system adopted lab to land programme where the farmers were mere beneficiaries of top-to-down dissemination of technologies.

Days have come to recognize bottom-up innovations and of equal partnership of farmers in agricultural research and development. Realizing this, there is a greater need to identify, recognize and promote those farmers who are actually grassroots innovators of agricultural technologies. These innovations have emerged out of farmers vast experience and wisdom-based on their analysis of their own micro-level farming situations. Such explorations have led to the development of numerous technologies including new varieties by the farmers. The farmers have demonstrated in many locations that use of Grassroots Innovation (GI) in their farming activities enhance their rural economics. Further, these innovations are being considered as social technologies and solution based science from the farming sector, it is expected that GI enhances the socio-economic well-being of rural people.

GIs are conceived as bottom-up innovations and solve the rural problems using the local resources, experience, skills and traditional knowledge. It function as alternative to modern technology. GI supplements rural development by generating sustainable livelihood opportunity for the farmers / peasants.

Grassroots Innovation in India

Grassroots Knowledge

Sharing of knowledge between communities and innovators are important to seek intellectual property and for up-scaling and commercialization. It is now being increasingly felt that despite sharing their knowledge, the grassroots knowledge holders have remained poor whereas those who access their knowledge and develop products after seeking intellectual Property (IP) protection or otherwise, have become prosperous. Ironically, the very success of the commercial products developed through value addition in local knowledge many times becomes reason for the erosion of the knowledge itself. Realizing the gap between formal knowledge seekers and inform knowledge holders, the first initiative was taken by Padmasri Anil Gupta(source: www.iimahd).

Promoting Grassroots Innovations

Most of the Grassroots Innovators are not only promoters and provides of GI, but also entrepreneurs. They also generate employment and create opportunities for skill enhancement and develop local economics (Joshi, 2015). These innovators have been supported under Micro-Venture Innovation Fund (MVIF) in rural areas through HBN partner organizations. NIF has set-up a MVIF with the assistance of SIDBI to support the GIs which have market potential and repayment is expected and to support entrepreneurs and companies who are interested in investing these innovations. It generally provides a means of finance for innovation and traditional knowledge based micro and Financial Institutions (FIs). MVIF is not a grant, it is a loan and helps in extending risk capital to many ventures.

The GI is instrumental in providing a reorganization and a commercial platform for the innovations that have potential to succeed in the market. In India, GIs ecosystem involves innovator and institution that are part of National Innovation System (Joshi and Chelliah 2013).

GI system

In view of strengthening and promoting grassroots innovations in India, the Govt. of India declared 2010-20 as "Decade of Innovation". Further, a national innovation act was formulated to facilitate to have an inclusive innovation system and to develop a national integrated science, technology and innovation plan. It is being felt that a grassroots knowledge holders were have remained poor whereas those who access their knowledge and develop product after availing Intellectual Property Protection (IP) have become prosperous. Therefore, realizing the gap between formal knowledge seekers and informal knowledge holders the first initiative was started as 'Honey-Bee Network (Source: www.sdo.eea.).

Honey-Bee Network (HBN)

HBN was started in India in 1990 and now spread to 75 countries. This network strengthens people to people learning, links and networks by pooling the solutions developed by people. It involves innovators, farmers, policy makers, entrepreneurs and

NGOs etc. The network acknowledges, innovators, knowledge producers and communicators. It strongly believes in sharing knowledge among the providers among innovations in their own languages and also ensures that the fair share of benefits arising from commercial exploitation of local knowledge and innovation reaches the innovators and knowledge providers. The network has partner organizations namely SRISTI, NIF, GIAN, SEVA, Pritvi, PEDES, Innovation club, Network of Gram Vidyapeethas and Honey Bee AP. Besides these institutional collaborators, there are individual collaborators from Jammu and Kashmir, Rajasthan, HP and Punjab also build HBN. Using ICT, these institutes connect Grassroots people to mentors, engineers, designers, investors and entrepreneurs thereby shaping the development agenda for regional growth (Maurya et.al. 2014.)

Partners of HBN

It is reported that Innovations from grassroots not only empower people through local development, design ownership and control of technology, but it also helps to challenge mainstream innovation agendas and development pathways (Fressoli et al. 2014). Partners of HBN provided legal and marketing advice such as technology transfer, patenting, licensing, design rights, branding, prototype development, quality testing and positioning to Grassroots Innovators. These innovators are also transforming the socio-economic life and reducing inequities by providing affordable quality products and employment opportunities. In this process the GIs has transformed the rural lives.

Society for Research and Initiatives for Sustainable Technologies and Institutions (SRISTI)

SRISTI which means creation, was born in 1993, essentially to support the activities of the Honey Bee Network to recognize, respect and reward creativity at the grassroots. Based in Ahmedabad, Gujarat, SRISTI (Society for Research and Initiatives for Sustainable Technologies) is a registered charitable organization that is devoted to empowering the knowledge rich-economically poor people by adding value in their contemporary creativity as well as traditional knowledge. It has helped established Grassroots innovation Augmentation Network (GIAN), National Innovation Foundation (NIF) Micro Venture Innovation Fund (MVIF) and Academy for Augmenting Sustainable Technological Inventions, Innovations and Traditional knowledge (AASTIK). SRISTI was mainly set up to provide organizational support to the Honey Bee Network (Source:www.sristi).

The organization work on the principal that a synthesis of six Es i.e., Excellence, Equity, Environment, Efficiency, Ethics, Empathy and Education would provide the right chemistry for societies seeped in mediocrity to get over their inertia and move toward a compassionate, creative and competitive as well as collaborative society with the following objectives.

- To expand space in society for building upon sustainable technological, institutional and educational initiatives and innovations at the grassroots with special focus on women's knowledge.
- To document, analyse and disseminate innovations developed by people themselves.
- To validate and add value to local innovations through experiments (on farm and on-station) and laboratory research for generating nature-friendly sustainable technologies.
- To conserve local biodiversity through in-situ and ex-situ gene banks managed by local people.
- To protect the intellectual property rights of grassroots innovators and to generate incentive models for recognising, respecting and rewarding grassroots creativity and associated ethical values and norms.
- To provide venture support to grassroots innovators to scale up products and services based on grassroots innovations through commercial or non-commercial channels.
- To embed the insights learnt from grassroots innovations in the formal educational system in order to expand the conceptual and cognitive space available to these innovations.

It also manages the Honey Bee database of innovations, and supports the publication of the Network newsletter in three languages, English, Hindi and Gujarati. Lately SRISTI has being focused on more concerted ways of hitherto neglected domains like women knowledge systems, value addition through a natural product laboratory, and innovations in education.

Grassroots Innovations Augmentation Networks (GIAN)

Grassroots Innovations Augmentation Networks (GIANs) were established with the primary objective to link up innovations, investment and enterprise. Since the process of covering grass-root ideas / innovations into a prototype, product and enterprise is a critical and risky process, there was a felt need for a platform to facilitate the transition of an innovation into produce and later into enterprise. This transition requires input of formal science and technology, design, handholding support for project planning and management and finance with marketing intelligence. In addition, confidence building measures are needed since failure and set backs are not uncommon in this journey (Source: www.gian.org/).

The first model of GIAN was set up on March 1 1997. During these years of existence, GIAN has made steady progress in commercializing technologies, filing patents in India and abroad and building of a portfolio of new projects. It has received excellent support from TePP (Technopreneurial Promotion Programme) of Department of Science and Technology and Department Scientific and Industrial Research. Through national Innovation Fund (NFI), GIAN has been set up in different regions of the country in close collaboration with respective state governments and other institutions of excellence. For eg: GIAN-NORTH, GIAN-NORTH EAST. Currently the scope of GIAN-GUJARAT to Maharashtra and Goa was expanded and upgraded to GIAN-West. Rajasthan Government set up GIAN North as an independent society at Jaipur with the participation of NIF nominee on the board. GIAN Northeast is a cell set up by NIF in the campus of IIT, Guwahati. NIF took the decision without waiting for any formal communication from any of the state governments in the regions. Of late there has been some response from state government of Chattisgarh for forming GIAN cell at Raipur. Kerala IT Mission has already decided to set up Kerala Innovation Foundation. A GIAN cell was recently set up at TCE campus at Madurai in collaboration with SEVA for Tamil Nadu. Additionally the GIAN-Karnataka cell was established at SSIT (Sri Siddhartha institute of Technology), Tumkur for State of Karnataka.

National Innovation Foundation (NIF)

The National Innovation Foundation (NFI) has evolved to meet a long felt need for recognizing, respecting and rewarding innovation and outstanding traditional knowledge at the grassroots. Honey Bee Network had triggered a movement to scout spawn and sustain unaided creative and innovative urges in unorganized sector of our society. The events following this trigger led to many important developments leading to SRISTI, GIAN and other initiatives. A need for institutional framework was felt and voiced. In 2006, the Department of Science and Technology, Government of India, set up National Innovation Foundation (NIF). The foundation has been set up with the main goal of providing institutional support in scouting, spawning, sustaining and scaling up grassroots green innovations and helping their transition to self-supporting activities. Thus NIF provides an institutional platform for the knowledge-rich economically poor people. It is expected to help the unsung heroes /heroines of our society who have solved a technological problems through their own genius without any outside help. It is committed to making India innovative by documenting, adding value, protecting intellectual property rights of the contemporary unaided technological innovations, as well as outstanding examples of traditional knowledge on a commercial as well as non-commercial basis. NIF also seeks to develop a new model of poverty alleviation and employment generation by helping convert grassroots innovations into enterprises with or without value addition through institutional science and technology (Source: nif.org.in).

Challenges for Innovations

It is reported that, the agriculture research and technologies have enabled increased in overall global food production over the least fifty years. However, about one billion people remain hungry and impoverished and declining natural resources, our capacity to sustainably feed nine billion people by 2050 is a serious concern. An estimated increase in food production of 70 per cent globally is required to meet this demand upto 100 per cent in developing countries. (Ashley et.al. 2009). Therefore the global agri. food system has major challenges to respond to climate change soil degradation, pest and disease management besides production challenges. To address these challenges, a new generation of agriculturalists will need to develop with innovative ideas, ways of working, skills and leadership abilities. The innovation challenge is not only for the traditional researchers-extension workers-farmers but also for other actors like processors, marketers, transporters, input suppliers, policy makers and more. FAO defined Agricultural Innovation System is a network of individuals, organizations and enterprises focused on bringing new products, processes and forms of organization into social and economic use, together with the institutions and policies that affect their behavior and performance, to achieve food and nutrition security, economic development and sustainable natural resource management (Rajlahti, 2012).

Changing Context for Rural Innovation

Let us take an example of population growth. World population was 3 billion during 1950 and about 65 per cent lived in rural areas. By 2050 the population is expected to be 9 billion and 70 per cent will be in urban areas. Further, the consumption pattern of food also changed rapidly as per the need of growing middle class communities. Hence, there is a need for an integrated approach to innovation that manages production, social, environmental, market, political and economic issues.

It outlined six changes in the context for agricultural development that highlight the need to examine how innovation occurs in the agricultural sector (World Bank, 2008):

- Markets, not production, increasingly drive agricultural development.
- The production, trade, and consumption environment for agriculture and agricultural products is becoming more dynamic and evolving in unpredictable ways.
- Knowledge, information, and technology is increasingly generated, diffused, and applied through the private sector.
- Exponential growth in information and communications technology has transformed the ability to take advantage of knowledge developed for other purposes.

- The knowledge structure of the agricultural sector in many countries is changing markedly.
- Agricultural development increasingly takes place in a globalized setting.

Extension and Innovation systems: Innovation systems began to emerge as an overarching alternative concept. This was driven by a range of factors, including the critique of technology transfer, the emerging new ideas and practices that did not fit the 'extension' image, the growing lack of enthusiasm for traditional extension services, and the often large disconnect between research and real world problems. The innovation system concept was given prominence by a 2006 World Bank publication which defined it as "a network of organizations, enterprises, and individuals focused on bringing new products, new processes, and new forms of organization into economic use, together with the institutions and policies that affect their behavior and performance. The innovation systems concept embraces not only the science suppliers but the totality and interaction of actors involved in innovation. It extends beyond the creation of knowledge to encompass the factors affecting demand for and use of knowledge in novel and useful ways." (Hall et al., 2006; Hiroven, 2008).

Who innovates what, for whose benefit and at whose cost will always be an issue of power and control over resources. It remains critical to remember that small-scale producers and rural communities are often in a marginalized position where they are 'price takers' and often have limited and dwindling assets. Innovation aimed at improving the livelihoods and food security of these groups must look at issues of power.

Innovation systems approaches facilitates the application of particular knowledge or research outputs. It provide an entry point for individuals or organizations seeking to increase the impact of their work through using innovation systems approaches.Hence, building the capacity of a system to respond to emerging challenges and opportunities to innovate on a continuing basis is required.

The following principles are suggested to place innovation system approaches into practice.

- Innovation requires both scientific knowledge and dynamic learning networks between knowledge users.
- Technological innovation must be embedded in human processes: understanding the different people involved in those human processes is key. Innovation is an evolutionary process.
- An enabling environment is more than just policies, institutional frameworks and research and development programmes -supporting interaction and learning amongst actors is also key.
- Innovation is not just about bright ideas, but getting them into use.
- Learning is key to innovation processes and conflict can play an important role.
- The gender gap needs to be addressed in innovation system's thought and practice

In countries where innovation occurred, there will be an impact on income levels and poverty reduction (Ahmed & Al-Roubaie, 2012). Ahmed and Al-Roubaie further explain that "innovation empowers the economy by increasing productivity, enhancing technological learning and creating knowledge" (2012).

Process to promote Grassroots Innovations

Human responding hardship and difficulty in life can easily become accidental innovators. Grassroots innovation focuses on individuals as agent for innovation (Bhaduri & Kumar, 2009). Grassroots innovations are innovative product or process created at the bottom of the pyramid, usually due to necessity, hardship and challenges. Bhaduri and Kumar (2010) defined grassroots innovators as "...individual innovators, who often undertake innovative efforts to solve localized problems, and generally work outside the realm of formal organizations like business firms...". Mass poor from the grassroots level implements millions of solutions in facing their hardship (Gunu, 2010). In Malaysia, grassroots innovation is one of the high impact programs intended to empower the bottom 40 per cent of the income pyramid (Hashim, 2012). Malaysian's Ministry of Science, Technology and Innovation (MOSTI) initiated the Innovation Walk program to scout for grassroots innovation ("Innovation Walk," 2012). The Innovation Walk program is manage by the Malaysian Innovation Foundation and follow the framework and structure of similar programs established by Professor Anil Gupta who is an expert on grassroots innovation ("Grassroots Innovation Augmentation Network," 2012).

Innovative products discovered during these event are documented and highlighted as part of the National Grassroots Innovation Databank project. The products listed in the databank serve the purpose of recognizing, encouraging and celebrating innovation at the grassroots level. Furthermore, the database facilitates linkages and networks with investors, promote awareness and possible application of innovative products.

Innovation activities occurring at the rural grassroots-level are basically ingenious solutions developed locally to give simple solutions to system challenges for improving rural livelihoods and promote sustainability. These are novel bottom–up solutions which respond to the local situations and suit to the interests and values of the communities. These solutions may be innovative product or process which are created due to necessary, hardship and challenges at grassroots level in agriculture and allied sectors. These solutions/ innovations are termed as "Grassroots Innovations" (GRI).

In contrast to mainstream research and development (R&D), grassroots initiatives operate at both individual and community levels. Often, it involves committed activists helping and catalyze local people with social innovations. Innovations of this nature tend to take the shape of simple but rustic technologies which provide solutions to drudgery and solve the immediate problems faced by the user. A typical characteristic of these innovations is that these are limited to particular regions and often need a push to move forward to develop as technologies and later as business solutions.

Strategy and approach: Taking Grassroots Innovators towards validation, business incubation and IPR protection stream, following needs to addressed:

- Scouting of innovations products (varieties / breeds), processes, prototypes which can be up-scaled and commercialized towards agriculture development.
- Documenting ITKs, TKs or mere modification of existing innovation perse for further dissemination.
- Scientific validation of grassroots innovations at NARS laboratories/ research institutions.
- Support to develop prototypes for up scaling the proof of concepts through Agri-Business Incubation (ABI).
- Providing support for business incubation and commercialization through existing network ABI. Facilitate the GRIs to showcase in different platforms.
- Providing techno-legal and handholding support for addressing issues such as IP protection, marketing, negotiations etc., through ABI network.

Scouting of Innovations and Documentation

Scouting is to be done to discover and recognize grassroots innovations and traditional knowledge practices among various rural and urban communities through various methods namely campaign, organizing village meetings, organizing bio-diversity competitions, through literature, participation in traditional festival and using ICT modes etc. The scouted innovations has to be documented by entering innovators personal details, description of the innovation and details of traditional knowledge followed by verification of GRIs practices by the expert committee to go for secondary documentation. During the secondary documentation the experts can collect the samples of GRIs photographs, videos and other required details with respect to innovations.

Dissemination of Innovations

On completion of scouting and documenting rural innovations, it is necessary to develop a database which may be named as National Grassroots Innovation Databank as has been done in Malaysia (www.cuocient.com). The process involves various information technology and database applications for horizontal networking among innovators and traditional knowledge experts as well as other stakeholders besides following multi-lingual approach. This will help to have lateral linkages among innovators and traditional knowledge holders and also to promote sharing of public domain knowledge with local communities and individuals. Further it is also necessitated to develop and maintain a national register for innovations and traditional knowledge with multi languages. For dissemination of innovations in a larger scale there is a need to manage database, electronic networking, web based management of value chain for grassroots innovations coordination with regional language portals and management information systems.

Multimedia kiosks at various places, educational institution and local bodies will provide necessary support for dissemination of innovations.

The dissemination of technologies which are either already matured for technology transfer to potential entrepreneurs are which can be easily matures has been taken up quite earnestly through participation in various exhibitions and fairs for this purpose. For eg. Mobile Exhibition initiative by SGK, Pantnagar in Uttaranchal. SGK modified a traditional chakra using LPG kit and equipped it with modern IT features like a projector for presentation and named as 'SAKSHAM'.

Intellectual Property Management

On completion of scouting and documentation there is a need to empower the knowledge rich, economically poor grassroots innovations and traditional knowledge holders for protecting their Intellectual Property Rights. A separate cell for recognizing the true value of Intellectual Property Rights is to be created with the following activities.

- Help the innovator in prior art searchers.
- Drafting and filling patent application.
- Coordinating with various intellectual property institutions and attorney for mobilizing pr bono or paid support for grassroots innovators in filing patents trademarks and other means of IP protection on their behalf.
- Providing legal assistance to the innovators to deal with issues of infringement of the IP rights.
- Screening of patents and patent applications based on Indian traditional knowledge and grassroots innovations, so as to oppose the improper applications/ granted patents, particularly those dealing with practices entered in the National Register.
- Coordinating with National and international organizations / officers to secure IP Protectionof grassroots innovators globally.

Value Addition for GIs

Most of the GI products will exhibit low value due to its design, application process and outlook. Therefore, it is to optimize the design development and product formulation of the GIs through value addition. This will enable the GIs to have a marketable product. Value addition can be done by merging modern science and technology inputs. The research and development organizations like ICAR institutes, CSIR institutes, IITs etc. can also provide technical as well as financial support to the potential GIs. Many sources have indicated that these institutions have provided support to few potential GIs for its incubation activities like prototype development, testing the innovations, design optimization and development of concept proof model. The value added GIs can lead for commercialization either directly by innovator or through any other well established organizations on cost sharing basis. The following business models are suggested for value- addition and commercialization.

- Coordination with public and private sector R & D institutions, academic institutions, private associations, industry clusters and innovation intermediaries.
- Formulation of strategies, undertaking detailed product planning and prototype development while maintaining a WBS (work breakdown structure).
- Mobilization of support for product development proposals through the DST's TePP fund and other similar programs / projects either at local or international level.
- Creation of a contractual value chain through either benefit sharing through an up-front fee as suitable in the specific context.
- Establishment of grassroots incubation centers to coordinate regional product development activities jointly with other National Coordinators and local advisory committees.
- Recommendation of domain expert members for Research Advisory Committee (RAC) to scrutinize the innovations for their novelty value, cost effectiveness, social impact etc. Further the process of incubation is tracked through incubation Advisory Committee (this process needs considers strengthening) so as to increase the probability of commercial success of the innovations or traditional knowledge.

Business Development

As most of the grassroot innovators are not able to undertake entrepreneurial activity on their own, they should develop linkages with entrepreneur / industry associations, management institutions and incubator for mentoring and management support. The linkages and coordination with these agencies will support the grassroot innovators to undertake business activity. Further, to promote business opportunities for GIs, any public or private institutions can develop a business development cell for GIs with the following objectives.

- Identifying application domains, conduct market benchmark studies, modify the product and ascertain techno-commercial feasibility of innovations.
- Coordinating with various entrepreneur/industry associations, management institutions and incubators to mobilize mentoring and management support for grassroots innovators.
- Coordinating with private and public sector industrial and financial institutions and associations to link innovations with investment and enterprise wherever possible.
- Encouraging various industry associations and other developmental bodies to set up mechanisms for licensing innovations for business development. The benefits may be shared in an equitable manner with the innovators. In addition, a share may be contributed towards conservation of nature, innovators association, and for community development.
- Providing financial assistance to innovators in the supply chain in all activities related to technical and business incubation.

Upscaling of GIs

To commercialize the innovations, a framework has been developed by NIF (Fig. 27.1) to build the value chain around various innovations and traditional knowledge. It is found necessary to form mentoring teams around each product for which business development has to be done. One has to locate willing entrepreneurs or business managers who would help in market research, business planning, developing a proposal for raising micro venture finance and, eventually, help convert innovations or traditional knowledge into commercial ventures.

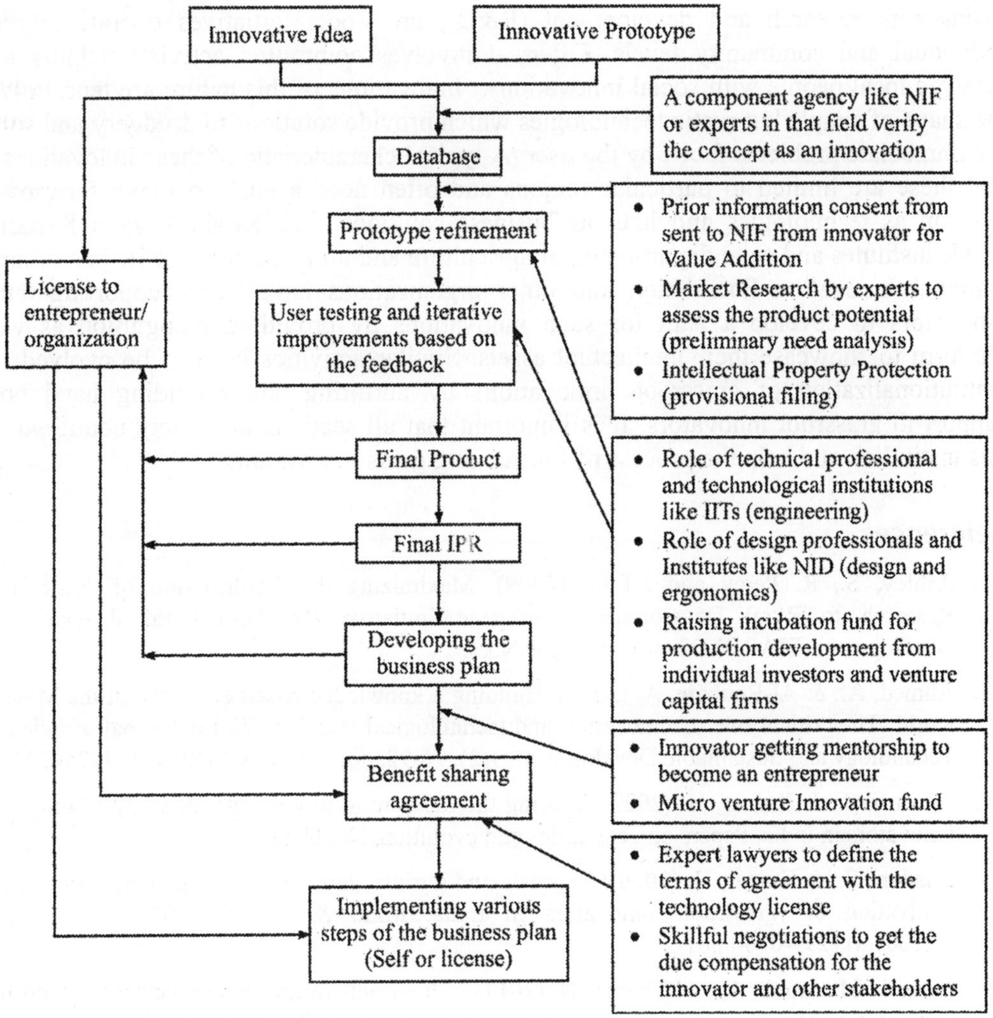

Fig. 27.1 Value chain of Upscaling Process for Grassroots Innovations - Process adopted by NIF

(Source: Presentation CD on Green Grassroots Innovation and outstanding traditional Knowledge Vol.1-3, NIF)

Conclusion

Innovation activities occurring at the rural grassroots-level are basically ingenious solutions developed locally to give simple solutions to system challenges for improving rural livelihoods and promote sustainability. These are novel bottom–up solutions which respond to the local situations and suit to the interests and values of the communities. These solutions may be innovative product or process which are created due to necessary, hardship and challenges at grassroots level in agriculture and allied sectors. These solutions/ innovations are termed as "Grassroots Innovations" (GRI). In contrast to mainstream research and development (R&D), grassroots initiatives operate at both individual and community levels. Often, it involves committed activists helping and catalyze local people with social innovations. Innovations of this nature are tend to take the shape of simple but rustic technologies which provide solutions to drudgery and solve the immediate problems faced by the user. A typical characteristic of these innovations is that these are limited to particular regions and often need a push to move forward to develop as technologies and later as business solutions. The Krishi Vigyan Kendras, ICAR institutes and state departments of agriculture and allied sectors, CSIR institutions, National Innovation Foundation and other organizations have been supporting these innovators to develop a path for such innovations by providing recognition as well platform to showcase their intellectual assets. Similar activities have to be evolved for institutionalization of grassroots innovations by nurturing and providing hand hold support to grassroot innovators. It is important that all sections of society contribute to this institution and help the process of formulizing grass innovations.

References

1. Ashley, S., R. Percy and J.Tsui. (2009). Maximizing the Contribution of Agriculture Research to Rural Development. European Initiative for Agricultural Research for Development (EIARD) Discussion Paper No.1.
2. Ahmed, A., & Al-Roubaie, A. (2012). Building a knowledge-based economy in the Muslim world: The critical role of innovation and technological learning. World Journal of Science, Technology and Sustainable Development, 9(2), 76-98. doi: 10.1108/20425941211244243
3. Bhaduri, S., & Kumar, H. (2009). Tracing the motivation to innovate: A study of grassroot innovators in India. Papers on economics and evolution, No. 0912.
4. Bhaduri, S., & Kumar, H. (2010). Extrinsic and intrinsic motivations to innovate: tracing the motivation of 'grassroot' innovators in India. Mind & Society, 10(1), 27-55. doi: 10.1007/s11299-010-0081-2
5. Fressoli, M., Smith, A, and Thomas H. (2014), "Grassroots Innovation movements: Enduring dilemmas as sources of knowledge production", Journal of Cleaner Production, vol. 63, no, 15, pp. 114-124.
6. Grassroots Innovation Augmentation Network. (2012) Retrieved 29 October 2012, from http://www.gian.org/

7. "Grassroots Innovation Augmentation Network," (2012). http://www.cuocient.com©2012Chimera Innova Group

8. Gunu, U. (2010). Entrepreneurship Development in Micro Enterprises as a Medium for Poverty Reduction in Kwara State, Nigeria. Interdisciplinary Journal of Contemporary Research in Business, 2(6), 235-252.

9. Hashim, H. (2012, 3 September 2012). Inclusive entrepreneurship for the rakyat, New Straits Times.

10. Hall, A., W. Janssen, E. Pėhu and R. Rajalahti. (2006). Enhancing Agricultural Innovation: How to go Beyond the Strengthening of Research Systems. The World Bank, Washington, DC.

11. Hirvonen, M. (2008). A Tourist Guide to Systems Studies of Rural Innovation. LINK Policy Studies Resources on Rural Innovation Series, No. 1. Learning Innovation Knowledge (LINK), Hyderabad.

12. Innovation Walk. (2012) Retrieved 6 October (2012), from http://www.yim.my/jejakinovasi/

13. Joshi, R.G. and Chelliah. J. 2013, "Sharing the benefits of commercialization of traditional knowledge: what are the key success factors?" The Intellectual Property Forum, No. 93, June, pp. 60-66.

14. Joshi R.G., 2015, Fostering Transformation through the mobilization of Grassroots Innovations, www.unescap.org/.

15. Maurya, N., Kumar V., Patel, R., Mahanta H., and Gupta, A. (2014), "ICTs in Support of Grassroots Innovation". Information Technologies & International Development, Vol. 10, No. 1, pp. 21–25

16. National Grassroot Innovation Datal ank. (2012) Retrieved 6 October 2012, from http://www.yim.my/databank/

17. Rajalahti, R.(2012). Sourcebook Overview and User Guide. In: World Bank. Agricultural

18. Innovation Systems: An Investment Sourcebook. The World Bank, Washington, DC.p.1-13

19. 'Rural Innovations' Course material of PGD-TMA 2015, NAARM, Hyderabad.

20. Schőn, D.A. (1983). The Reflective Practitioner: How Professionals Think in Action. Basic Books.

21. World Bank (2008). World development report: Agriculture for development. World Bank, Washington DC

28 Role of Extension in Developing Skills and Entrepreneurship to Attract and Retain Rural Youth in Agriculture

- Dr. Devi Prasad Juvvadi

Preamble

Youth are the present and the future of any country and it is important that youth are both viewed as an investment opportunity and are treated as partners in the development process. The future of agriculture and food security in India will depend upon the contributions of rural youth both men and women in shaping its agriculture. Present trend of youth moving away from agriculture and migrating to cities in search of better employment and better life is posing severe problem to agricultural development in India..Evidence from Nigeria (Akpan2010) points four major problems largely accounted for youths' lack of interest in agriculture, namely: drudgery in farm operations; lack of competitive market for agricultural products; lack of credit and; and lack of Buy Back Scheme (BBC) from the government which can be attributed to many developing countries including India. Gandhi etal (2001) reported that migration is due to drudgery, limited resources, weak infrastructure, and unemployment in rural areas and youth mainly don't see it is as an attractive profession in the long run. However, it is likely that the youth will select agriculture as a profession only if it is profitable and made interesting with innovation and use of information and communication technology. Supporting youth in agribusiness with incentives further helps to attract and retain youth in agriculture. Globally there is an increasing interest in finding ways of engaging youth in agriculture (IFAD, 2012; Paisley, 2013). The population in the age-group of 15-34 in has India increased from 353 million in 2001 to 430 million in 2011 with more than half of population under the age of 25 and 65 percent of the population under 35 years. The rural population is about 70%, and the indications are that the migration of rural youth to cities is around 45% in the country, which is quite alarming. India's decisive advantage of demographic dividend needs to be leveraged for agricultural development. Considering the fact that every year about 800,000 farmers are quitting agriculture and youth are not interested to enter farming, it is of even more relevance to meet food requirements for increasing population which is estimated to reach 1.60 billion by 2050.

Challenges for Youth in Agriculture

Attracting youth to and retaining them in the agriculture sector remains a global challenge because all over the world agriculture has an image problem (Mukemboetal 2014). Retaining youth in agriculture has been a prominent topic recently and has risen up the development agenda, as there is growing concern worldwide that young people have become disenchanted with agriculture. World over there is a decreasing interest among youth in entering agriculture due to persistent perception that agriculture is risky and non remunerative because of minimal financial returns. This is attributed to "farm problem' meaning economic difficulties faced by farmers as a result of low farm incomes and great instability and variability in the income from farming (Sharma and Bhaduri, 2006).

Sharma (2007) reported rising trend in withdrawal among youth from farming in India, particularly in the regions with low value of agricultural production per capita and in villages close to towns. The trend is stronger among higher caste, better educated and youth with non-farm skills. Interestingly, both the small and marginal landholding class and the large landholding class showed a trend towards withdrawal. However, the small and marginal farmers are largely being pushed out of farming; the big farmers are moving to tap better opportunities outside the farm sector being better off in terms of education and other resources such as capital. It is well known that as economies mature, the percentage of people involved in agriculture undergoes a significant decline particularly in developing countries. But India is losing more than 2,000 farmers every single day and that since 1991, the overall number of farmers has dropped by 15 million (Sainath, 2013). This has several implications for the future of Indian agriculture and food security. According to National Sample Survey Organization (NSSO) survey (2005) as much as 40 % farmers expressed their bitterness with farming and would like to quit agriculture if they have an alternative.

New Approaches to Agricultural Extension

Agricultural extension is the function of providing need and demand based knowledge in agronomic techniques and skills to farmers in a systematic, participatory manner, with the objective of improving their production, income leading to improved quality of life. Extension is essentially education and it aims to bring about positive behavioural changes among farmers.

Agriculture is changing, and with it, a revised set of skills is needed to address new challenges in agriculture (Blackie et al, 2009).. As attitudes, expectations and employment in agriculture have changed, there is evidence that the skills and competencies of farmers do not meet the needs of today's agricultural sector. Till date, the agricultural extension focused only on transfer of technology with production as main accomplishment and not considered agribusiness, value addition, private sector orientation and market-driven systems etc to make agriculture not only profitable but to attract youth in farming. Extension in agriculture should focus on skill development and entrepreneurship to improve competitiveness of the farmer leading to sustainability of agriculture. Augmenting knowledge and skill levels of the agricultural workforce and youth in particular is essential to enhance productivity, boost innovation, manage finance, mitigate risks and improve decision making ability. Agricultural extension services should focus on rural youth to tap their energy and creativity to transform agricultural sector. Addressing new and growing challenges in agriculture requires extension to play an expanded role with diverse objectives. New approaches to extension need to emphasize three elements: strategies to develop Agricultural Innovation Systems, pluralism of service providers, and extension services should be demand-driven.

Focusing on Youth in Agricultural Extension

Literacy rate was only 12 % at the time of independence, as a result earlier agricultural extension programmes, obviously were mostly designed focusing on interventions aimed at improving the functional literacy among rural adults. However, over the years, the literacy rates have gone up and in 2011, it stands at 74.04%. Literacy in rural youth has increased substantially and they are anxious in employment opportunities in nearby towns and urban areas.

Today India is the youngest nation in the world, with nearly 70 percent of population below 35 years, and interestingly 70 percent of them in rural areas. In 2020, the average Indians will be only 29-year-old; whereas in the US and China, the average age is estimated to be 37 years. It is this group to be tapped for extension programmes and capacity building with focus on agribusiness and entrepreneurial development to attract and retain them in farming. This offers India the possibility to utilise its demographic dividend, and can take Indian agriculture to further hights. Encouraging rural youth to take up farming would be an appropriate way of harnessing youths' potentials. This would not only help enhance agricultural productivity and deliver food security but also reduce rural-to-urban migration. It will be young farmers, with greater capacity for innovation and risk-taking, who can help address the emerging requirements of agriculture and thus, give a boost to the rural economy.

Youth are more techno-savvy and they could access information and knowledge promoted through the new ICTs which uses computer, internet and mobiles. Young farmers often have greater capacity for innovation, imagination, initiative and entrepreneurship than older adults and these characteristics should be effectively harnessed by extension services to provide better livelihood opportunities for youth in agriculture (Mahesh Chander, 2014).

In recent years young farmers are taking up innovative farming activities and agri-ventures like protected agriculture, precision farming, organic agriculture, floriculture, medicinal and aromatic plants cultivation etc. which were avoided by the earlier generation farmers due to high risks involved. Government and financial institutions should encourage youth to venture in to these kinds of activities with skill training, financing and marketing support.

The agriculture extension services are still ignoring youth as dynamic partners of extension work. A vibrant agriculture sector is a must for a country to attain economic stability and the youth must be encouraged imbibing farming as a noble profession. One of the things missing in agriculture today is imagination and the poor image of people involved in agriculture needs to be changed. Agricultural advisory services including extension functionaries, agricultural scientists, policy makers and even media specialists, have collectively failed to trigger youths' imagination (Lyocksetal 2013). Generally youth are willing to adopt new ideas and technologies and therefore agricultural extension services should target youth to transform agriculture. The youth could be the ideal catalyst to change the poor image of persons involved in agriculture due to their greater possibility to adapt new ideas, concept and technology which are all important to changing the way agriculture is practiced and perceived. Agricultural extension services can effectively address these issues by encouraging and supporting youth participation in agriculture. Improving their capacities and increasing their involvement will also help in changing the negative perception about farmers as "uneducated and unskilled, physical labourers engaged in a glamour less vocation with extremely low economic returns" (Bhatetal, 2015).

Ongoing Programmes for Youth

There are number of youth policies, initiatives and skill development schemes in the country but they often do not cater for poor rural youth but tend to be biased towards urban areas. Consequently, there is growing disenchantment among rural youth towards farming and they find it much lucrative to do even menial jobs in cities. The resultant large-scale migration of rural youth from farming to urban areas has caused concern among the agricultural policy makers because, if the trend is not checked, it is likely to affect agricultural activities severely. Thus, checking migration and retaining youth in agricultural sector is currently a big challenge. Some of the ongoing initiatives and programmes to enhance capacities of rural youth and some of them include:

KrishiVigyan Kendra (KVKs): Imparting need based vocational training to farmers, farm women and rural youth to change their knowledge, skill & attitude with the main mandate of uplifting their standard of living.

Nehru Yuva Kendra Sangathan (NYKS): For nearly three decades, they have been channelizing the power of youth between 13-35 years on the principles of voluntarism, self-help and community participation. Over the years, NYKS has established a network of youth clubs in villages, where Nehru Yuva Kendras have been set up for harnessing youth power for development by forming Youth Clubs, which are village level voluntary action groups of youth at the grassroots level to involve them in nation building activities, working for community development and youth empowerment. However, the scope of all NYKS has to be expanded in skilling rural youth in agricultural activities because presently only a few of them are involved in agricultural development.

NABARD Farmers' Clubs: Mainly for linking technologies with farmers while facilitating market access through capacity building of members in leadership training; linkage with technology and markets; Self Help Groups (SHGs)/Joint Liability Groups (JLGs) formation and forming Federations of Farmers' Clubs, Producers' Groups, Companies etc. It is essential that the amount of Rs.10,000 given to each club should be increased with compulsory 50% membership to youth under 21 years. Further, hand holding of groups is more important than formation. The model of women SHGs can be mimicked with thrift concept imbibed.

ARYA (Attracting and Retaining Youth in Agriculture): The Indian Council of Agricultural Research (ICAR) has constituted a seven-member expert committee, to suggest ways of attracting youth to agriculture. The committee recommended a Rs.200 crore pilot project to implemented in 30 different places across the country during 12 th Plan to motivate rural youth to remain in agriculture. This would involve a series of activities, including exploring opportunities in secondary agricultural operations such as value addition of crops and hiring, and servicing of mechanized farm implements.

National Rural Livelihood Mission (NRLM): NRLM aims at creation of opportunities for both wage employment and skill development for the rural youth, who lack skills in many areas of agricultural production and processing. National Skill Development Mission and the National Skill Qualification Framework are, thus, aggressively pushing the agenda of skill development to build the capacity of rural youth so that they are meaningfully employed in rural areas itself. The need based experiential skill learning supported by public sector banks/organizations in rural areas is the key to strengthen the Rural Self Employment Training Institutes (RSETIs) being set up in all districts under NRLM to assist such youth (Likhi, 2013). NRLM requires suitable decentralized convergence of skill development programs run by multiple central ministries including the National Skill Development Corporation (NSDC).

Agricultural Skill Council of India (ASCI): Considering the need for skilling the work force in agricultural sector, the Agricultural Skills Council of India (ASCI) has been recently proposed by National Skill Development Corporation (NSDC), which could be one ideal institution to train rural youth. The ASCI proposes to train, certify and accredit 56.5 million workforce comprising of farmers, wage workers, entrepreneurs and extension workers, over 10 years through its training partners.

However, most of the initiatives are not functioning to expectations. Agricultural extension programmes should focus on these programmes.

Skill Development in Agriculture for Rural Youth

A skilled workforce is the key to a country's overall growth and it is one of the determining factors for attracting domestic and foreign investments in different sectors including agriculture. Skill development and training are essential to face agricultural challenges of 21st century like climate change, rising input costs, urbanization and changing consumer preferences for strategic decision making, market awareness and optimum utilization of resources. The skills to improve productivity, increase adaptability to deal with change and crisis, and facilitate the diversification of livelihoods to manage risks are essential for rural India. The renewed significance of skill development in agriculture sector is requisite as the sector formsthe backbone of Indian economy, contributing around 14% to the nation's GDP, 11% to the export basket and is the primary source of income for almost half the country's population.

Skill development contributes to agricultural and rural development. However, close attention is not paid for training and training needs of youth in this. Efforts to provide more training for rural people are frequently part of more general commitments to expand skills development for marginalized groups and fail to recognize the specific needs of rural areas.

Like most other sectors, the Indian agriculture sector is also in dire need of skill development because of increased use of technology in the sector. There is notion that agriculture requires very limited or no skill and thus labour wages in the sector are low compared to other sectors. Because of increased use technology and innovation in agriculture, increased productivity relies on improved skills in efficient usage of all resources. This highlights the need not only for skills development to enhance agricultural productivity but also for more and better employment in the rural non-farm economy. Augmenting knowledge and skill levels of the agricultural workforce and youth in particular is essential to enhance productivity, boost innovation, manage finance, mitigate risks and improve decision making ability.

Horticulture and dairy sectors produce the high value commodities which can give good returns to farmers. However, these sectors face numerous challenges in terms of availability of skilled manpower. Skill development interventions in these sectors are relevant with respect to technical skills of operating machines, better farm practices like integrated nutrient management, nursery management, breeding techniques, harvesting techniques etc. There are other segments as well where the rural youth can be skilled and effectively employed, some of which include activities like seed production, providing para veterinary services, repairing and maintenance of farm equipments, retailing, providing crop advisory services by using information and communication technology etc. There is a need to focus on skill development in agriculture extension programmes to harness the advantage of 'demographic dividend'.

Youth and Entrepreneurship in Agriculture

The issue of educated unemployment of youth in villages and increase in farm income needs much concentration by the extension functionaries working in the field of agriculture and allied sectors. Extension functionaries should encourage unemployed rural youth to attempt towards entrepreneurship development. A serious attempt was made in India towards massive creation of self-employment opportunities through entrepreneurship development programmes. Recent trends in agri-business sector and emerging ICT has substantially improved Indian agriculture system. In the modern era agriculture is not only restricted to farm but it has expanded to the global market system. Hence, the farmer now is not only restricted to farm and domestic market, but also has opportunity to reach global markets as an agri-business entrepreneur.

As farmers and youth lack entrepreneurship, the opportunities in agriculture are highjacked by outsiders, particularly the urban businessmen and traders, leading to exploitation and deprivation of employment for the farmers. Considering the growing unemployment among rural youth and slow growth of the agricultural sector, it is necessary to tap the opportunities for promoting entrepreneurship in agriculture, particularly for youth, which in turn can address the present problems related to migration, agricultural production and profitability.

Entrepreneurship in agriculture is not only an opportunity for additional income but also a necessity for improving the production and profitability. Farmer entrepreneurs should consider farming as a business and as a means of earning profits. Apart from level production enterprise, there is a need to promote entrepreneurship and agribusiness among rural youth and there are different types of enterprises in agri-business, that include;

1. **Service Providers:** For optimizing agriculture by every family enterprise, there are different types of services required at the village level. These include the input procurement and distribution, hiring of implements and equipment like tractors, seed drills, sprayers, harvesters, threshers, dryers and technical services such as installation of irrigation facilities, weed control, plant protection, harvesting, threshing, transportation, storage, maintenance of information kiosks with ICT use etc. Similar opportunities exist in the livestock husbandry sector for providing breeding, vaccination, disease diagnostic and treatment services, apart from distribution of cattle feed, mineral mixture, forage seeds, etc.
2. **Input Producers:** There are many prosperous enterprises, which require critical inputs. Some such inputs which can be produced by the local entrepreneurs at the village level are biofertilizers, biopesticides, vermicompost, fruits and vegetable nurseries and other plant material, seed production, seed banks, root media for raising plants in pots, agricultural tools, small irrigation, production of cattle feed concentrate, mineral mixture and complete feed. There are good opportunities to

support sericulture, fishery and poultry as well, through promotion of critical service facilities in rural areas.

3. **Processing and Marketing of Farm Produce:** Efficient management of post-production operations requires higher scale of technology as well as investment. Such enterprises can be handled by people's organizations, either in the form of cooperatives, service societies or joint stock companies. The most successful examples are the sugar cooperatives, dairy cooperatives and fruit growers' cooperatives in many States. However, the success of such ventures is solely dependent on the integrity and competence of the leaders involved. Such ventures need skilled professional support for managing the activities as a competitive business and to compete well with other players in the market, particularly the retail traders and middlemen.

Pandey (2015) identified youth's difficulties to become agro-entrepreneur and presented the scope of agro-based entrepreneurship development in agricultural extension programs as following;

1. Identify the steps required for wider involvement of young professionals in promoting sustainable intensification and profitable entrepreneurship in the agricultural sector;
2. Chart the way out for mentorship programmes in order to bridge research and knowledge gaps in agricultural system;
3. Generate conceptual tools needed for wider engagement and youth's contribution in agricultural development.

Conclusion

New approaches in agricultural extension are required with focus on rural youth to face the emerging challenges in agriculture. Skill development and Agribusiness and entrepreneurship is solution to many problems in agriculture particularly in checking migration of youths and retaining them in farming leading sustainability. Agricultural extension should play a major role in motivating rural youth to promote small enterprises. The extension programmes should enhance the knowledge and skills of farmers and youth in prediction of the future demand, introduction of modern technologies, cost control and business expansion etc. The agricultural extension agencies should popularize the services of the entrepreneurs to enhance the credibility of the services provided by youth (Bairwa et al, 2014). While it is the responsibility of government to promote agro-based enterprises in rural areas, extension should furnish information, enhance knowledge and develop skills of local people. The Government schemes to entrepreneurs and various concessions and incentives should be part of extension services. Networks of entrepreneurs may be established to share their experiences. These networks can also establish a close link with research institutions and universities to become acquainted with the latest research findings and seek solutions for their field problems. Extension should also promote direct marketing by establishing close interaction between producers and consumers to increase the benefits, while encouraging a large number of unemployed rural youth to turn into microentrepreneurs and traders leading to agriculture sustainability.

References

1. Akpan, S. B. (2010). Encouraging youth's involvement in agricultural production and processing. International Food Policy Research Institute. Policy Note 29, pp. 1-4.
2. Bhat, Prashanth., Swathi Bhat and Anantha Shayana. (2015).Retaining Youth in Agriculture – Opportunities and Challenges. International Journal in Management and Social Science. Vol.03 Issue-02, (February, 2015) ISSN: 2321-1784
3. Blackie, M., Mutema, M., & Ward, A. (2009). A study of Agricultural Graduates in Eastern, Central and Southern Africa: Demand, Quality and Job Performance issues. *A report to ASARECA and RUFORUM.*
4. Bairwa, S.L., Lakra, K. and Kushwaha, S. (2014) Agripreneurship for Rural Development. Edited by Rai and Goyal "Agriculture and Rural Development in India". M.R.F. Publication, Varanasi. PP 422 – 425. ISBN 978-81-926935-6-9
5. Gandhi, V., Kumar, G. and Marsh, R. (2001) Agro industry for Rural and Small Farmer Development: Issues and Lessons from India. International Food and Agribusines Management Review, 2(3/4): 331-344. In Family *farming and Rural Development..* CHAPTER - 38 *Agribusiness: The Way of Attracting and Retaining Rural Youth in Farming Bairwa, S.L.*, Lakra, K. and *Kushwaha, S. (2014c)* Agripreneurship for Rural Development.
6. IFAD. 2012. Youth in agriculture: Special session of the Farmers' Forum Global Meeting, 18 February 2012, IFAD, Rome.
7. Likhi, Abhilaksh 2013. Challenges for India's Livelihood: Youth Skill Development in Rural Areas. http://blogs.worldbank.org/publicsphere/challenges-india-s-livelihood-youth-skill-developmentrural-areas
8. Lyocks, J. S., Lyocks, S. W. J., & Kagbu, J. H. (2013).Mobilizing Youth for Participation in Nigerian Agricultural Transformation Agenda: A Grassroots' Approach. *Journal of Agricultural Extension, 17*(2), 78-87.
9. Mahesh Chander, (2014). Youth: Potential Target for Agricultural Extension http://www.aesa-gfras.net/images/youth.pdf
10. Mukembo, S.C., Craig Edwards, M., Ramsey, Jon W., and Henneberry, S.R. (2014). Attracting Youth to Agriculture: The Career Interests of Young Farmers Club Members in Uganda. *Journal of Agricultural Education, 55(5), 155-172. doi: 10.5032/jae.2014.05155*

11. NSSO (2005). National Sample Survey Organization.https://data.gov.in/dataset-group-name/national-sample-survey
12. Pandey, Dinesh (2015). Promoting agro-entrepreneurship among the Nepalese youth. 19th June 2015. Young Professionals for Agricultural Development. http://www.ypard.net/2015-january-19/promoting-agro-entrepreneurship-among-nepalese-youth.
13. Paisley, Courtney. 2013. Engaging youth in agriculture: Investing in our future. Global Food for thought. The official blog of the global Agricultural Development Initiative.
14. http://globalfoodforthought.typepad.com/global-food-for thought/2013/02/ commentaryengaging-youth-in-agriculture-investing-in-our-future.html
15. Sainath, P. (2013). Over 2,000 fewer farmers every day. The Hindu, May 2. http://www.thehindu.com/opinion/columns/sainath/over-2000-fewer-farmers-everyday/article4674190.ece>
16. Sharma, A. (2007). The Changing agricultural demography of India. Evidence from a rural youth perception survey. *International Journal of Rural Management, 3*(1), 27-41.
17. Sharma, A., & Bhaduri, A. (2006). The "tipping point" in Indian agriculture. Understanding the withdrawal of Indian rural youth. Draft prepared for the IWMI-CPWF project on "Strategic Analysis of National River Linking Project of India. In Colombo, Sri Lanka: International Water Management Institute.

Focused Areas in Extension

29 Convergence of Agriculture and Allied Sector Programmes

- Dr. V. V. Sadamate

Convergence defined

Convergence is the process that results in the achievement of common objectives through value addition and targeted and efficient use of financial and human resources. Coordinated planning and service delivery ensures timely inputs from multiple sources, simultaneously avoiding duplication and redundancies. The planning process drawing in from mutually agreed programmes, underlines clarity regarding targets, timeframes, shared responsibilities and monitoring parameters. Specific convergence initiatives could be of a complementary or supplementary nature, aimed at either more comprehensive treatment, adding productive value to assets created, ensuring sustainability or up-scaling successful initiatives.

In other words Convergence process brings in

(i) shared values and responsibilities,

(ii) on supplementary and complementary mode,

(iii) to achieve common objectives and mutual benefits to the converging partners,

(iv) around the targeted programmes.

Why Convergence?

- Convergence focuses on synergies required to move towards a more integrated delivery approach, using the comparative strengths of different partners to address the specific challenges of production and livelihood systems. There is now far greater need that the programmes are required to be implemented through a cluster approach integrating broader technological and social concerns as the basis for planning and for consolidation of impacts. The convergence approach provides a dynamic framework that can benefit from the activities undertaken through certain programmes while providing a platform to other programmes that can use the potential created through the projects, to realize incremental outcomes.

- Certain other linkages could ensure community empowerment through capacity building and skills development and overall sustainability of impacts for both programmes. To elucidate:

Elucidation-I: Comprehensive treatment – Natural Resources Management (NRM) works

- Mahatma Gandhi National Rural Employment Guarantee Act (MGNEGA),Rural Development (RD) – for land treatment especially earthworks such as farm and contour bunds, clearing drainage lines, preparing pits for plantation
- Repair, Renovation & Restoration (RR&R), Water Resources (WR) – to restore and augment storage capacities of water bodies
- Forest Department (FD) programme – for treatment of ridge areas in the upper reaches – especially in the Reserve Forest areas, afforestation through the Green India Mission in common lands, farm bunds, etc.

Elucidation-II : Programmes/ Initiatives that can benefit from potential created (indicative)

- Rural Drinking water supply (RD) – for Improved water availability from water conservation measures
- Agriculture - Different Programmes including National Mission for Sustainable Agriculture (NMSA), National Food Security Mission (NFSM), National Horticulture Mission (NHM), Rashtriya Krishi Vikas Yojana (RKVY), etc. – for Improved productivity potential through assured water availability, also addresses climate change concerns from more frequent droughts, reduction in soil degradation due to erosion, etc.
- Animal husbandry – Improved fodder availability through treatment of commons, agricultural residues and third fodder crop. Self Help Group (SHGs) and SHG federations can take up dairying as in income generating option
- Fisheries – Perennial water bodies created through IWMP ensure fisheries as a livelihood option
- National Rural Livelihood Mission (NRLM) – SHGs are already created as an integral part of the livelihoods component
- Credit support – Better outlook for loan repayments as farmers are less susceptible to weather fluctuations and women's SHGs have reached a certain level of maturity
- Public Private Partnership (PPP) – assured farm produce provides an impetus for value addition and marketing. Farmers' groups and Commodity Interest Groups (CIGs) already formed to facilitate the process.

These broadly align with the objectives addressed through the Production Systems and Livelihood Components.

Convergence for Integrated Development & wider impact

Given the magnitude of the investments envisaged, it becomes imperative to ensure Convergence of various area development and livelihood improvement programmes. This would bring in integrated development in a given micro situation, enhance income of the farming communities and provide wider impacts across the sectors. To achieve this, convergence becomes a necessary instrument that is internalized right from the planning to the consolidation phase through an inbuilt institutional mechanism at various levels. Programme delivery would need to be organized and reoriented accordingly, enhancing institutional capacities of the agencies and communities involved.

Critical levels for Convergence

State specific Convergence policies and strategies would need to be formulated in consultation with the concerned line Departments. At the national level, discussions with concerned Ministrieswould be necessary to assess and identify broader convergence potentials. At the State level, the leaddepartment should take the initiative to discuss convergence with other State Departments for both, Central and State schemes and issue necessary guidelines and instructions. Given the decentralized planning framework, the critical level for convergence planning should be a district. However, it matters most at the level below the District / Block, at the cluster level, where actual implementation is envisaged to address common concerns and realize mutual benefits.

Who should lead the process of Convergence?

- Initial meetings with the State line Departments & decision makers to explore specific convergence potential and kick start the process is necessary, followed by dialogue with relevant Central Ministries.
- Concerned Ministries/Departments may then provide separate guidelines to their State formations on relevance and modalities of Convergence. Joint Guidelines issued by Ministry of Rural Development (MORD) as in the case of MGNREGA is one such example.
- The lead Department & the converging State Line Departments may bring out the instructions in securing co-ordination and convergence of line departments/other implementing agencies at the State, District and Project level. This would enable the Collectors, District Planning Committees (DPCs) or Chief Executive Officer (CEO)/ District Development Officer (DDO) - Zilla Parishad/Panchayat (ZP), as applicable in a State, to take decisions for convergence. This key coordinating authority at the district level has an important decision making role in bringing in convergence at the district level. Functional responsibilities of the line departments need to be clearly defined and included under the convergence process.
- On the strength of above instructions & in consultation with the concerned authority at the district level would facilitate linkages with relevant programmes of

agriculture, horticulture, animal husbandry, rural development, etc. for enhancement of productivity and livelihoods at the district level. The States would have liberty to fix this district level responsibility as per administrative arrangements in the State.

- The Convergence Potential and Modalities would need to be clearly spelt out in the Convergence and resultant matrix which would be an integral part of the Detailed Implementation Strategies
- Project Implementing Agencies (PIAs)and field functionaries would facilitate the implementation of important programmes through convergence of other Departments such as MGNREGA, Backward Region Grant Fund (BRGF), NFSM, NHM, Tribal Welfare Schemes, Artificial Ground Water Recharge, Green India, IWMP etc. on priority in collaboration with their field functionaries.

Possible Departments to converge with at various levels

The Departments that are potential convergence partners include: Rural Development, Panchayati Raj, Water Resources, Minor Irrigation, Ground Water, Drinking Water Supply & Sanitation, Agriculture, Horticulture, Animal Husbandry, Dairy, Fisheries, Agriculture Engineering, Soil and Water Conservation, Environment & Forests, Sericulture, Marketing, Cooperation, Social Welfare, Tribal Development, Adult Education, etc.

Possible Institutions / Organizations to work for Convergence?

Training, research, extension and other facilitating institutions/ organizations that could be considered to promote Convergence would include: State Administrative Training Institutes, State Agricultural/ Horticultural / Veterinary Universities, State Marketing Boards, National Bank for Agriculture & Rural Development (NABARD), Indian Council of Agricultural Research (ICAR) institutes, State Agricultural Management and Training Institute (SAMETIs), Forestry Research and Training Centres, Watershed Training Institutes/Centres, Water & Land Management Institutes (WALMIs), Irrigation Management Training Institutes, National Institute of Rural Development (NIRD), State Institutes of Rural Development (SIRDs), National Institute of Agricultural Extension Management (MANAGE), State level Co-operative Training Institutes, National Remote Sensing Centre (NRSC), State Remote Sensing Agencies, Krishi Vigyan Kendras (KVKs), Farmers Training Centres (FTCs), Training Centres of Input Support Agencies like Indian Farmers Fertilizer Cooperative (IFFCO), Krishak Bharati Cooperative (KRIBHCO), Training Institutes of Banks, etc.

Existing Planning Instruments/Mechanisms facilitating Convergence

A number of mechanisms are already available through various programmes and schemes to facilitate convergence. Some of these are:

(i) Comprehensive District Agriculture Plans (C-DAPs)& State Agriculture Plans(SAPs) of RKVY,

(ii) Strategic Research and Extension Plans (SREPs) and Block Action Plans (BAPs) of Agricultural Technology Management Agencies (ATMAs) of State Agriculture Departments,

(iii) District Level Potential linked Credit Plans (PLPs) of NABARD,

(iv) Micro Level Farming Situation Analysis Reports of KVKs(in District Profile),

(v) Detailed Project Reports and State Perspective Plans of IWMP and

(vi) Work Plans of Forest Departments. There should be a clear focus on capture of priorities to strategize and facilitate convergence process.

Brief description of selected programmes of various ministries/organizations wherein convergence possibilities could be explored:

Agriculture

Rashtriya Krishi Vikas Yojana (RKVY)

RKVY is a multi-sectoral programme covering agriculture and allied sectors. This is an important scheme for convergence and many innovative programmes can be promoted across the sectors. It provides flexibility to the States in formulating their development priorities for the State, region and district. The programmes are designed broadly as per the priorities reflected in Comprehensive District Agricultural Plans(C-DAPs). State Department (Lead Programme) may develop a system of convergence of agriculture, horticulture, animal husbandry, dairy, fisheries, and agro-forestry and sericulture programmatic interventions for selected districts/regions as per potential. It may then approach the concerned line department for their support in implementation. The interventions could be with or without funding support.

National Food Security Mission (NFSM)

Production support for Rice, Wheat, Pulses, Coarse Cereals and Commercial crops (Jute, Cotton and Sugarcane) is extended under this Mission. The scheme provides for field demonstrations on production technologies, inter-cropping and cluster demonstrations to be conducted by the States in collaboration with the ICAR institutes and State Agricultural Universities. The scheme also provides for cropping systems based training support and assistance on High Yielding Varieties (HYV) seed distribution of certified seeds of pulses (arhar, moong, urad, field pea, gram and moth). NFSM interventions could be suitably converged with lead programmes.

National Mission for Oilseeds and Oil Palms (NMOOP)

Diverse agro-climatic conditions are favourable for promoting cultivation of 7 edible oilseeds (groundnut, rapeseed & mustard, soybean, sunflower, sesame, safflower, niger and 2 non-edible oils (castor and linseed). NMOOP provides assistance on distribution of certified seeds of major oilseed crops, supply of soil amendments, plant protection chemicals and field extension support. Oilseed cultivation covers about 27 million hectares mainly on marginal lands of which 70% are confined to rainfed farming. Oil Palm is highest vegetable oil yielding perennial crop. With quality planting material, irrigation and proper management there is potential to achieve higher yields. The NMOOP supports distribution of Oil Palm sprouts, drip irrigation system for oil palm, construction of farm ponds and water harvesting structures and bore well for oil palm cultivators. There is scope to expand Oil Palm cultivation in the coastal States and a few NE States like Assam, Mizoram and Tripura Further, common tree borne oilseeds like Karanj, Neem, etc., could be taken up in rainfeds. There is scope for promotion of these crops in areas having such potential. It requires joint strategies, linkages and interface between field functionaries of the State Department of Agriculture, especially in planning and conducting training programmesand demonstrations.

Integrated Nutrient Management (INM) as a part of NMSA

INM is focused on promoting soil test based balanced and judicious and timely application of chemical fertilizers, bio-fertilizers and locally available organic manures such as farm yard manure, vermi-compost and green manure to maintain soil health and soil productivity. District Soil Testing Laboratories (STLs), Mobile Soil Testing Laboratories and Soil Testing Laboratories of Krishi Vigyan Kendras (KVKs) are available in the public domain. This facility is also provided by Private Input Support providers (mainly seed, fertilizers and plant protection chemicals) in potential areas. Besides, private and co-operative sugar factories are also extending this support to the farmers in their catchments. Farmers are encouraged to go for Soil Health Cards indicating status of major and micro-nutrients. This is far more important for the farmers in rainfed ecologies for balanced and timely use of fertilizers and for making the best use of available moisture regime and scarce water resources. An assistance for soil improvement (in the form of soil ameliorants, micro-nutrients, vermin-compost, bio-fertilizers, organic inputs, etc.) is extended through various Missions such as NFSM, NHM, NMOOP & NMSA. Farmers training and field demonstrations may be organized by the agriculture department and KVK scientist in the selected areas. KVKs and ATMAs may focus on training of the farmers in relevant technologies.

Integrated Pest Management (IPM) as a part of National Mission on Agricultural Extension & Technology (NMAET)

IPM is an eco-friendly approach which aims at keeping the pest below Economic Threshold Level (ETL) by employing all available pest control methods and techniques such as cultural, mechanical and biological controls. Greater emphasis is laid on use of bio-pesticides and use of pesticides of plant origin such as neem formulations. The use of chemical pesticides is advised as a last resort when the pest population crosses the threshold limits. IPM activities are promoted through Farmers Field Schools (FFSs) wherein farmer to farmer learning is promoted through field observations and real farm situations. The Agriculture Department functionaries and KVK scientists may emphasize IPM approach in the watershed areas through training, demonstrations and exposure visits. Farmers Field Schools may also be encouraged and their capacities built to undertake improved IMP practices. Sub- Mission on Agricultural Mechanization of NMAET provides for plant protection equipment like sprayers. NFSM & MIDH have a component for providing plant protection chemicals, bio pesticides, IPM etc., these could be suitably extended to the potential areas through their field programmes like training, demonstrations, fields schools and exposure visits.

Horticulture

Mission for Integrated Development of Horticulture (MIDH)

National Horticulture Mission (NHM) was launched in 2005-6 as a Centrally Sponsored scheme to promote holistic growth of horticulture sector through area based regionally differentiated strategies. The Scheme has been subsumed as a part of Mission for Integrated Development of Horticulture (MIDH) from 2014-15. The MIDH continues to be Centrally Sponsored Scheme covering fruits, vegetables, root & tuber crops, mushrooms, spices, flowers, aromatic plants, coconut, cashew, cocoa and bamboo. While Government of India (GOI) contribution is 85%, the State's contribution is 15 %(except North Eastern States where GOI contribution is 100%). Major interventions operated under the scheme are: Research & Development (R&D) Support, Production and productivity improvement, Production and distribution of planting material, Rejuvenation of Senile plantations, Creation of Water resources, Protected Cultivation, Precision farming, Human Resource Development (HRD), Technology Dissemination, Post-Harvest Management, Processing & Value addition, etc. Most of these interventions are applicable to the areas suitable for promotion of horticulture related activities. Field functionaries may obtain quality planting material (as per potential /requirement) from MIDH credited outlets and seek technological back up from MIDH as per horticultural development needs. The convergence with this sector could be strengthened further by:

(i) providing representation to the field formations on the District Horticulture Mission,

(ii) Line departments and KVKs may interact for development and distribution of quality planting material,

(iii) IWMP and MGNREGA may develop farm ponds; conservation structures and polythene lining, and drip system required for horticulture development may come from other programmes such as NMSA. Activities like production of planting material for potential areas and pest and disease control of horticultural plantations there may be proposed for convergence with MIDH.

Natural Resources Management and Climate Change

National Mission on Sustainable Agriculture (NMSA)

Developed under the National Action Plan for Climate Change (NAPCC), the NMSA aims at bringing in a systems approach to rainfed farming. The major components are Rainfed Area Development Programme (RADP), Soil Health Management, on Farm Water Management and Climate Change pilots. Intensive activities like soil resource mapping, promoting soil testing in the given watershed through static and mobile laboratories, distribution of portable soil testing kits, making the soil fertility maps available at field level, conducting soil health awareness campaigns at the village/Panchayat levels, etc., could be undertaken on the convergence mode combining the efforts of both public and private facilities. The field formations would need to orient farmers about soil health concerns. Capacities of field agencies would need to be enhanced through systematically developed training modules.

Prime Minister's Krishi Sinchaai Yojana(PMKSY)

Micro Irrigation Component of PMKSY focuses on enhancing water use efficiency by promoting efficient on farm water management technologies and equipment. This would include promotion of drip and sprinkler irrigation. These systems would need to be extended around water bodies created or renovated through watershed & MGNREGA works with focus on judicious and timely use of available water. Training and MI technology support (installation, maintenance, etc.) may also be provided to the farmers adopting drip/sprinkler practices. Emphasis is being given on effective harvesting and management of rain water (through IWMP Component of PMKSY) and assistance is also extended for adopting water conservation technologies. Further, efficient delivery and distribution systems are promoted through AIBP Component of PMKSY.

National Initiative for Climate Resilience Agriculture (NICRA)

NICRA is an ICAR initiative that covers: cropping systems, water management, weather-based agro-advisories, risk management and enhanced extension efforts. These climate resilient initiatives are implemented through selected KVKs, wherein Village Climate Management Committees are promoted. Learnings from NICRA villages could be captured by field programme teams and up-scaled if found advantageous.

Rural Development

Mahatma Gandhi National Rural Employment Guarantee Act (MGNREGA)

The IWMP (now a part of PMKSY) has stipulated per hectare cost norms which at times proves inadequate to undertake comprehensive soil and water conservation measures. Convergence would provide opportunities and flexibility to undertake more labour intensive earthworks under MGNREGA, while IWMP resources could fund the permanent structures. Concerned officials of Mahatma Gandhi National Rural Employment Guarantee Scheme (MNREGS) should actively participate in DPR preparation and ensure that the activities identified are included in the shelf of the works of the concerned Gram Panchayat. Considering the commonality of the objectives, works, target areas and beneficiaries between IWMP & MGNREGS, it would be appropriate to take up MGNREGS works in villages under watershed projects. Activities on the ground would include construction of farm ponds, field building, water harvesting structures and plantations.

Convergence would further ensure targeted planning on a wider canvas. What IWMP brings to the table is a more scientific approach to the planning of soil and water conservation works as a whole with specific activities apportioned to MGNREGS and IWMP. Such works can be taken up under MNREGA after exhausting IWMP funds.

Further, convergence with MGNREGS have been detailed in the recently issued MORD circulars No. 11017/17/2008/NREGA(UN), dated 11th August,2014 (Watershed management works independently under MGNREGS or in convergence with IWMP) & 9th December, 2014 (Advisory on Water Harvesting Structures (WHS) & desilting /repair & renovation of existing structures). These may be referred appropriately by the field agencies for locally relevant convergence.

Aajeevika-National Rural Livelihood Mission (NRLM)

The National Rural Livelihoods Mission currently under implementation in selected blocks in the Country was launched in June, 2011. The Mission aims at creating efficient and effective institutional platforms of the rural poor, enabling them to increase household income through sustainable livelihood enhancements and improved access to financial services. NRLM is complementing rural poor groups with knowledge, information, skills, tools, finances and collectivization. Wherever feasible, synergies should be explored with watershed projects. As NRLM expands, these groups including women SHGs already created could be involved in skill development initiatives of NRLM. Income generating and livelihood promotion activities of NRLM may be promoted in potential areas and joint training strategies could be taken up accordingly. Also, there is need for promoting more SHGs and Credit and Thrift Group linkages on convergence mode.

Panchayati Raj

Backward Region Grant Fund (BRGF)

This scheme provides for Panchayats as institutions for planning and implementation of need based programmes to address the intra district variations. 250 of the most backward districts are targeted for the purpose. The BRGF allows a high degree of flexibility to the community in the choice of activities. These could well include soil and water conservation and other activities that are typically considered under development projects. Further, there is scope for involvement of Panchayats in local priority settings for managementof development programmes.

Drinking Water Supply & Sanitation

National Rural Drinking Water Supply Programme

The scheme provides for taking up of conservation measures for sustained supply of water through rainwater harvesting and ground water recharge structures. The projects include taking up water conservation and recharge measures for source strengthening for drinking water. The scheme could be taken up in the selected rainfed areas. Joint actions could facilitate in identification of such sites and aligning development porgramme plans to ensure adequate water availability.

Swachh Bharat Mission Campaigns may be aligned to the needs of the field programmes, targeted population(farming and non-farming) and their involvement.

Water Resources

Repair, Renovation and Restoration (RRR) of Water Bodies

This Pilot Scheme for "National Project for Repair, Renovation & Restoration (RRR) of Water Bodies" is directly aligned to the watershed objectives of improving water availability and creating the potential for Agriculture. The objectives of the Scheme are to restore and augment storage capacities of water bodies, and also to recover and extend their lost irrigation potential. The RRR Scheme, as a part of Accelerated Irrigation Benefit Programme (AIBP, now a part of PMKSY) has been approved for 26 district projects in 15 States. Water bodies having original irrigation culturable command area up to 2000 hectare or less are covered under the scheme. There is scope for convergence around such bodies if they fall in a given watershed. There is strong need for watershed teams to interface with irrigation/minor irrigation department for convergence purpose.

Central Ground Water Board (CGWB)

CGWB monitors ground water levels four times in a year. In addition, there is a bimonthly water level monitoring of selected observation wells by the States too. These activities would need to be dovetailed in the rainfed areas so as to assess the surface and ground water availability. This analysis would facilitate crop diversification and promotion of other livelihood activities there and converging schemes of agriculture and line departments.

Animal Husbandry, Dairy & Fisheries

National Dairy Plan (NDP)

NDP is under implementation through National Dairy Development Board (NDDB) with World Bank assistance since 2012 with an objective to meet the projected national demand of 150 million tonnes of milk by 2016-17. The scheme has focused components of increasing productivity of milch animals and providing greater access to the rural milk producers with organized milk processing sector. Also, it has a focus on promoting Dairy Farmer Producing Companies and Dairy Farmer Producer Organizations. These could be taken up in areas having such potential. The District level milk unions and the field functionaries of line departments may collaborate on use of primary milk cooperatives for programme delivery. SHGs, CIGs in a given programme can be part of this process. Further, SHGs could also be milk collection agents. State Animal Husbandry Department in the project States may provide necessary instructions to their district level formations for convergence of schemes/activities related to livestock development.

Livestock Health and Disease Control Programmes

This is one of the major schemes of the Department of Animal Husbandry. Animal Health Camps could be organized in collaboration with the departmental functionaries in the irrigated andwatershed areas. Watershed development teams may focus on small ruminants and backyard poultry depending on the potential. Para vets/Barefoot vets and Non-Government Organizations (NGOs) trained through the programme, could play an important role in preventive and primary health care of animals. Network of *Gopals* and *Pranibandhus* promoted by various States may need to be adequately trained and their capacities built for addressing feed fodder and cattle health care including of small ruminants.

National Fisheries Development Board (NFDB) Programmes

NFDB programmes are available both for inland and marine fishery initiatives. Water availability in perennial surface water bodies and fish seeds in such bodies may come from the Fisheries Department. NFDB's scheme of Aquaculture in Ponds and Tanks could be extended for this purpose. Community participation for water management, fish seed production and auction may be promoted through village level PRIs. Training of fish farmers may be taken up through KVKs/ATMAs and outlets of State Agricultural Universities (SAUs) and ICAR institutes. Further, involvement of the Department of Fisheries is necessary for providing fingerlings to farm ponds with lining developed under watershed/MIDH programme.

Credit and Insurance Support

Crop Insurance for Risk Coverage in Rainfed Areas

Small and marginal farmers in rainfed areas face partial or total crop loss due to risks associated with weather fluctuation. While crop diversification and supplemental irrigation measures help in coping with risks to a large extent, rainfed farming would still be vulnerable to longer drought spells. Crop insurance is an important tool for risk mitigation for small and marginal farmers. The principle crop insurance scheme – National Agricultural Insurance Scheme presently covers less than 10% of farmers. Climate risks are often highly spatially correlated and therefore call for expansion of Weather Based Insurance Scheme. Crop Insurance as a risk mitigation measure is effective only in combination with risk reduction measures like soil and water conservation, adoption of sustainable practices, inter-cropping and diversification, use of appropriate seed varieties, etc. Hence, rainfed agriculture needs: location specific insurance products for crops and livestock, insurance education of small and marginal farmers and adoption of comprehensive risk reduction agricultural packages. The field formations of development progrmmes (Agriculture, etc.) may arrange interface between farming communities and insurance agencies for educating the farmers on the risks and insurance coverage.

National Bank for Agriculture and Rural Development (NABARD)

NABARD has a dedicated district level functionary, District Development Manager (DDM) who could be associated with watershed programmes especially for providing credit support linkages. The much required farming systems - input support and credit linkages equilibrium could be obtained in this manner. The field formations may look into Potential Linked Credit Plans (PLPs) as one of the instruments of convergence. NABARD/ Bank linkages would help in mobilizing credit support for SHGs/entrepreneurs in watershed areas. Further, issues like financial inclusion, sensitization of the bankers, activating priority sector lending, providing credit on lower interest rates to the watershed farmers and linkages with the SHGs would also be undertaken by the Agriculture Department in collaboration with project level watershed functionaries. DDM could play an important role in this process. Besides credit, the NABARD also supports watershed programmes primarily promoted through NGOs. Successful watershed programmes can be used for exposure visits of the field functionaries.

Technology Validation and Dissemination in Agriculture & Allied Areas

Krishi Vigyan Kendras (KVKs) Programme of ICAR

Krishi Vigyan Kendras(now over 650) are constituted as front line extension institutions and activities focus on technology validation and refinement at the local level. In the process, KVKs are involved in technology demonstrations and training of farmers/field extension personnel. Other extension activities taken up by the KVKs include: organizing technology weeks, production of information material and their dissemination, farmer-scientist interactions, *Kisan Melas*, Mobile Advisory Services and providing technological backstopping to the field programmes of the agriculture and line departments. KVK – Agriculture and line department linkages could be improved further by:

(i) having Memorandum of Understanding (MOU) based linkages for technology testing and training,

(ii) involving KVK Subject Matter Specialists (SMSs) for technology backstopping to ATMAs,

(iii) drawing learnings from micro level farming situation reports of KVKs. The process of convergence could be strengthened by having jointly developed training modules as per specific training requirement of a particular areas and developing joint strategies for dissemination of technologies. Every KVK is headed by the Programme Coordinator who in turn is assisted by six Subject Matter Specialists covering various disciplines such as agronomy, plant protection, soil science, home science, agricultural extension, etc. Most State Agricultural Universities (SAUs) have outreach programmes and field specialists in position in the districts (like Districts Agriculture Technology & Training Centres of Andhra Pradesh University, Extension Agronomist in Maharashtra Universities, District SMSs of Dharwad University, Farm Advisory Services of Punjab Agriculture University, etc). They are involved in dissemination of farm advisories. What really matters is the linkage of these SMSs with the field extension functionaries. There is a strong possibility of collaboration of these SMSs with field level formations in the given area.

Agricultural Technology Management Agency (ATMA) as a part of NMAET

The ATMA concept focuses on extension reforms and programme delivery. There is a broad menu of extension reforms for State and district levels. The extension reforms are centered around training, demonstrations, exposure visits, farmer empowerment, farmer field schools and field extension activities. Farmer Interest Groups and Commodity Interest Groups that are formed at the field level may converge with other programmes. Most of the district developments departments are suitably represented on the Governing Board of ATMA to achieve synergy between the extension interventions. Further synergy could be obtained through:

(i) ATMA field progammes are jointly organized through Block Technology Teams of ATMAs,

(ii) Field formations of development programmes to look into the Strategic Research and Extension Plans (SREPs) and Block Action Plans (BAPs) of ATMA for convergence modalities,

(iii) The Farmer Friends (FFs), Village Extension Workers (VEWs) and SMSs of KVKs and ATMAs be oriented on project requirements, and

(iv) ATMA/KVK training strategy may be sharply focused on requirements of a particular area.

The faculty members of State Agricultural Management and Extension Training Institutes (SAMETIs) be promoted as master trainers. The subject matter areas that would need to be stressed in these model training programmes may include: Programme Planning& implementation, Monitoring & Evaluation (M&E), Application of Remote

Sensing (RS) & Geographical Information System (GIS), Sustainability, Convergence & programme delivery, etc.

The KVK & ATMA Convergence Guidelines brought out jointly by the DAC & DARE in 2012 is an excellent example of convergence process, though some of the stakeholders are yet to fully firm up the actions proposed and roll them out in the field.

Environment and Forests

Programmes of Forest Department

Convergence with Forest Department programmes could be worked out at the State level between the two departments. In Reserved Forest areas joint planning for the treatment of the upper reaches/ ridge areas needs to be promoted. Forest Department should treat upper reaches so as to ensure comprehensive treatment with adequate soil, water conservation and afforestation measures. As no other department is permitted to work in the reserved forest areas the inadequate treatment of the upper areas defeats the purpose of watershed treatment or at best offers only fractional benefits. For Community and Social forestry in rainfed commons the planting material may come from the Forest Department. Accordingly, there is need for a joint capacity building strategy. Convergence with Agro-Forestry and Non Timber Forest Produce (NTFP) collection can be facilitated through field functionaries. This will ensure better integration of activities of Joint Forest Management Committee (JFMCs). Convergence of specific NRM activities would need to be identified for forest and non-forest lands.

Green India Mission

The Mission is focused on enhancing eco-system services through afforestation on degraded lands. Possibility of massive plantation drives maybe considered in watersheds, especially wastelands. Agro/Farm forestry linkages need to be strengthened especially for providing good quality seedlings from accredited nurseries.

Promoting a Convergence Matrix Approach

Every Development programme, in the Planning Stage itself, should mandatorily include a Convergence Matrix indicating Lead Programme & resource inputs and roles of the converging partners at various levels. This matrix could specifically indicate the GOI and State programmes, separately.

The logical steps followed in developing Convergence Matrix should include:

Sr. No	Sector/ Schemes & Lead Progra-mmes (a)GOI & (b)State	Description of Possible Activities (of Other Schemes/ Progra-mmes) proposed to be Converged with Lead programme	Specific activity(s) identified for conver-gence	Quantifiable Estimated amount of Conver-gence at Planning Stage (DPR)	Time Period / Phase for Convergence	Quantifiable expected Outcomes from proposed convergence	Remarks of M & E mechanism	Actual Outcomes /Benefits accrued from conver-gence

Convergence modalities could be initially piloted in a couple of districts where convergence possibilities are under process or in those which have just shown interest in the convergence process. There appears to be a great scope for convergence under sharing of Human Resources (HR), capacity building and techno-managerial services of agriculture &line departments. The Convergence Matrix could be an addendum to the DPR of the development programmes.

Existing Convergence instruments like C-DAPs of RKVY, SREPs of ATMAs, PLPs of NABARD, Micro Agro-Eco Situation Analysis Reports of KVKs, DPRs of Watershed programmes, schemes under RD and Forest Departments, etc., should be intensively reviewed / used in formulating convergence matrix and priority setting.

Activities felt necessary for integrated and comprehensive treatment that are reflected in the DPRs, but not covered through programme funding, may be supported through MGNREGA, BRGF, Integrated Action Plan (IAP), Special Component Plan (SCP), Tribal Sub Plan(TSP), DPC and other State sponsored schemes.

The Way Forward – Action Points

Orientation of Decision-makers at State/ National level

State Nodal Department seeking convergence should hold initial discussions with state level decision-makers from other converging Departments to create awareness regarding the opportunities and benefits of such convergence at the State, district and project level.

Convergence Arrangements & Issue of Necessary Instructions

State nodal department (seeking convergence) is required to interface with other Departments at the State level to work out suitable convergence arrangements with the State and District level agencies of various Departments. Convergence would be effective if concerned Government departments at the State level issue necessary instructions to their respective implementing agencies and field formations.

Documentation of existing good practices

Some good examples already exist in States. These may be documented and explored further to better understand processes and modalities for upscaling or customization as per needs.

MOU-based Convergence Arrangements

There is need for having a draft MoUs that can be customized to State and project specific requirements. The task specific convergence MoUs may be considered with ICAR institutes, SAUs, SAMAETI, ATMA, KVKs and other related institutes in converging Departments. These if need be, may be endorsed by the respective Departments at appropriate levels. The MOUs so agreed upon should specifically highlight technological support and participation required for intended convergence.

Making Convergence Matrix mandatory at the Planning Stage itself

All the development programmes should have a Convergence Matrix approach built into the process right in the Planning Stage itself. This should identify the gaps in the ongoing programmes and possible fill up through other related programmes. It has already been detailed in Para-III above.

Proposal for Line Department Support

The nodal department seeking convergence would develop broad outlines for converging other departments' programmes for each agro-climatic zone in the State, indicating scope and opportunitiesthat the nodal department can offer them and how they could make best use of it.

Further,location specific convergence support would be worked out by the district and block formations in collaboration with such functionaries of the department(s) proposed for convergence

As a strong convergence requirement the district/block level functionaries may be represented on the district level structures of the other important programmes such as ATMA Governing Board and ATMA management committees, District Horticulture Mission, District NFSM mechanism, Scientific Advisory Committee of KVKs, and on district and block level formations of other converging core Departments like Rural Development, Panchayati Raj,Forestry, Water Resources, Ground Water, Minor Irrigation, Marketing, Animal Husbandry/Dairy, Fisheries, Tribal Development,etc

Training Strategies Supportive to Convergence

The nodal department seeking convergence of other departments would need to develop a broad training strategy in collaboration with SAMETI, ZPD (ICAR), SIRD, Rural Development & Self Employment Training Institutes (RUDSETIs), Skill and

Entrepreneurial Development Institutes and training establishments of other Departments for need based involvement of line departments.

Further, location specific training modalities would be worked out by an inter-disciplinary expert group under the guidance of nodal department in collaboration with the training institutes at the district/ block level like KVKs/ATMAs/NGOs/Farmer Organizations (FOs) and training establishments & field formations of various schemes in converging sectors.

The field agencies like KVKs, ATMAs and WDTs would need to be geared up accordingly for training and capacity building of farmers/ farmer groups and other resource poor for providing them with better production and market linkages.

Promote Demand Driven Convergence Approaches including Indigenous Technical Knowledge (ITKs)

Concerted efforts are required in selecting the locally relevant priorities and assigning deliverables to the identified converging agencies, public/private, as applicable.

Separate accountability indicators would need to be set for various converging partners.

Documentation, validation and promotion of ITKs as applicable to the rainfed areas may be undertaken, with systematic involvement of ICAR institutes, agricultural universities, relevant Consultative Group on International Agricultural Research(CGIAR) institutes, SIRDs and rainfed networks (both national /international).

Credit back up – to be tracked by DDM, NABARD

Emerging entrepreneurial initiatives in a given watershed, covering agricultural and allied areas would need credit back up for sustenance and up-scaling. There should be appropriate tie up of such efforts with identified credit institutions. The DDM, NABARD in a given district could facilitate this process.

Community involvement for Convergence and up-scaling

- Farmers and farmer groups would be encouraged to successfully adopt convergence practices at the field level. Field functionaries could supplement such efforts.
- Farmer Field Schools (FFSs) may be promoted in large numbers so as to replicate convergence successes. Agriculture and line departments may promote FFS in collaboration with district agriculture department as a part of Capacity Building (CB)/Training activities under various development programmes.
- Successful convergence experiences may be widely shared and disseminated, clearly delineating converging schemes/partners, both public and private.

Assessment of Convergence Impact

- There should be third party evaluation of the impact of initiatives undertaken through convergence. It should lead to identification of strengths and weaknesses of the convergence processes and should be able to qualify and quantify the impact in terms of higher adoption, better economic gains accrued to the farming communities and sustenance of the impact.
- Given the magnitude of investment envisaged in agriculture and allied sector programmes, it becomes imperative to ensure convergence of IWMP with other development programmes in sectors such as Agriculture, Rural Development, Panchayati Raj, Water Resources, etc. This would bring in an integrated approach to the development in a given area, enhance incomes of the farmers, provide wider impact across the sectors and make such efforts more sustainable. Accordingly, a Convergence Matrix (CM) approach is suggested, capturing the process right from the planning to implementation stage.
- The Convergence Matrix broadly indicates: The possible areas for convergence, Specific areas identified for convergence, Estimated amount of convergence (at DPR Stage), Time period/ Phase of convergence, Quantifiable expected outcomes, Remarks of M&E System on Convergence and finally Actual convergence outcomes.
- There is need to design a mechanism to set the Benchmarks (Performance levels) of Convergence Outcomes including indicators and benchmarks for convergence process. It would be taken up by the States in collaboration with identified institutions.

Monitoring & Evaluation (M&E) of Convergence

- Institutional M&E arrangements of IWMP would take care of the Convergence aspects as well. As indicated, third party M&E arrangements may be put in place by the States to assess the impact of IWMP including convergence. There should be adequate scope for mid-course corrections as a follow up on the recommendations of M&E findings.
- SLNAs, as a part of M&E process, may assess social impacts of convergence strategies in terms of better community partnerships, sharing of the watershed gains by the communities, social participation, development exposure, awareness of programmatic interventions, etc.

Management Information System (MIS)

MIS for collating, processing and reporting on progress *vis a vis* convergence actions envisaged in the DPR would need to be established at the project, SLNA and DoLR levels. MIS of relevant line departments may also have a provision to monitor the convergence component promoted under IWMP.

Conclusions

- The process of convergence is far more essential to make the maximum returns to the farming communities by converging various development programmes at various levels, defining suitable converging arrangements at various levels.
- Broad outlines have been suggested as above, however, specific detailing needs to be worked out by the respective State and the converging departments.
- Convergence Matrix Approach (as a mandatory component of the DPR) right from planning, implementation to evaluation stage is suggested and may be prioritized for action.
- State Nodal Department (that seeking convergence) may work out Convergence Mechanism at various levels obtain State Government approvals and issue necessary Government Orders for convergence. State-wise convergence arrangements/ Options would vary but the common thread of arguments in this paper may be suitably piloted first.
- Each Nodal Department would need to organize Convergence Orientation Workshops for the project functionaries to make them understand the concept and operation of Convergence.
- These Operational issues brought out in this paper would need to be analyzed, discussed and internalized first by the lead department and then shared with other Departments dealing with the identified sectors to orient them on the mutual benefits to be obtained by this convergence process.

References

1. Central Research Institute for Dryland Agriculture (2013-14), Annual Report-NICRA Chapter
2. Department of Agriculture and Co-operation, MoA(2013-14), Annual Report
3. Department of Agriculture Research and Education, MoA (2013-14), Annual Report
4. Department of Animal Husbandry, MoA (2013-14), Annual Report
5. Department of Land Resources, MoRD (2013-14), Annual Report
6. Extension Division of DAC (2012), Research Extension Linkage Guidelines .
7. FAO (2008), Development Challenges of Indian Agriculture, Background Technical Papers
8. Ministry of Rural Development, Department of Land Resources, GOI(2015), Operational Guidelines For Convergence of Programmes Under Various Ministries with IWMP
9. Ministry of Agriculture (2009) Proceedings of National Seminar on Agricultural Extension
10. Planning Commission (2011-12), XIIth Plan Working Group Report, Agricultural Extension

11. Planning Commission (2011-12), XIIth Plan Working Group Report, Natural Resource Mgment
12. Planning Commission (2011-12), XII th Plan Working Group Report, Agricultural Res.& Edn.
13. PlanningCommission (2011-12), XII th Plan Working Group Report, Decentralized Planning
14. Sadamate, V.V. (2003), PPPs in Extension Management, Report, Fulbright Supported Study
15. Sadamate, V.V. et.al (2012), KVK QRT Report, Zone-V , Indian Council of Agricultural Research, New Delhi
16. World Bank (2012), The Community of Practice in Art of Knowledge Exchange

30 Rethinking Agriculture Development:

The Value Chain Way!

- *Dr. Hemnath Rao Hanumankar*

Introduction

Having set a very impressive record of agricultural production in regard to food grains as well as commercial crops during the last few decades, the country is keen to move beyond productivity and sustainability gains in agriculture towards a market based farming systems approach to help enhance income levels of the farmers. In keeping with the spirit of globalisation of agri-business markets around the world and the accompanying escalation in competition for procuring agricultural inputs as well as marketing agricultural produce, it is important to ensure the competitiveness of Indian agricultural produce, even as we seek productivity gains. If the farm production is not in tune with the shifting patterns of market demand, valuable opportunities for value addition and value realisation by farmers can be lost to competing countries.

To embrace the new paradigm of market orientation and competitiveness, bold new approaches to agricultural development that are no less revolutionary than the 'Green Revolution' of the yesteryears are needed. This chapter attempts discussion of the value chain approach based on envisioning any given crop production system as a 'giant enterprise' seeking to integrate forwards and backwards as well as diversify and exploit the fullest economic potential from the crop, on and off the 'farm from seed to plate'. Drawing from the strategic management literature developed by industrial economists and business school academics, the chapter dwells on the genesis of the value chain thinking from the industrial economy and its growing influence on agriculture development policies and models. Obviously, the value chain approach calls for convergence and extension of traditional agriculture development models and the product level competitiveness and firm level competitive advantage based strategic thinking that dominates the industrial economy.

Goal Setting in Agriculture

Typically, central and state governments have been articulating their goals for agriculture development as follows;

- To increase farm incomes in a sustainable and environment friendly manner.
- To improve efficiency of agriculture production by improving yields, reducing costs, developing export markets, building infrastructure and investing on research, development and extension etc.
- Ensure adequate and timely supply of quality inputs – seeds, pesticides, and fertilizers to the farmers.
- Bring all eligible farmers into the banking net by restructuring existing credit delivery institutions.
- Institutionalize Integrated Pest Management (IPM) for forecasting and managing outbreaks of pests and diseases.
- Promote private sector participation in agriculture through contract farming, buy-back arrangements, private investment in storage facilities and marketing.
- To develop the state as a leader in agriculture/ horticulture/ fisheries/ etc.
- Balanced regional development in Agriculture.

While the above goals and objectives sound clear and compelling, the strategic framework to guide and achieve the same received little or no attention as the policy makers seemed to believe that the articulation of the goals was an end in itself on the assumption that the functional research and extension machinery at the national and state levels possessed all that was needed to realize the stated objectives. This assumption produced results to an extent as the states chased with support from the Government of India (GoI) production and productivity targets through enhanced major, medium and micro irrigation projects; seed improvement and input delivery systems; intensive and inclusive credit coverage; strengthening extension systems for technology diffusion; setting up marketing and processing facilities for agricultural produce and a host of other development interventions.

Even the celebrated green revolution of the 1970s and 1980s that bolstered the country's food security is attributable to the laudable programmes and projects that were followed up on the above assumptions and interventions. Yet the production and productivity gains have not translated into livelihood security for millions of farm households across the country who form the bedrock of the agriculture production system that is at the heart of India's rural economy nor has the country been able to carve out a leadership position in regard to a particular product or product mix in the global marketplace. The lack of a strategic focus on holistic planning and goal setting that should have nurtured crop specific competitiveness and secured livelihoods for all those involved with the crop value chain stood exposed as a missing link in the agriculture development system that has been built over the decades. Recognising this gap, the country's agriculture research and development wing led by the Indian Council of Agricultural Research (ICAR) launched the National Agricultural Innovation Project (NAIP) in September, 2006 with funding support by the World Bank. Research on crop based production to consumption systems through value chains approach constituted one of the four pivots of the NAIP.

NAIP's Value Chain Component

With an overall objective of facilitating an accelerated and sustainable transformation of Indian agriculture towards supporting poverty alleviation and income enhancement through collaborative partnerships for promoting agricultural innovation, the NAIP rested on the following four components:

1. The role of ICAR as a catalyst in managing change in the National Agriculture Research System.
2. Research on production to consumption systems (PCS).
3. Research on sustainable rural livelihood security.
4. Basic and strategic research in frontier areas of agriculture science.

It is the second component of NAIP that was designed to capture integration and economies of scale benefits across select sub-sectors of Indian agriculture through market-oriented collaborative research centres for sustainable improvement of selected agricultural production to consumption systems. The concept (World Bank, 2006) behind this component was that around 15 well financed consortia comprising public, private, civil society and farmers' groups will be "able to galvanize greater interest from different value chain partners" involved in producing, harvesting, processing and marketing a particular product, capturing value through increased collaboration and openness.With this, a new paradigm of realigning research and development activities with the emerging global market scenario through a value chain based PCS was ushered in. It is important at the same time for academics, research and extension specialists to appreciate the genesis of the concept of value chain in the literature of business management so that agribusiness value chains can potentially be configured in more creative ways to harvest enhanced value in favour of the primary producers without hurting the interests of any [particular vale chain partner.

Porter's Value Chain

The concept of value chain primarily has its genesis in the industrial economy linked strategic management literature of 1980s that evolved in the context of seeking out firm level competitive advantage in a given industry. While strategy consulting groups such as Mc Kinsey had their own version of an industrial firm's business system, Porter, M.E.

(1985)posited the argument that competitive advantage cannot be understood by firms in any given industry without considering in great detail, the many discrete activities that a firm performs in designing, producing, marketing, delivering and supporting its product. Therefore, Porter believed that a systematic way of examining all the activities performed by a firm and how the activities interacted was the basis for analyzing the sources of competitive advantage and propounded what has popularly come to be known as Porter's Value chain **(Figure 1)**. The value chain disaggregates a firm into its strategically relevant activities so that one can understand the cost behaviour of the firm as well as identify the current and potential sources of differentiation which he considered were the principal sources of competitive advantage in an industry.

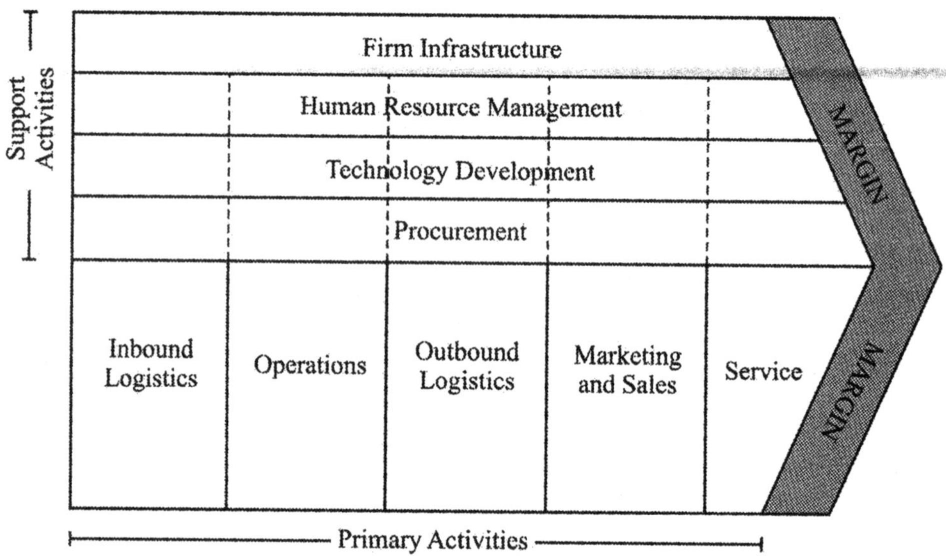

Fig. 30.1 Porter's Generic Value Chain

As can be seen from the Figure 1, Porter's value chain displays total value formation in a firm with value activities and the margin that they contribute. Value activities are broadly split into two categories- primary activities and support activities. The five primary activities listed in the posterior half of the value chain are those that are associated with the physical creation of the product, its sale and transfer of ownership to the buyer with necessary after sale assistance. The four support activities shown in the upper half of the value chain are the firm wide functions performed to support the primary activities, with the dotted lines reflecting the fact that three support activities namely procurement, technology development and human resource management can be associated with specific primary activities as well as firm wide chain of operations. The fourth support activity described as firm infrastructure, however, supports the entire firm value chain seamlessly and hence cannot be linked to any discrete primary activity.

Examples of firm infrastructure include activities such as compliance with regulatory provisions, firm level strategic planning or management information systems that aid corporate decision making.

A wider concept that accompanied Porter's Value Chain and would be of deeper interest for the agriculture sector is the configuration of a firm's value chain within a larger stream of activities that Porter termed as the Value System. The value system views suppliers' value chains as impacting upstream value through delivery of purchased inputs used in a firm's value chain. Similarly, as the firm's products pass through the value chains of the channel partners (distributors) on their way to the end buyers, downstream channel value is generated in various ways that contribute to the firm's product becoming eventually a part of the buyers' value chain. Hence, it was suggested by Porter that creating and sustaining the competitive advantage depended not merely on understanding a firm's value chain but also on how the firm's value chain was aligned with the overall value system.

Exploring Value System of Indian Agriculture

To impart increased gravitas to Indian agriculture and to combat the current agrarian crisis, it is necessary to ensure that no point in the value chain across the major agriculture crops/ commodities is allowed to go unexplored and unexploited in terms of economic value. While the value chain serves as a conceptual tool to strategise and operationalise secondary processing initiatives, the outcomes in terms of increased returns from farming, employment opportunities, checking migration and above all improving the sagging share of agriculture's share in the country's GDP cannot be exaggerated. As gains from secondary processing become evident, price buoyancy of the primary produce will motivate farmers to strive for higher yields and productivity gains. In the current situation of distress enveloping farmers in most geographical parts of the country and segments of agriculture, the value chain based approach to crop planning, production and processing seems to be the way forward for sustainable growth. As the NAIP envisaged, there is a need for multiple stakeholders to converge and explore the value system across the major crops which undoubtedly involves massive capacity building of all the involved actors, particularly extension staff, to promote a shared understanding of the concept.

Globally, the industrial models of value chain based business strategies have appealed to the imagination of agribusinesses and farming groups such as the soyabean industry, corn and cotton growers who have been deeply influenced and motivated to deploy value chain based approaches for enhancing returns from farm production. The results have been quite positive as certain industries like the wineries, tea and coffee processors have realized, which is prompting the farmers to adopt value chain based strategies across all crop and animal husbandry produce. As discussed earlier in this chapter, the value chain based approach to planning crop production and secondary processing is very nascent to the Indian context. In fact, many African and Latin American economies seem far ahead of India on the learning curve. A beginning has been made, thanks to the NAIP as value chain based studies have been carried out on selected arable crops, livestock and fisheries related products.

Need for Sectoral Restructuring

A strategic initiative such as the value chain approach cannot progress entirely on the steam of a single project like the NAIP. In the absence of well researched generic guidelines, most product studies supported under the NAIP reflected the understanding and perception of individual researchers and research teams which explains the diversity of value chain models that have emerged. This could in fact be deemed as a healthy manifestation of the creativity of the researchers and it is now time to analyse the available literature and encourage more research to cover more agricultural products and product groups and uncover the hidden value in the value system. While research is needed to initiate the value chain based strategic approach to agriculture development, there is a need for buy in by other institutions and agencies including the extension machinery with responsibility for driving agriculture development policies and programmes.

To disseminate the concept of value chain based agriculture development widely across all other players in the policy space, a strategy-structure alignment assumes critical importance (Hanumankar, H.R. 2008). Any strategic initiative can be only as effective as the robustness of the structure supporting the strategy. As Hanumankar H.R. (2008) argued, a sub-sector linked restructuring of the entire agriculture development linked organizational machinery in the state and central government could mark a significant baby step towards aligning the institutional structures with the value chain based strategy. **Figure 2** depicts the sectoral value chain for Indian agriculture which can potentially guide any restructuring exercise aimed at achieving the structural fit with a value chain based strategy. Once the restructuring is initiated along the sub-sectoral value chain at the apex levels, fieldstructures befitting product value chains will evolve as a logical corollary to the higher level organizational design and staffing patterns. It must also be reiterated that a well planned change management programme accompanied by an adequate learning and capacity building component along the lines of the first component of the NAIP will hold the key to the success of the value chain based approach to agriculture development.

Agricultural Inputs Delivery System	Agricultural Production System	Agricultural Value Addition System (Processing and Marketing)
☞ Crop Nutrition ☞ Plant Protection ☞ Quality seeds and planting material ☞ Credit ☞ Irrigation ☞ Agricultural Input Delivery Information System	☞ Crop Yields ☞ Soil Conservation ☞ Farm Equipment and Mechanisation ☞ Farm Agronomy and management ☞ Animal Husbandry ☞ Sericulture ☞ Fisheries ☞ Sugar ☞ Extension ☞ Information System	☞ Storage Commodity ☞ Marketing Agro Processing ☞ Product mechanizing ☞ Export promotion ☞ Co-operatives/SHGs ☞ Marketing Information System

VALUE

Sector Value Chain for Agriculture

Value Chain Agriculture: Implications for Agricultural Extension

Since the strategies for agriculture development and agribusiness would increasingly be influenced by value chain based approaches, there is a need for reorienting the agriculture extension machinery towards the new paradigm. An extension system, originally designed to promote technology diffusion and improved crop/ livestock husbandry practices for primarily increasing agricultural production may not readily embrace the value chain approach and there could even be some resistance in this regard as it often happens with the challenge of any change that involves new learning.

To overcome the challenge of change in promoting a market led agriculture extension system, two major initiatives could prove helpful. First, a massive capacity building programme would need to be designed for empowering the available extension personnel with the knowledge of value chain based agriculture. The capacity building process would mean that all the available extension training institutions in the country including the apex training institutions will have to be equipped adequately to implement a series of trainers' training programmes across the country. Secondly, nothing changes until behaviour changes and this can be very true of the extension personnel who are themselves adept at understanding behavioural issues. Hence, incentivising the extension personnel, both monetarily and non-monetarily could be another lever for making the change happen.

Conclusion

Indian agriculture has the potential to escape the pangs of the current distress impacting the rural economy in the country, if a new approach to agriculture development based on crop specific value chain and a commitment to harvest the value system across all agricultural products and product groups could be pursued. Globally, agribusinesses have demonstrated the utility of a value chain based strategy in creating competitive advantage across product lines such as soybean, corn, tea and coffee etc. In the process, agribusiness value chains are inching closer to the industry based generic value chain model advocated by Porter and the pre-cursor business system model popularized by the strategy consulting firm, Mc Kinsey. Even as the learning curve of other developing countries in Africa and Latin America in recasting their agriculture development programmes around the value chain based strategy could be ahead of India, an earnest beginning was made with a significant component of the NAIP implemented by the ICAR oriented towards production to consumption systems along product based value chains. Indeed, the success of any strategy is predicated on the strength of the supporting organizational structure and an initial restructuring of the entire agriculture development related institutional machinery around the sectoral value chain alongside a mammoth change management and capacity building exercise, could make a huge difference to the future of India's agriculture.

References

1. The World Bank. (2006). *Project Appraisal Document for National Agricultural Innovation Project.* India Country Management Unit, South Asia Region.
2. Porter, M.E.(1985). *Competitive Advantage: Creating and Sustaining Superior Performance,* New York: The Free Press.
3. Hanumankar, H.R. (2008). Bridging Strategy and Structure for Agricultural Development. *ICFAI University Journal of Rural Management, 1.1, 55-59.*

31 Participatory Irrigation Management : A Successful Agri Extension Approach

- *Dr. R. Ratnakar*

Introduction

Historically minor irrigation structures have an important place in Indian agriculture and rural life. Among the minor irrigation structures tanks have a predominant position. Such tanks have been the lifelines of several thousands of villages in the dry regions of the country in general and in the states of Telangana State and Andhra Pradesh in particular. They have a long history in the country dating back to two millennia or even more years. The rainfall and terrain conditions determined the number and size of tanks in various regions. In scanty rainfall regions with undulating terrain, more tanks are seen as that was the strategy to harvest and store water to meet the scarcity conditions. The tanks with their traditional technology, eco-friendly structures and sizes were mostly community managed and to a large extent able to address the needs of non complex traditional societies. In addition, these tanks have been directly or indirectly contributing to the livelihood of a few lakhs of households by way of facilitating fisheries, animal husbandry, sheep rearing or traditional industries like pottery and brick making.

The need for suitable MI structures was long realized, and a large number of tanks have been built throughout history by local rulers, temple authorities and the wealthy elite. Tanks, as useful storage facilities, supported not just agriculture and various other livelihoods in the village, but also played a crucial role in the rural ecology by serving as water-harvesting structures. While the aristocracy undertook the construction and further extension work of the tanks, it was the community that always had a role in managing them irrespective of their size and ayacut (command area).

Water management is an essential element to utilize the available surface, tank and ground water equitably and efficiently. Participatory irrigation management is the solution, the answer. Institution building is a social process, motivating the civil society of farming community to form collectives for expression of common will. Community involvement and institution building is essential to ensure that the actions are planned with community in a consultative manner and their active participation is sought so as to ensure future operations and maintenance by them for long term sustainability

Agricultural Extension Approaches

Agricultural Livelihoods Support Systems (ALSS) in the AP TS CBTM project include intervention areas under Agriculture, Animal Husbandry, Fisheries and Agri-Business. Final End Project Impact Assessment (FEPIA) for the period spreading from April 2014 to July 2014, was conducted covering a total of 200 selected tanks across 21 districts in different agro climatic zones of AP and Telangana.

The major objectives of agricultural livelihoods support services component in the project are (i) to improve production and productivity of tank system, and (ii) to enhance returns to farmers and other tank users through development of better market linkages and promotion of business development services. The Project Implementation Plan (PIP) included the following livelihood activities for improvement in agriculture, horticulture, livestock, fisheries and agri-business:

- Agriculture: technology demonstrations for traditional and non-traditional crops
- Horticulture: technology demonstration for vegetables and medicinal/aromatic plants
- Livestock: breed improvement, nutrition management and health management
- Fisheries: technology demonstrations and market linkages
- Foreshore plantation
- Marketing and Agri-Business: collective marketing, market led production, developing marketing information system, agri-business development.

Agriculture/Horticulture Development: Technology Demonstrations

The project has taken up agriculture and horticulture interventions with the objectives of improving productivity of crops by technology demonstration, training and backstopping, influencing cropping pattern and cropping system changes (crop diversification) through market led extension, to facilitate cost reduction through demonstration of appropriate technologies, agronomic practices and collective inputs procurement and to promote water saving technologies in crops to enhance water use efficiency (WUE) and water productivity.

The project envisaged increase in area of economic crops cultivation by infrastructure development and water management; to increase productivity through demonstrations linked to adoption, adoption support and capacity building; and also to increase cropping intensity by market led production, crop diversification and water audit based crop planning.

Demonstration and adoption plots

A Demonstration Plot (DP) of about half an acre was to be selected. Each DP was to be provided with inputs from the project covering seed, fertilizer, pesticide and incremental operational costs. Lead farmer's contribution has to be in the protection of DP. The agriculture / horticulture department at the state and district level are expected to play an important role in supervision and facilitation of the overall process. ATMA in the project district have to be actively involved in the planning and implementation of project activities. A well laid out field demonstration in a tank ayacut area is the best tool to spread awareness amongst large number of farmers on the best package of practices. In the project it was planned to conduct at least - one demo in each tank each year.

Demo themes

The project planned to promote the adoption of the System of Rice Intensification (SRI) method in the tank command areas by building the capacities of the farmer in technical aspects through training and field demonstrations as paddy is the most predominant crop in majority of tank areas. Adoption of SRI and MSRI methods in paddy would improve the productivity by better utilization of water for maximization of output. It was projected in PIP that 40 per cent of the demonstrations would be under SRI method. The SRI requires two specific new implements, i.e., a rotary weeder and line marker for planting. Hence, the project planned to supply one set of implements to each tank wherever required for adopting this method. The project also envisaged to conduct demos on integrated crop management (ICM) for major crops under the tanks in the project. This covers "seed to seed" package of practices in an intensive way. This includes pre planting agronomic operations, Integrated Pest Management (IPM), Integrated Nutrient Management (INM), post-harvest technology practices *etc.* In addition to this, DPUs recently planned to conduct Mechanized System of Rice Intensification (MSRI) with the help of Agricultural Research Station Scientists nearby.

The interaction with the farmers at focused group discussions indicated that the farmers were not inclined to adopt SRI method because special care has to be taken to nursery raising and transplanting, difficulty in maintaining of water level at initial stages, difficulty in getting labour for running rotary weeders, though there was yield improvement and water saving from 10-28 per cent over farmers practice.

Field Level Trainings

As per PIP, the farmers were to be trained at demonstration sites by experienced resource persons on different themes. The trainings were to be spread over the entire demonstration period. Nearly 3-4 training sessions are linked to each demonstration plot, in addition to the field days. The agricultural coordinator of the SO/NSO has to coordinate the resource persons and facilitate farmers' mobilization with the help of the lead farmer. The documentation and co-ordination has to be done by the APD (Agrl.), DPU at respective districts.

Field level trainings at the demo plot are required to spread the improved/proven practices undertaken at the demo plot to other progressive farmers so that a spread effect through adoption by other farmers is to be achieved.

Field days

As per PIP, it is proposed to organize at least one field day for each demonstration at an appropriate time before harvesting in order to demonstrate the improved situation on demo plot compared to present farming practice. The field day is open for participation of farmers from the adjoining villages also. This is expected to encourage the farmers to adopt the demonstrated new practices and the trainer is supposed to visit the demo plot of lead farmer as well as the progressive farmers in other tank areas and provide feedback to the DPU. The agri. extension specialist of DPU and the M&E team has to randomly check the conduct of field days.

NADEP pits/ Vermi-Compost Demos

NADEP composting pits and vermi compost units are part of INM strategy and need to be established and demonstrated at each and every tank area. The selection of the beneficiary is to be transparent and objective. Other activities include promotion of green manuring and bio-fertilizers as part of integrated plant nutrient management. Uses of vermi compost / FYM reduce the usage of in-organic fertilizers as well as enrich the soil fertility and enhance water holding capacity. As per plan, it was envisaged that in each tank one vermi compost pit was to be maintained as a demo, so that the farmers will get exposed benefit by observing and replicating this practice.

Kisan Melas

Kisan mela is an opportunity to expose and orient the farmers on new seed varieties, bio-fertilizers, bio-pesticides and improved farm implements etc. It is a focused area of project intervention. The project aims to organize Kisan Melas on seasonal basis in the tank areas besides conducting two melas every year in each tank. Animal health care and animal vaccination services were also to be provided in the melas. The DPU and NSOs were supposed to coordinate with the line departments for participation and the lead farmer was to play an active role at the tank level in organizing the kisan melas. The inputs required for animal deworming and vaccination were to be procured from the department on payment of cost from the project. The vaccines which were not available with the department would be procured from the open market. Kisan Mela is one of the best agriculture extension methods to spread technology and awareness amongst vast number of stakeholders. The farmers normally prefer to participate in the kisan melas as experts and technologies to be demonstrated are both available at one place.

Exposure Visits

As per PIP, WUA members were to be taken on exposure visits both within and outside the state. Exposure visits help the farmers to improve knowledge on production related technologies practiced within the same agro-climatic regions. Exposure visits also help farmers to have firsthand interaction with practicing farmers, clear their doubts and improve their level of understanding to stimulate their desire for adoption of new techniques/practices.

The project planned to select three farmers every year from each tank for exposure visits within the state. Similarly, one farmer from about 25 per cent of total tanks is to be selected for outside state exposure visit. Each batch for visit outside of the state consists of 30 farmers, to expose them on improved agricultural technologies so as to replicate the same in their villages. Farmers are also to be taken for exposure visits within the state to explain the potential of the vegetable cultivation and motivate them to cultivate vegetables through interaction with practicing fellow vegetable farmers and sharing of experiences in production and marketing processes. Two visits per district per year for each batch with 30 farmers are to be organized during the five years of project implementation.

Vegetable cultivation

The objective of conducting demonstrations is to improve the productivity of the existing vegetable crops and to increase the area under vegetables. Vegetable nurseries / poly tunnels are to be promoted. These are to be linked to the demonstrations of vegetable cultivation. The criteria to be used for selection of tanks for vegetable promotion are proximity to the urban area and availability of water. About 10 per cent of tanks are to be selected for promoting vegetables towards diversification from paddy. It has been observed that the diversification from paddy to vegetables was not observed and vegetable nurseries/poly tunnels were not promoted in the project during the survey period. This aspect has to be taken care in future implementation of the project.

Seed production

The project plans to introduce seed production on a business mode. The major constraint found in the project tanks is the availability of the pure seed, at the right time. The strategy involves building the capacities of the command area farmer for seed production in their own village. The seed production activity is to be taken up in two ways depending on the crop. In case of the crops where high yielding varieties are popular, the breeder/foundation seed is to be supplied to the Commodity Interest Group (CIG) for demonstration and further multiplication. In case of the crops where hybrids are highly productive, introducing the seed companies for demonstrative production in the tank area as seed production area is to be planned.

It has been found from the MTR and FEPIA survey that there was no activity of seed production under the tanks selected for the study. These aspects are to be looked into for future implementation of the project by strengthening WUA linkages with Department of Agriculture for inclusion in seed village programme.

Custom hiring centers

Lack of access to the improved implements and inability to procure improved implements were the reasons for non-adoption of new implements in farming as well as non-adoption of the improved agronomic practices or technology. Project desires to provide improved implements at community level (WUA) with contribution from community. The WUAs are expected to manage the implements and earn service charges from their use.

Monitoring of demonstrations by community, neighboring WUAs, line department staff and DPU

As per PIP monitoring and evaluation, visits by community members, WUA members, agriculture coordinator from SO/NSO and other stakeholders are to be organized to review and monitor the ongoing progress, methodology and results. The opinions and suggestions from the stakeholders would be recorded in a fact sheet by the resource person for improving the demonstrations in future.

Farmers Adopting Improved Cultivation practices

Efforts were made by the Government of India to improve agriculture technology and transfer the same to the field through various extension techniques. One of them is technology transfer which assumed greater importance especially after the green revolution. The primary objective of this process is to create greater awareness on improved technology for various crops and convince the farmer to practice them in order to achieve higher productivity and higher income. In the present study, an attempt was made to analyze the same by comparing with the baseline in order to study changes in adoption of different cultivation practices being followed by the households in the selected project and control tank areas.

Livestock - Breed and Milk Improvement

Livestock plays a complementary and supplementary role in rural earning and provides livelihood especially to small, marginal farmers and it's a part and parcel of farming system. The main objective of the livestock component is to improve the productivity of livestock in an integrated manner to ensure increased contribution to family income. This was to be achieved by Breed Improvement through doorstep Artificial Insemination (AI), improved nutrition through fodder development and improved animal health care. In addition to this, the project planned for development of sheep and ram-lamb rearing activities for poor and vulnerable groups in the tank areas.

The main focus of the project is on providing technology support, management of inputs and establishing market linkages to maximize the returns through strengthening the existing Gopal Mitras and introduce new Gopal Mitras in the new dairy potential areas, organizing animal health care camps, creating facilities to access fodder seeds/saplings and trainings to cattle owners with the technical support of the Animal Husbandry department. After MTR evaluation, the programmes on livestock interventions were discontinued as progress was slow and to focus on fewer sub-components under ALSS. It was also felt that there were sufficient departmental programmes for farmers interested in animal husbandry. The data collected from household surveys were presented to represent the key performance indicator on animal husbandry.

Fisheries Development

Fisheries Development is an integral part of the project objectives for enhancing the economic returns from fishery in the tanks, with intervention in technology promotion, inputs, credit, linkages, post production management and extension. The project interventions are being implemented in close coordination with the Fisheries Department (FD) as per project implementation plan. The support from fisheries department is specifically in the areas of selection of tanks, planning, training support on technical aspects relating to production and technical backup for supervision and monitoring.

Fingerlings of Rohu, Kaatla and Mrigal were introduced in selected tanks. The practices adopted by the Fisheries Co-operative Societies were use of disease free fingerlings, missing of different varieties of fingerlings (Rohu, Katla and Mrigal). Application of lime, FYM, maintaining adequate population and supplementary feeding. The project has provision for co-option of two FCS representatives as special invitees for the WUA Management Committee meetings. This improves involvement of fisher folk in TIMP preparation water audit, crop planning and plans for fisheries development, as well resolve any conflict arising in view of the overlapping seasonality of operations related to agriculture and fish culture.

Technology introduction for production improvement includes promotion of fingerlings and provision of supplementary feed to fish. The project proposes both extensive and intensive approaches for fisheries development. But in practice the approach was same in all the selected tanks, i.e., fingerlings supply and subsidy for supplementary feeding.

Capacity building of fisher folk including orientation training on technical aspects and skill trainings was conducted in selected tanks. The members trained were six (6) per FCS and on an average three members were taken on exposure visits to fish production and marketing centre of Kaikalur, Bheemavaram and Kakinada. Film shows and folk shows were organized in the selected tanks. Women are mostly involved in local retail marketing of fish. Encouraging and supporting the retail marketing of fish by women results in direct improvement of their income levels. The project proposed to provide ice boxes to the women for retail marketing of fish and was procured at different selected fisheries tanks. The existing contractors' agents in the market were to be identified and made partners in the marketing process. They would be linked to the FCS on commission basis for smooth process of transportation and marketing.

Agricultural Marketing and Business Development

Marketing of the produce plays a very critical role in the farm economics and livelihoods. Lack of supply chain integration and lack of integration of the farmer with the other end of the supply chain, i.e., choice of the consumer are the core issues. There is a need to integrate the farmer with the supply chain and reduce the length of the supply chain and the best possible way is public-private partnership. The project aims to intervene in the agricultural marketing system of the tank area to enable the command area farmers to utilize emerging market opportunities. The agri-business component of the project envisages to bridge these gaps through various targeted interventions. The first step towards planning and introduction of the Agri business interventions under the project is to understand demand projections of various commodities, gain knowledge about existing and emerging opportunities, insights into different business models present and its suitability to the tank command area farmers, and identifying gaps in the present market system at village level. This will then lead to designing the implementation strategies for field level support and identification of the appropriate partners.

The project proposes different processes for development of agri-business: Identification of the Commodities and training to WUA properly and guiding/helping them in the identification of commodities for marketing, Value chain analysis and identification of gaps, Identification of Tanks for business development, Orientation training to all the WUA members, Exposure visits, Formation of CIGs, Training of CIG business committee and lead farmers, Preparation of Business plans, Preparation of Marketing calendar, Estimation of working capital requirement, Arranging Working capital and other basic requirements for procurement, and Execution of the Business. The project also plans to support business interventions in Collective Marketing, Market driven production (Seed production of Jowar, Bajra, Maize and Sunflower) and Specific crop production (Sweet sorghum and vegetables). The project also proposes to intervene in the identification of new business opportunities, updating market information (through Buyers sellers meets, Public Private Partnership building and Market intelligence), and Dissemination of market information to users.

In the selected tanks the CIG farmers/producers followed the following channel in order to sell their produce, i.e., paddy.

Product –> CIG. –> Rice mill –> Consumer

From these tanks all the CIGs were analyzed for value chain analysis of paddy and the results indicated that the product average sale price was Rs.1375/qtl. The final consumers' price was Rs.3100/qtl of rice after attending to value addition but from one qtl. The farmer obtained only 65 kgs of rice after milling. The average final sale price of rice was Rs.31/kg. Hence the total income realized from one quintal of paddy after value addition was Rs.2015. The producer's share in consumers' rupee through this agri-business was 68.24% and the same without CIG was 64.50%. Thus on an average the farmer realized Rs.75/quintal in the selected tanks where agri-business is being operated.

Conclusion

The participatory irrigation management process in the AP TS CBTMP project have shown that farmer participation in irrigation management is viable and relevant. Women farmers have been especially empowered to participate fully in irrigated agriculture. Significant impacts are seen in the field and various inputs to increase productivity are being provided in response to continuous adoption of participatory approaches.

References

1. External Monitoring & Evaluation Team reports of AP TS CBTMP. 2007- 2014
2. Participatory Hydrological Monitoring: An effective tool for community managed Ground Water. S V Govardhan Das. December 2000.

32 Gender Concerns Mainstreamed in Agricultural Extension – Learning from India

- *Dr. Geetha Kutty P.S.*

Introduction

A country that is signatory of the Convention on Elimination of All forms of Discrimination against Women (CEDAW), one of the major international human right treaties, is supposed to" (1) take into account the particular problems faced by rural women and the significant roles which rural women play in the economic survival of their families in rural areas--- (2) take all appropriate measures to eliminate discrimination against women in rural areas in order to ensure, on a basis of equality of men and women, that they participate in and benefit from rural development----, shall ensure such women the right: to participate in the elaboration and implementation of development planning--, to obtain all types of training and education to increase their technical proficiency; to organize self-help groups and co-operatives in order to obtain equal access to economic opportunities---; to have access to agricultural credit and loans, marketing facilities, appropriate technology and equal treatment in land and agrarian reform as well as in land resettlement scheme". It is easy to locate from the above clauses what kind of opportunities are available for the policy makers and agricultural R&D system of a country which is CEDAW signatory, towards the goal of alleviating gender inequalities in rural areas! Gender inequality in the farming sector has been a major development issue in India, which had signed the CEDAW quite early, in 1980. Hence, to start a reading on the scope of integrating gender concerns in agricultural extension, a perusal of the promises under the Article 14 of the CEDAW will be highly useful and hence the above quote. Could the policies and R&D interventions in the farming sector of India, utilize the clarity CEDAW had provided on rural women's human rights and address the gender inequalities in farming sector effectively? With this question, a glance through the trajectory of the various women empowerment and gender integration efforts undertaken in the farming sector of India is attempted in this paper.

Macro Level Policies and Commitments Towards Gender Equality

" Towards Equality" was the title of the Report of the Committee on the Status of Women in India (Sharma et al 1974) and it is considered as the historic benchmark of women's status in India, which had revealed the landscape of the gender based atrocities, discriminations and women's disadvantageous status in the country and provided useful guidelines for the formulation of social policies. Agriculture being the major livelihood for millions of men and women in the country, it is very important to note that, more than thirty five years down the line, the situation is not much different, particularly with respect to the inequalities under the strategic gender needs of the farming sector i.e.the gender equality issues in the accessibility and ownership of land- the basic and primary issue in the context of farm based livelihood - is still remaining unresolved for a lrge majority of women in the country. Land provides the critical productive resource, asset, and title deed for getting recognized as farmer and accessing all other productive resources including credit and water ;hence in accessing land most of the women in agriculture are still victims of gender discriminations in the country!

Apart from the CEDAW agreement, in the Beijing Declaration of 1995 also, as a part of the 4th International Conference on Women's Rights, India was a party and had agreed for gender mainstreaming and gender budgeting as strategies for attaining gender equality. The recently completed UN Combat on Millennium Development Goals also had urged the member countries to focus on curbing gender inequalities. As follow up on these efforts for women empowerment and gender equality initiated since the seventies and strengthened in the nineties by the various international agencies (UN, FAO, IFAD, IFPRI, World Bank, UN Women, etc), there were regional and country level units of such agencies also to initiate and monitor interventions towards gender equality. Thus, various international projects piloted through government and non- governmental and community based organizations in the country also had tried introduction of gender integration in rural development and agriculture in India. In this context, the observation of the UN Women Expert Group Committee on Gender Mainstreaming Approaches in Development Programming (Cohen et al.2013) is important and demand concerted attention -Although high-level global policies and corporate procedures on gender mainstreaming are in place, recent assessments suggested that broad international commitments to gender equality and its mainstreaming in all development policies and programmes have not translated into sustained development cooperation and scaled-up programme implementation at country level".

Gender Concerns in Agricultural Extension

To the men and women actors of the farming sector, the socio- economic, demographic, market and technological changes occurring in the internal and external environment of the farming sector present increasingly varied challenges as well as opportunities. One of the important outcomes of the socio economic transformation and consequent male out migration from rural areas is the increasing rate of feminization of agriculture. This has in fact increased the burden and exploitations, the rural women as livelihood earners are subjected to. Despite the fact that women are major partners of farming along with men, a large percentage of women in agriculture are still landless, not included as farmers and also deprived of the accessibility to farm resources and farmer support services.

Though the women are running farms and rural enterprises in the capacity as heads of the female headed households, co- farmers of male headed households, farm workers, farmer groups and entrepreneurs, most of the farmer organizations are functioning as male dominated, thus denying the women their opportunities as the decision makers of their livelihood issues. Together with the associated issue of the absence of land ownership and the non inclusion in farmer organizations, illiteracy, cultural barriers and technology divide, even now, most of the women farmers are not reached by the development functionaries; thus not accessing available information and extension support services as well. Mostly, the landless women groups, the new entrants in farming, are forced to take up hired land farming without proper documents and thus face the denial of subsidized inputs, low interest credit and crop insurance. These exclusions and inequalities are in fact critical deciding factors of their performance. The landless nature of the women farmers is the key reason causing a chain of deprivation for them, as the land ownership is the base for farmer registration, accessing credit and market, to avail technologies, insurance claims and inputs from development agencies (Geethakutty, 2011). Does the agricultural extension system in the country identified this issue face, and formulated a gender responsive definition of farmers to recognize the women engaged in varied types of farming as farmers?

The SOFA Report (FAO, 2011) has clearly highlighted how these gender discriminations are adversely affecting the rural women's livelihood and the global food security at large. The report indicates, "yield gap between men and women averages around 20–30 percent, and most research finds that the gap is due to differences in resource use. Bringing yields on the land farmed by women up to the levels achieved by men would increase agricultural output in developing countries between 2.5 and 4 percent. Increasing production by this amount could reduce the number of undernourished people in the world in the order of 12–17 percent". India has a population of 1210.6 million out of which 833.5 million are living in the rural sector. But only 14% is the GDP contribution of the agricultural sector in the country, though 49 % of the total population and 55 % of the working population (Census, 2011) are engaged in agriculture based livelihoods in the country. Though the impacts of accelerated growth of industry and services, globalization, climate change, rural urban divides and requirement of updated skill, resources, technologies and infrastructure facilities are often identified as reasons for the low economic share of the agricultural sector in the plan and policy documents, no efforts to correlate the low GDP contribution of agriculture with the gender divide that exists in the sector are observed (Geethakutty, 2015 a). The national level data generation on the economic impacts of the gender inequalities existing in the farming sector of the country and timely remedial action plan are of utmost importance in this context .A fully fledged gender disaggregated database system on the livelihoods of the farming sector is an essential requirement for effective gender budgeting in this context.

Women in Agricultural Extension Programmes

The Pilot project "Women in Agriculture" of the Department of Agriculture and Co-operation (DAC), Ministry of Agriculture(MoA), Govt of India implemented in selected states (TANWA (TamilNadu), WYTEP(Karnataka), TEWA (Odisha), MAPWA (Madhya Pradesh), Central Sector WIA (Kerala) TWA (Gujarat), ANTWA (Andhra Pradesh) etc., during nineties was the first of its kind in the history of agricultural extension programmes in the country, wherein women's role in agriculture was recognized and promoted. The programmes implemented under the Central Sector Scheme were pilot projects to test the potential of bringing special focus on the women in agriculture. Though evaluation studies of such pilot projects had reported success and significant positive impacts, the project on women in agriculture was not up scaled as a comprehensive field programme of the DAC. In some of the States, the programme had built up separate cadre of women extension officers as well to work among farm women, which was effective in overcoming cultural barriers. It was by then, the Ministry of Agriculture had declared gender mainstreaming as a national strategy of agricultural policy and implemented for farmer support during the 10th Five Year Plan. The standing guidelines for ensuring 30 percent participation of women in all central sector programmes (DAC, MoA 2011& 2014) and the gender budgeting approach of checking what percentage of funds and benefits are allocated for women are availed by women beneficiaries under each sector are some of the gender budget practices then introduced under MOA. The 2014-15 annual report of the DAC also reveals that the 30 percent quota for women beneficiaries of all Central Sector Programmes in Agriculture is the major gender mainstreaming strategy still in practice. In an operational environment, where the definition of farmers are not made gender responsive, can the all category of women in agriculture get included as beneficiary of this thirty percent quota? Moreover, the question of not insisting similar measures of gender mainstreaming in the various State level development plans of farming sector is also relevant here.

The provision of earmarking budget not less than thirty percent of funds available under budget to women was the approach of Women Component Plan (WCP) initiated under the Eight Five Year Plan. The introduction of the Women Component Plan as part of gender budgeting and its utilization in selected local governments was one area where the needs and problems of women in agriculture could gain some focus in the field. The entry and participation of women in local planning and grassroot level collectivization of the women are observed to be enabling the rural women to function as pressure groups in this regard in some states.

It is important to note that, Gender mainstreaming is "the process of assessing the implications for women and men of any planned action, including legislation, making women's as well as men's concerns and experiences integral dimensions in the design, implementation, monitoring and evaluation of policies and programmes in all political, economic and social spheres so that women and men benefit equally and inequality is not perpetuated. The ultimate goal is to achieve gender equality" (United Nations, 1997).The lack of a holistic approach to evaluate the benefits and costs of the existing women focused programmes and not using gender analysis and related methodologies on adopting the gender mainstreaming as the strategy of agricultural extension by the MOA had been reported in earlier review reports (Sujaya, 2006). Rigorous capacity building among the functionaries for gender analysis and gender responsive development planning is an essential pre requisite for gender mainstreaming and the system is yet to attain this competence which is also affecting effective policy translation (Geethakutty.2015.b). The relevance of building gender competence in the present functionaries through rigorous training and introducing gender perspective as mandatory component of the educational curriculum of farm courses in the nation for moulding gender sensitive human resource is getting reiterated here.

With the objective of mainstreaming gender in agricultural research and extension, the Indian Council of Agricultural Research initiated a National Centre for Women in Agriculture in 1998 (presently the Central Institute of Women in Agriculture (CIWA); but so far the institute has not identified the need of engendering agricultural curriculum as a means of effective gender mainstreaming in research and extension! In the same period itself, the Kerala Agricultural University had established its Centre for Gender Studies in Agriculture for engendering education, research and extension in farming sector and in collaboration with FAO (Rome), ICAR, Ministry of Agriculture, Govt of India and M.S Swaminathan Research Foundation (MSSRF) had organized a series of national and state level capacity building programmes gender analysis competencies for the scientists and extension personnel and developed resource materials on gender perspectives of undergraduate education in farm universities. As a result, in 2007 the Deans Committee had introduced gender perspectives as part of an introductory course of BSc(Ag) honors programme in India (Geethakutty,2015a) . But among the faculty of the 42 SAUs across the country the required capacity building on teaching on gender perspectives in farming sector is still awaited. The lack of gender analysis capacity among the scientists is a limiting factor which demands a comprehensive scheme of capacity building in the system right from the decision makers. The requirement of building research capacities on gender analysis and related methodologies among the social scientists of SAUs is to be recognized as a priority for effective gender mainstreaming in agricultural extension and attempted through introducing mandatory courses on gender analysis as part of post graduation.

The National Institute of Agricultural Extension Management (MANAGE) had introduced gender perspectives as a mandatory component of its training syllabus. Of late, the various State Institutes of Extension Management (SAMETI) are also observed including gender as a topic in the various courses offered. The Ministry of Agriculture, Govt. of India had initiated a National Gender Resource Centre (NGRC) to take care of gender integration in the sector. Apart from the extension activities of the MoA, the women empowerment focus adopted by the Krishi Vignan Kendras (KVK) under ICAR and the extension research and development activities of the State Agricultural Universities (SAU) have brought good number of location based women empowerment models across the country. Most of them were on technology promotion through training of women groups on food processing, farm technologies and SHG interventions. The focus introduced by these systems on the promotion of women friendly technologies for the women organized farm operations like mechanized paddy transplanter, winnower, seed drill, maize decorticator, cono weeder, garden tillers, new varieties of crops and animals is worth highlighting. Unfortunately, such success cases and models are not getting mainstreamed . Limited utilization of the scope of social science research with focus on gender analysis and data generation on the basic gender issues of the farming sector and extension functionaries only have happened. Periodic stock taking and mainstreaming of the extension models and field research observations of the SAUs and KVKs on women empowerment and gender integration into the National and State level agricultural programmes is a much wanted strategy. The CIWA, NAARM, MANAGE and NGRC like national agencies of the agricultural R&D have a greater role to be played in this context.

At this point, if a serious audit of the agricultural extension strategies and women focus adopted in various phases of planning in India is done, it may leave one to wonder whether the Agricultural Extension System in India has missed these opportunities of mainstreaming to learn from these models in the right perspective or not? If a review of the Agricultural Extension Programmes right from the Community Development Programmes and National Extension Services era of India from 1952 is attempted, it can be noticed that no special focus on women in these programmes was given initially; but later on isolated training components under the rural development programmes were implemented for the upliftment of the women with focus on activities of household involvement. All such programmes such as Applied Nutrition Programmes (ANP), Development of Women and Children in Rural Areas (DWCRA) and promotion of cottage industries with home management related avenues like tailoring, food processing, etc had placed a welfare approach of considering women as weaker section, confined to household management and not as major contributors in the farm management. Hence, it will not be incorrect to state that, till late eighties, no consideration of women in farming sector was observed to be given under the planned development programmes of the farming sector like High Yielding Variety Programmes (HYVP), Intensive Agricultural Development Programmes (IADP), Intensive Cattle Development Programmes (ICDP), Training and Visit System, etc. Then the basic assumption adopted in development strategies was that men are the major actors of farming sector and women are the managers of the household and hence there is no need of focus on the activities performed by women in the farming sector and improved provision of technologies, resources and support services among women. On comparing the gender inclusive efforts in agricultural sector with other development sectors like rural development, women and child development, etc one may have to doubt that even after establishing the successful impact of adopting gender perspective as a cross cutting competency of functionaries in research and extension in the farm sectors, the agricultural extension system per se in the country is still struggling to shrug off the stereotyped mindset of keeping water tight compartmentalization between men in farm and women in home !

Rural Development Programmes and Gender Integration

An interesting observation on the progress of women in agriculture in the country is that, they are mainly benefited from the women empowerment interventions of the departments like Rural Development, Women and Child Development, Panchayathi raj etc and not much from the Department of Agriculture. There are evidences from various parts of the country that changes could be brought in the lives of rural women through policies and programmes largely targeted on practical gender needs of rural women. Mostly, the components of such women oriented interventions were meeting the training needs and increased access of micro credit among women of low income groups. Women were capacitated to initiate their own organizations through self help group networks. The grassroot level political empowerment of women and opportunity for equal participation in the local level government was a remarkable achievement of strategic nature towards gender equality. The collective power and the leadership women could gain through SHG networks are empowering women's entry and effective participation as elected representatives in the three tier Panchayati raj system at least in some of the states. The cumulative accessibility which the rural women of poor rural families could gain through these opportunities – access to productive resources such as credit, hired land, training, market etc, have largely contributed to their increased participation in farming sector. In other words , the development achieved among rural women through their engagement in microcredit based self help group networks , increased political participation and livelihood opportunities have largely benefited the agriculture sector. Now one interesting twist is slowly evolving- more than , women depending on farming sector , the sustainability of farming sector is observed to be depending on women's participation in agriculture. The commercial level farming enterprises taken up on hired farm lands by the women farmer groups of the Kudumbashree Mission (Poverty Alleviation mission of Kerala) is an example in this context. About four lakhs women are reportedly operating joint farming enterprises under the Kudumbashree programme on lands hired from traditional farmers who have abandoned cultivation and kept the lands fallow. Another example of similar impact among rural women is of the Society for Elimination of Rural Poverty (SERP) in Andhra Pradesh (now the States of Andhra and Telangana). Such programmes of the Departments of Rural Development enable the rural women to emerge as farmer groups of their own crops, livestock, fisheries, poultry and other farming based enterprises in Assam, Odisha and Tamil Nadu (Geethakutty et al, 2011).

The grassroot level political empowerment of women has brought active participation and leadership of women in local planning and governance in some of the States like Kerala, Karnataka, Tamil Nadu etc. The trust and confidence women SHGs have created with banking agencies through timely pay back of micro credit has persuaded the banking sectors to come up with women friendly schemes like Joint Liability Groups (JLG) from NABARD which are providing credit support to landless also to do farming on leased land. The currently implemented Mahila Kissan Swasshaktheekaran Pariyojana (MKSP) implemented under the NRLM Scheme of the Department of Rural Development is another example of gender integration done in rural development programmes of the selected states. The Mahila Samakhya Programme of the Ministry of Women and Child Development (MWCD) has made wide impact in promoting landless women to take up farming in Andhra and Telangana through the Andhra Pradesh Mahila Society (APMS) . Most of the SHG members are participants of the Mahatma Gandhi National Employment Guarantee Scheme (MGNREGS) and reported to be making utilization of the MGNREGS component for the land development work of their group farming. An examination of the project framework applied in the MGREGS reveals a typical example of gender mainstreaming process. The promotion of duck farming among the poor women in Assam by introducing better varieties of duck and collective marketing was reported as a success story of technology promotion by the State Institute of Rural Development (SIRD), Assam under the Ministry of Rural Development (MoRD) (Geethakutty et al., 2011.). Thus , the outcomes of the women targeted programmes of the MoRD, viz., the self help group (SHG) promotion, the micro credit based poverty alleviation programmes of the Department of Rural Development among rural women under the SGSY and NRLM schemes, MGNREGS, etc are found contributing to their livelihood generation in the farming sector.

Thus, a comparison of the visible changes of women empowerment focus and the passive gender mainstreaming strategy of the Ministry of Agriculture (MoA) will be revealing the need of a combined strategy. To have a long term focus, gender mainstreaming may guide and be helpful, but the sector should keep the priority of comprehensive women exclusive and women targeted programmes as well. Hence, concerted attention of the policy makers and planners of the Ministry of Agriculture is invited here to make note of the fact that the current visibility of the women in agriculture is mainly due to interventions of the MoRD, MWCD and MoPR etc. The focus of women empowerment through science and technologies promoted by the Department of Science and Technology (DST) in the country deserves a special mention. A major chunk of the women focused projects taken up by non- governmental and governmental agencies with the support of DST and the Department of Biotechnology (DBT) had focus on transfer of technologies for the farm based livelihoods of rural women. The impacts of the DST projects on technology promotion and group based interventions of NGOs among rural women also are important in this context.

Implications and Wayford

The MoA seems to have missed the opportunity of introducing a combined approach of gender inclusion wherein programmes of women focus along with gender mainstreaming could be effectively implemented for gender equality in farming sector. Since,limiting to gender mainstreaming mainly to the thirty percent quota as the field level intervention the MoA was not able to bring in effective integration of gender concerns in agriculture and address gender inequalities; hence it is felt that revamping of gender integration in agricultural extension with amulti pronged approach is the need of the hour. Another opportunity the MoA can utilize is the potential linkage of convergence with the women development programmes of the various departments such as MoRD, MWCD etc., which is remaining as untapped. The Report of the Expert Group Committee of UNWomen (Cohen et al .2013) had examined the status of gender mainstreaming in development across the globe and observed "- the prominence of parallel cross-cutting issues and new modalities in development assistance and in sector policy-making, also called for - revisiting gender equality priorities and actions" While observing trend and shortfalls of relying on 'gender mainstreaming' as the means of gender equality programming , observations of the above Committee of UN women in this context were:

(i) Gender mainstreaming has been interpreted as making gender equality programming "everyone's business, by which gender considerations were diluted, became "invisible" resulted in ineffective gender-responsive national policies and strategic planning processes

(ii) the meaning of gender issues resulted in gender mainstreaming content being dependent on political will and power relations among bureaucrats rather than deriving from sound gender analysis,

(iii) The absence of accurate sex and age disaggregated data and weak monitoring and evaluation link in mainstreaming of gender issues in development programming were shortfalls "The conclusion of UNWomen committee is very relevant for the agricultural extension in India – "Adopt a more pragmatic, strategic and synergistic vision to gender equality programming".

References

1. CEDAW (1979) Convention on the Elimination of All forms of Discrimination against Women,WWW.un.org/womenwatch/daw/cedaw

2. Cohen Sylvie I, Neena Sachdeva, Sharon J. Taylor and Patricia Cortes (2013) Expert group meeting gender mainstreaming approaches in development programming: being strategic and achieving results in an evolving development context, UN System Coordination

Division, United Nations Entity for Gender Equality and Empowerment of Women, UN Women, New York .

3. Census, 2011. Population Census 2011. New Delhi, Government of India
4. Beijing Declaration- Fourth world Conference September,1995.www.un.org/womenwatch/daw/Beijing/index.html
5. DAC, MOA(2015) Annual Report 2014-15 of the Department of Agriculture, Govt. of India, Krishi Bhavan, New Delhi, pp126-134
6. FAO- SOFA Report (FAO, 2011) The State of Food and Agriculture-Women in Agriculture -Closing the Gender Gap for Development. http://www.fao.org/catalog/inter-e.htm
7. Geethakutty, 2011.Can government programmes enable women farmers to access credit and land ? Case Study presented in the IFAD –MSSRF Workshop on Gender Resource Book, 28 Nov-2nd December,2011, MSSRF, Chennai
8. Geethakutty, 2015.Gender Inclusive S&T and Agricultural Education in India in Vol 3. India Science and Technology, CSIR-NISTADS, Cambridge University Press, India
9. Geethakutty.2013. Lessons Learned from Advocating for a Women inclusive Agricultural Policy in Kerala State, India http:// www.g-fras.org/en/policy-compendium.html
10. Geethakutty.P.S, Jiju P Alex, R.M.Prasad. 2011. Case Studies on opportunities and Challenges of Women in agriculture', NCAP, ICAR Consultancy Report. New Delhi and CGSAFED, KAU
11. Sharma et al(1974 Sharma Kumud, C.P. Sujaya & Vina MazumdarTowards Gender Equality -Report of the Committee on the Status of Women in India (Edited) http://www.cwds.ac.in/towards-report-of-the-committee.htm
12. Sujaya. C.P. 2006. Climbing a Long Road-Women in Agriculture in India – Ten Years after Beijing. M.S. Swaminathan Research Foundation, Chennai pp144

33 Climate Change Adaption and Mitigation – Role of Extension

- AVR Kesava Rao, Suhas P Wani and Mukund D Patil

Introduction

Evidences over the past few decades show that significant changes in climate are taking place all over the world as a result of enhanced human activities in deforestation, emission of various greenhouse gases, and indiscriminate use of fossil fuels. Global atmospheric concentration of CO_2 has increased from pre-industrial level of 280 parts per million (ppm) to 400 ppm in 2014. Global projections indicate higher temperature of 1.5 to 4.5°C by the year 2050, as a result of enhanced greenhouse gases. Climate change predictions for India indicate that warming is likely to be above the global mean and fewer very cold days are very likely. Frequency of intense rainfall events and winds associated with tropical cyclones are likely to increase.

The global average surface temperature in 2015 broke all previous records by a strikingly wide margin, at 0.76 ± 0.1°C above the 1961-1990 average. For the first time on record, temperatures in 2015 were about 1°C above the pre-industrial era, according to a consolidated analysis from the World Meteorological Organization (WMO, 2016).

Under the threat of increased greenhouse gases and resultant higher temperatures and uncertainty in rainfall regimes, there is a critical need to communicate climate change scenarios, adaptation and mitigation strategies to all stakeholders particularly farmers and agricultural extension personnel to enhance resilience and also to reduce greenhouse gas emissions.

Climate Variability and Change in India

Various studies show that climate change in India is real and it is one of the major challenges faced by Indian Agriculture, more so in the semi-arid tropics (SAT) of the country. India ranks first among the countries that practice rainfed agriculture in terms of both extent and value of production. Rainfed agriculture is practiced under a wide variety of soil types, agro-climatic and rainfall conditions. Rainfed agriculture supports nearly

40% of India's estimated population of 1.21 billion in 2011 (Sharma, 2011). The rainfed agro-ecologies cover about 60 per cent of the net sown area of 141 million ha and are widely distributed in the country (DOAC, 2011). Even after achieving the full irrigation potential, nearly 50% of the net cultivated area may remain dependent on rainfall. Changes in climate would affect agriculture directly through abiotic stresses and indirectly through biotic stresses.

Climate change is seen as changes in temperature, increased variability in rainfall, enhanced carbon dioxide concentrations. Climate change is likely to make changes in the length of the rainfed crop-growing period. Rainfed agriculture in India plays a crucial role in ensuring food security for the larger and poorer segment of the population but often it coincides with a high incidence of poverty and malnutrition. Reduction in yields due to climate change is likely to be more prominent in rainfed agriculture and under limited water supply situations. Crop yields in dryland areas of the country are quite low (1-1.5 t ha^{-1}) which are lower by two to five folds of the yields from researchers' managed plots (Bhatia et al., 2006). Current rainwater use efficiency in dryland agriculture varies between 35-45% and vast potential of rainfed agriculture could be unlocked by using available scientific technologies including improved cultivars.

The International Crops Research Institute for the Semi-Arid Tropics (ICRISAT) is one of the 15 Future Harvest Centers of the Consultative Group on International Agricultural Research (CGIAR). It recognizes that opportunities for sustainable productivity increases in the SAT will be firmly anchored on Integrated Genetic and Natural Resource Management (IGNRM) strategies, improved input-output market delivery systems for agricultural produce, and knowledge dissemination through capacity building.

Due to anthropogenic activities, a steady increase in atmospheric turbidity is observed in India. Indian annual mean (average of maximum and minimum), maximum and minimum temperatures showed significant warming trends of 0.51, 0.72 and 0.27 °C 100 y^{-1}, respectively, during the period 1901–2007 (Kothawale et al., 2010). However, accelerated warming was observed in the period 1971–2007, mainly due to intense warming in the recent decade 1998–2007. Mean annual temperature of India in 2010 was + 0.93°C above the 1961-1990 average and the India Meteorological Department (IMD) declared that 2010 was the warmest year on record since 1901. Mean temperature in the pre-monsoon season (March-May) was 1.8°C above normal during the year 2010.

At the country scale, no long-term trend in the southwest monsoon rainfall was observed, although an increasing trend in intense rainfall events was reported. Goswami et al., (2006) analysed gridded rainfall data for the period 1951-2000 and found significant rising trends in the frequency and the magnitude of extreme rain events, and a significant decreasing trend in the frequency of moderate events over central India during the monsoon seasons. The seasonal mean rainfall does not show a significant trend, because the contribution from increasing heavy events is offset by decreasing moderate events. They concluded that a substantial increase in hazards related to heavy rain is expected over central India in the future. Increased frequency and intensity of extreme weather events in the past 15 years were reported (Samra et al., 2003 and 2006).

A study carried out by ICRISAT under the National Initiative on Climate Resilient Agriculture (NICRA) project described a net reduction in the dry sub-humid area (10.7 m ha) in the country, of which about 5.1 Million ha (47%) shifted towards the drier side and about 5.6 Million ha (53%) became wetter, comparing the periods 1971-1990 and 1991-2004 (Kesava Rao et al., 2013a). Results for Madhya Pradesh have shown the largest increase in semi-arid area (about 3.82 Million ha) followed by Bihar (2.66 Million ha) and Uttar Pradesh (1.57 Million ha). Relatively little changes occurred in AP; semi-arid areas decreased by 0.24 Million ha, which were shifted to both towards drier side (0.13 Million ha under arid type) and wetter side (0.11 Million ha under dry sub-humid type). Results indicated that dryness and wetness are increasing in different parts of the country in the place of moderate climates existing earlier in these regions.

Climate Change Impacts on Agriculture

Due to global warming, length of the growing period (LGP) is likely to increase, however due to increase in day and night temperatures, physiological development is accelerated resulting in hastened maturation and reduced yields. Increased nighttime respiration may also reduce potential yields. With global climate change, rainfall variability is expected to further increase. When decrease in rainfall coupled with higher atmospheric requirements due to elevated temperatures, the LGP is likely to shorten. At Nemmikal watershed in the Nalgonda district of Telangana, the LGP has decreased by about 15 days since 1978 and the climate has shifted to more aridity from semi-arid (Wani et al., 2012). Shift in the length of growing period, if not understood by the farmers, generally results in more crop failures due to late season drought (Fig. 33.1). Present popular varieties of maize and pigeonpea are likely to produce lower yields more often in future.

In the Eastern Dry Agroclimatic Zone of Karnataka (consisting of Bangalore and Kolar districts and parts of Tumkur district), there is a perceptible shift in rainfall pattern from July to August and also from September to October (Rajegowda *et al.,* 2000). If sowing is done in July, crops would suffer from moisture stress due to the reduction in rainfall during September and also the crop grown would be caught in October rains causing considerable loss in the grain yield. Thus, sowing of crops (long duration variety crops of about 115 days) could be done during August preparing the land using June and July rains. In the years of early onset of southwest monsoon, sowing can be recommended during last week of July also. Crops sown during August would reach the grand growth period during October. As October receives higher rainfall the crop in its grand-growth period would not suffer for want of moisture and higher crop yields are expected.

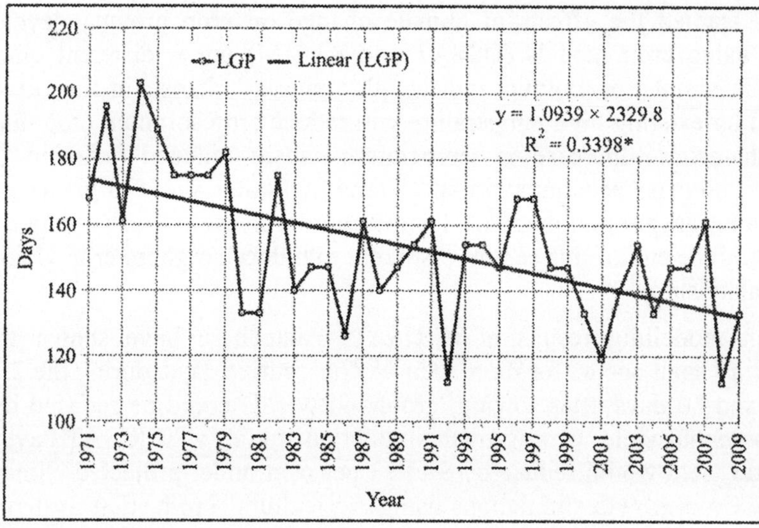

Fig. 33.1 Shift in Length of Rainfed Growing Period at Nemmikal, Nalgonda district

Rise in the mean temperature above a threshold level will cause a reduction in agricultural yields. A change in the minimum temperature is more crucial than a change in the maximum temperature. Grain yield of rice, for example, declined by 10% for each 1 °C increase in the growing season minimum temperature above 32 °C (Pathak et al., 2003). Climate change impact on the productivity of rice in Punjab (India) has shown that with all other climatic variables remaining constant, temperature increases of 1 °C, 2 °C and 3 °C, would reduce the rice grain yields by 5.4%, 7.4% and 25.1%, respectively (Aggarwal et al., 2009).

Field experiments and lab analyses were conducted at IARI, New Delhi in 2005, with five high-yielding rice varieties including aromatic and non-aromatic types, exposed to twelve different diurnal temperature (day/night) and radiation regimes to ascertain the impact of diurnal temperature and radiation changes on yield and yield components of aromatic and non-aromatic rice varieties in the field conditions and to document their effect on the grain and seed quality. Salient results indicate that the grain yield of all the five varieties was most significantly influenced by MNT ($P < 0.001$), followed by radiation ($P < 0.001$), explaining 87% and 77% of the yield variation respectively (Anand et al., 2015). Highest yields were recorded around a very narrow optimum temperature of 23°C to 24°C, with subsequent increase in temperature even by 1°C or 2°C, significantly reducing the grain yield.

Surface air temperature and diurnal temperature ranges are likely to increase along the high-ranges of the Western Ghats region of India and under such conditions; there is a threat to thermo-sensitive crops like black pepper, cardamom, tea, coffee, cashew and other plantation crops.

ICRISAT studied the effects of climate change on crop growth, development and productivity using crop models (DSSAT and APSIM) under different climate change scenarios. The simulation outputs indicate that climate change in the dryland regions characterized by existing high temperature will reduce crop sorghum crop duration by 15 days in Maharashtra. Increase in temperature causes reduced radiation interception, harvest index, biomass accumulation and increasing water stress in plants as a result of increased evapotranspiration demand. Temperature increase of 3.3°C, which is expected to take place by the end of this century, is likely to reduce sorghum crop yields by 27% at Parbhani, Maharashtra.

Groundnut modelling results at ICRISAT, Patancheru have shown that in SAT Alfisols, if a dry spell occurs for more than 15 consecutive days, during the 25-day period between 35 and 60 days after sowing, groundnut yields could be reduced by 35-38 per cent of the potential yield. Crop-growth simulation studies at ICRISAT have shown that groundnut pod yield would reduce by 9 to 13 per cent under projected climate scenarios (Table 33.1). Crop growth simulations using Agricultural Production Systems Simulator (APSIM) showed that in Gulbarga, Karnataka increase in temperature by 2°C could reduce pigeonpea yields by about 16% (Kesava Rao et al., 2013b). Rainfall decrease of 10% from present coupled with 2°C increase in temperature could reduce yields further by 4%, making the total reduction to be at 20%. Crop-growth simulationhas shown that pigeonpea yields would reduce by about 11 per cent under projected climate scenarios. In both crops, more runoff is likely which will lead to more soil erosion and nutrient loss.

Table 33.1 Impacts of projected climate on groundnut yields at ICRISAT

CC Scenario	Pod / Seed yield ($kg\ ha^{-1}$)	Change in Pod / Seed yield (%)	Total Dry Matter production ($kg\ ha^{-1}$)
Groundnut on Alfisols			
Current	2000	–	5430
HadGEM2-ES	1820	– 9	5410
GFDL-CM3	1830	– 9	5350
CNRM-CM5	1750	– 13	5250

At Guna, Madhya Pradesh, grain yield of soybean would reduce by 14% with increase in temperature by 2 °C and further reduce by 5% when coupled with reduced rainfall of 20%. At Guna, wheat crop duration would reduce by 10 days with increase in temperature of 2 °C. The increase in temperature by 2 °C would reduce the grain yield by 29%.

Climate change affects dynamics and interaction among species and it will affect and change the pattern of pest damage and pest control strategies. Increase in temperature may increase the need for application of pesticides and may reduce pesticide effectiveness and increase residues. Using the 'Rice FACE' facility in northern Japan,

Kobayashi et al., (2006) studied the effect of 200–280 ppm above-ambient CO_2 on rice blast and sheath blight disease for three seasons. Severity of leaf blast (*Magnaportheoryzae*) was consistently higher at the elevated CO_2 levels in all the three years assessed at two different stages of rice growth. Global warming will lead to earlier infestation by *Helicoverpa armigera* (Hub.) in North India (Sharma, 2010), resulting in increased crop loss. Rising temperatures are likely to result in availability of new niches for insect pests. Climate change is likely to make sleeper weeds to become invasive and favours expansion of weeds into higher latitudes and altitudes.

Pest-warning systems are key elements of Integrated Pest Management (IPM) efforts to reduce excessive use of chemical pesticides. The five components of an IPM program are prevention, monitoring, correct disease and pest diagnosis, development and use of acceptable thresholds, and optimum selection of management tools (Das et al., 2011). The management strategies available include genetic control, cultural control, biological control, and chemical control. Weather based pest and disease forewarning systems help farmers to avoid the risk of outbreaks of economically damaging crop pests and diseases by applying pesticides and fungicides, only when it is absolutely essential.

Climate-smart Agriculture (CSA)

Climate-smart agriculture (CSA), as defined and presented by the FAO at the Hague Conference on Agriculture, Food Security and Climate Change in 2010, contributes to the achievement of sustainable development goals. Climate-smart agriculture is a way to achieve short-and-long-term agricultural development priorities in the face of climate change and serves as a bridge to other development priorities. It seeks to support countries and other actors in securing the necessary policy, technical and financial conditions to enable them to:

1. Sustainably increase agricultural productivity and incomes in order to meet national food security and development goals.
2. Build resilience and the capacity of agricultural and food systems to adapt to climate change.
3. Seek opportunities to mitigate emissions of greenhouse gases and increase carbon sequestration.

These three conditions (food security, adaptation and mitigation) are referred to as the "triple win" of climate-smart agriculture. Climate-smart agriculture includes practices and technologies that sustainably increase productivity, support farmers' adaptation to climate change, and reduce levels of greenhouse gases. Climate-smart approaches can include many diverse components from farm-level techniques to policy and finance mechanisms.

Climate adaptation refers to the ability of a system to adjust to climate change (including climate variability and extremes) to moderate potential damage, to take advantage of opportunities, or to cope with the consequences. Adaptation to climate change refers to adjustment in natural or human systems in response to actual or expected climatic stimuli or their effects, which moderates harm or exploits beneficial opportunities. Various types of adaptation can be distinguished, including anticipatory and reactive adaptation, private and public adaptation, and autonomous and planned adaptation.

Climate Adaptation

Adaptation strategies need to be identified properly for increasing resilience of agricultural production to climate change. Several improved agricultural practices are evolved over time in various regions of the country. Management practices that are being followed under conditions of weather aberrations could also become potential adaptation strategies for climate change.

Resilience to climate change requires identifying climate smart crops and management practices and degree of awareness of community. Intercropping with grain legumes is one of the key strategies to improve productivity and sustainability of rainfed agriculture. Productive intercropping options identified to intensify and diversify rainfed cropping systems are:

- Groundnut with maize
- Pigeonpea with maize
- Pigeonpea with soybean

Some of the other initiatives are ridge planting systems; seed treatment; Integrated Pest Management (IPM); adoption of improved crop varieties and production technologies; promoting community-based seed production groups and market linkages. Farmers need to be encouraged to practice seed treatment with Trichoderma spp. and fungicides for managing seedling diseases and IPM options for controlling pod borer in chickpea and pigeonpea. Improved water use efficiency through IWM is the key in rainfed agriculture. Alternative sources of irrigation water are the carefully planned reuse of municipal wastewater and drainage water.

Climate Mitigation

Strategies for mitigating methane emission from rice cultivation could be alteration in water management, particularly promoting mid-season aeration by short-term drainage; improving organic matter management by promoting aerobic degradation through composting or incorporating it into soil during off-season drained period; use of rice cultivars with few unproductive tillers, high root oxidative activity and high harvest index; and application of fermented manures like biogas slurry in place of unfermented farmyard manure.

Methane emission from ruminants can be reduced by altering the feed composition, either to reduce the percentage which is converted into methane or to improve the milk and meat yield. The most efficient management practice to reduce nitrous oxide emission is site-specific, efficient nutrient management. The emission could also be reduced by nitrification inhibitors such as nitrapyrin and dicyandiamide (DCD).

Direct Seeded Rice (DSR) is an alternative method that can reduce the labour and irrigation water requirements. In the face of increasing population and growing demand for food, the upgrading of rainfed areas through DSR can help in soil and water conservation and deal with risks arising from climate change. Conservation agriculture technology helps to cope up with climate change impacts.

Legume-based systems are more sustainable than cereal only systems on Vertisols. Several soil and crop management practices affect C sequestration in the soil. Among them, conservation tillage, regular application of organic matter at high rates, integrated nutrient management, restoration of eroded soils, and soil and water conservation practices have a relatively high potential for sequestering C and enhancing and restoring soil fertility in the longer-term.

Leaf Colour Chart (LCC) is an easy-to-use and inexpensive tool for determining nitrogen status in plants. Use of the LCC promotes timely and efficient use of N fertilizer in rice and wheat to save costly fertilizer and minimize the fertilizer related pollution of surface water and groundwater. It is a promising eco-friendly and inexpensive tool in the hands of the farmers.

Renewable energy and farming are a winning combination. Wind, solar, and biomass energy can be harvested forever. Among various renewable sources of energy, biomass, which is produced right in the villages, offers ample scope for its efficient use to carry out domestic, production agriculture, livestock rising and agro-processing activities through thermal and bio-conversion routes. Usage of solar energy is slowly increasing in rural India for solar cookers for cooking, solar drier for drying agriculture produce, solar water heaters and solar photovoltaic systems for pumping devices which are used for irrigation and drinking water.Farmers can lease land to wind developers, use the wind to generate power for their farms, or become wind power producers themselves.

ICRISAT's Hypothesis of Hope to Address Climate Variability and Change

ICRISAT's research findings showed that Integrated Genetic and Natural Resources Management (IGNRM) through participatory watershed management is the key for improving rural livelihoods in the SAT (Wani et al., 2002, 2003 and 2011). Even under a climate change regime, crop yield gaps can still be significantly narrowed down with improved management practices and using Germplasm adapted for warmer temperatures (Wani et al., 2003, 2009 and Cooper et al., 2009). Some of the climate resilient crops are short-duration chickpea cultivars ICC 96029 (Super early), ICCV 2 (Extra-early) and KAK 2 (Early maturing); wilt resistant pigeonpea hybrid (ICPH 2671) with a potential to give 80% higher yields than traditional varieties and short-duration groundnut cultivar ICGV 91114 that escapes terminal drought.

Integrated Watershed Management (IWM) comprises improvement of land and water management, integrated nutrient management including application of micronutrients, improved varieties and integrated ·pest and disease management for substantial productivity gains and economic returns by farmers. The goal of watershed management is to improve livelihood security by mitigating the negative effects of climatic variability while protecting or enhancing the sustainability of the environment and the agricultural resource base. Greater resilience of crop income in Kothapally, Ranga Reddy district (Telangana) during the drought year 2002 was indeed due to watershed interventions. While the share of crops in household income declined from 44% to 12% in the non-watershed project villages, crop income remained largely unchanged from 36% to 37% in the watershed village (Wani et al., 2009).

Agroclimatic analysis at watershed level (Rao AVRK et al., 2008) coupled with crop-simulation models, and better seasonal and medium duration weather forecasts, help build resilience to climate variability / change. Farmers need to be encouraged to enhance soil quality and fertility through composting of organic wastes; and to promote cultivation of Leucaena, *Hardwickiabianta* and Glyricidia on farm bunds. Governments may also consider promoting and incentivizing the soil and water conservation measures taken by farmers. An improved agromet advisory service at the local level along with associated weather insurance packages is a sure way to enhance the resilience of poor farmers in the context of climate change. Policy interventions are needed to mitigate the climate change effects and Governments have to be proactive in developing adaptation strategies for those sectors like agriculture, water resources, forestry and biodiversity which are highly exposed to the future climate changes and have a significant impact on livelihoods.

ICRISAT's principle to improve the livelihoods of small-holder farmers even under future climate change scenario is built on the concept of Inclusive Market Oriented Development (IMDO), which is a Dynamic Development Pathway consisting of innovative environment, inclusive and market oriented.

Innovative Extension Systems for Climate Resilient Agriculture

Climate change adaptation involves adjustments to decrease the vulnerability of agriculture to current climate variability and to future changes. Farmers have traditional knowledge on sowing and harvest times and crop management practices. Farmers may not have the necessary information of possible climate adaptation options without an effective network of extension services that can filter knowledge gained through science to grass roots. In addition, it is necessary that farmers possess the necessary skills to implement an alternative production technique. Under the climate change scenario, due to uncertainty in rainfall conditions and occurrence of extreme weather, farmers need to be supported with both climate and weather information for sustainable crop production.

Agriculture extension system plays very important role in enhancing the knowledge and skills of farmers for improving agricultural productivity as lack of awareness among farmers about good agricultural practices is always been a key limiting factor for improving productivity levels. Thus, there is a clear and distinct role for strengthening extension services in agriculture to enhance farmer awareness of potential adaptation response options. Agricultural extension personnel are the main stakeholders to communicate with the farmers on how to cope with climate change through adaptation strategies. There is a need to develop appropriate training modules for the agricultural extension staff on the science of climate change and the various adaptation and mitigation strategies available in the universities, research institutes and the government departments. In this way, extension personnel will be acquainted with knowledge on climate change. Seminars / Workshops need to be organized frequently on climate change and extension personnel should be given the privilege to attend so that they can acquire required skills to help farmers.

The other dimension of extension system is knowledge delivery pathways (KDP). The tradition ways to delivery information are through announcements, info-graphics (wall writing or banners), and scheduled programs on television and radio, which still are effective option for mass communication. Often, farmers require information about weather, good agricultural practices, insect/pest identification and their control, where to purchase the input and where to sell the produce. But, this information should come at the time when it is actually needed, for which the traditional KDP are inadequate to provide this solution. Therefore, it is important to rejuvenate the existing Agricultural Extension Systems with innovative ICT models for knowledge generation and dissemination to make them truly innovative.

Information communication tools have provided the wide range of options to assist extension agents as well as farmer for getting up-to-date knowledge. Government of India and private companies are transforming the AES. There are several technologies are being used for information dissemination. For example, Government of India has Kisan Call Center (KCC) facility to satisfy information request as per farmers demand in 22 local languages. Karnataka State Natural Disaster Management Center is provider service to its subscribers in state to receive daily weather update including alert about abnormalities in weather. In addition to Government, private companies are also providing innovative solutions for agriculture extension. For example, IFFCO Kisan Sanchar Limited has introduced voice messages for agro-advisory system and Thomson Reuters introduced mobile based integrated agro-advisory system 'Reuters Market Light'. Information updates obtained from such advisory system allow farmers to take decision regarding various farm operations, which eventually help farmers to cope to climate variability and change.

Thus, the key strategies required for climate smart agriculture include training of extension staff to acquire the new knowledge and skills in climate risk management, setting up of emergency management unit by extension agencies, dissemination of innovations strategic research on best practices and building resilience capacities of vulnerable people in climate risk management, providing feedback to government and interested agencies with situation reports on various causes of climate change and its effects (Iqbal Singh and Jagdish Grover, 2013). Adaptation to climate variability and change must become an important policy priority to the Government and effectively be mainstreamed into national, provincial, local and sectoral development agendas (IPCC, 2011). There is a clear and distinct role for strengthening extension services in agriculture to enhance farmer awareness of potential adaptation response options. For an effective extension system a combination of on-field demonstration through farmers' participation, dissemination of results through conventional delivery pathways, and advance ICT for faster and interactive service is required so that gap between available information on climate change mitigation option and farmer impacted by climate change will be reduced. Enabling the farmers on the options to adapt also requires that other factors first be in place. In particular, investment in institutional support to promote the dissemination of knowledge through extension is important (Kurukulasuriya and Rosenthal, 2013).

Paris Agreement to Combat Climate Change

The United Nations Framework Convention on Climate Change came to a landmark agreement on December 12, 2015 in Paris. The Paris Agreement was signed by 196 nations and is the first comprehensive global treaty to combat climate change, and will follow on from the Kyoto Protocol when it ends in 2020. It will enter into force once it is ratified by at least 55 countries, covering at least 55% of global greenhouse gas emissions.

The agreement commits nations to keep temperatures well below 2 °C above the pre-industrial levels and to pursue efforts to limit the temperature increase to 1.5 °C. The French foreign minister Laurent Fabius, President of the COP21 summit has stated that "This agreement is differentiated, fair, durable, dynamic, balanced, and legally binding".

In order to create more vibrant and resilient communities and natural resources-dependent economic sectors able to mitigate and adapt to the risks associated with climate variability and change, there is a critical need to develop a "climate-literate society." People must gain a better appreciation of how a changing climate is likely to impact their own lives, local ecosystems, regional industries and society at large (Sea Grant Climate Extension Summit Report, 2013).

Let us join together to strengthen the Extension System to help manage the risks posed by climate variability and change for achieving sustainable agricultural production and better livelihoods.

References

1. Aggarwal PK, Singh AK, Samra JS, Singh G, Gogoi AK, Rao GGSN and Ramakrishna YS 2009 Introduction. In *Global Climate Change and Indian Agriculture*, Ed: PK Aggarwal, ICAR, New Delhi, pp. 1-5.
2. Bhatia VS, Singh Piara, Wani SP, Kesava Rao AVR and Srinivas K. 2006. Yield Gap Analysis of Soybean, Groundnut, Pigeonpea and Chickpea in India Using Simulation Modeling. Global Theme on Agroecosystems Report no. 31. Patancheru 502 324, Andhra Pradesh, India: International Crops Research Institute for the Semi-Arid Tropics (ICRISAT). 156 pp.
3. Cooper P, Rao KPC, Singh P, Dimes J, Traore PS, Rao AVRK, Dixit P and Twomlow SJ. 2009. Farming with current and future climate risk: Advancing a "Hypothesis of Hope" for rainfed agriculture in the semi-arid tropics. Journal of SAT Agricultural Research 7.An open access journal published by ICRISAT, Patancheru.
4. Das DK, Jitendra Singh and Vennila S. 2011.Emerging Crop Pest Scenario under the Impact of Climate Change – A Brief Review.Journal of Agricultural Physics. Vol. 11, pp. 13-20.
5. DOAC. 2011. Annual Report 2010-11. Department of Agriculture and Cooperation, Ministry of Agriculture, Government of India.http://agricoop.nic.in/Annual report2010-11/AR.pdf[accessed: 08 February 2016].
6. Goswami BN, Venugopal V, Sengupta D, Madhusoodanan MS and Xavier PK. 2006.Increasing trend of extreme rain events over India in a warming environment.*Science* 314: 1442–1445.
7. Iqbal Singh and Jagdish Grover. 2013. Role of extension agencies in climate change related adaptation strategies. *International Journal of Farm Sciences* 3(1): 144-155.
8. IPCC (Intergovernmental Panel on Climate Change). 2011. Managing the risks of extreme events and disasters to advance climate change adaptation, A special report on working group I and working group II of the intergovernmental panel on climate change. http://www.ipcc.ch/ipccreports/ar4-syr.htm [accessed: 08 February 2016].
9. Kesava Rao AVR, Suhas P Wani, KK Singh, M Irshad Ahmed, K Srinivas, Snehal D Bairagi and O Ramadevi. 2013a. Increased arid and semi-arid areas in India with associated shifts during 1971-2004. Journal of Agrometeorology 15 (1): 11- 18 (June 2013).
10. Kesava Rao AVR, Suhas P Wani, K Srinivas, Pushparaj Singh, Snehal D Bairagi and O Ramadevi. 2013b. Assessing impacts of projected climate on pigeonpea crop at Gulbarga. Journal of Agrometeorology, 15 (Special Issue II):32-37 (December 2013)
11. Kobayashi T, Ishiguro K, Nakajima T, Kim HY, Okada M and Kobayashi K. 2006. Effects of elevated atmospheric CO_2 concentration on the infection of rice blast and sheath blight. *Phytopathology*, **96**: 425-31.
12. Kothawale DR, Munot AA and Krishna Kumar K. 2010.Surface air temperature variability over India during 1901–2007, and its association with ENSO. Climate Research 42:89-104.

13. Kurukulasuriya, Pradeep; Rosenthal, Shane. 2013. Climate change and agriculture : a review of impacts and adaptations. Environment department papers; no.91.Climate change series. Washington DC ; World Bank.

14. Pathak H, Ladha JK, Aggarwal PK, Peng S, Das S, Yadvinder Singh, Bijay-Singh, Kamra SK, Mishra B, Sastri ASRAS, Aggarwal HP, Das DK and Gupta RK (2003) Climatic potential and on-farm yield trends of rice and wheat in the Indo-Gangetic Plains. *Field Crops Res.*, **80**(3): 223-234.

15. Rajegowda MB, Muralidhara KS, Murali NM and Ashok kumar TN. 2000.Rainfall Shift and its Influence on Crop Sowing Period.*Journal of Agrometeorology* 2(1), 2002.

16. Rao AVRK, Wani SP, Piara Singh, Rao GGSN, Rathore LS and Sreedevi TK. 2008. Agroclimatic assessment of watersheds for crop planning and water harvesting. Journal of Agrometeorology. Vol. 10, No. 1, 1-8.

17. Samra JS, Singh G and Ramakrishna YS. 2003. Cold wave of 2002-03: Impact on agriculture. NRM Division, ICAR, Krishi Anusandhan Bhawan-II, Pusa, New Delhi, India.49 pp.

18. Samra JS, Ramakrishna YS, Desai S, Subba Rao AVM, Rama Rao CAYVR, Rao GG SN, Victor US, Kumar Vijaya, Lawande KE, Srivastava KL and Krishna Prasad VSR. 2006. Impact of excess rains on yield, market availability and prices of onion. Central Research Institute for Dryland Agriculture (ICAR).52 pp.

19. Sea Grant Climate Extension Summit Report. 2013. The Role of Extension in Climate Adaptation in the United States. *Report from the* Land Grant - Sea Grant Climate Extension Summit. June 2013. http://seagrant.noaa.gov/Portals/0/ Documents/what_we_do/climate/SeaGrant_Climate Extension Summit Report.pdf [accessed: 09 February 2016].

20. Sharma, HC. 2010. Effect of climate change on IPM in grain legumes. In: *5th International Food Legumes Research Conference (IFLRC V), and the 7th European Conference on Grain Legumes (AEP VII),* 26 – 30 th April 2010, Anatalaya, Turkey.

21. Sharma KD. 2011. Rain-fed agriculture could meet the challenges of food security in India. Current Science 100 (11).

22. Suhas P Wani, William D Dar, Dileep K Guntuku, Kaushal K Garg and AVR Kesava Rao. 2012.Improved livelihoods and building resilience in the semi-arid tropics: science-led, knowledge-based watershed management, Climate Exchange, The Global Framework for Climate Services, World Meteorological Organization, October 2012, Pages 69-71.

23. Wani SP, Rego TJ, Pathak P and Singh Piara. 2002. Integrated Watershed Management for Sustaining Natural Resources in the SAT. Pages 227-236 in Proceedings of International Conference on Hydrology and Watershed Management.18-20 December 2002, Hyderabad, India, Jawaharlal Nehru Technological University (JNTU).

24. Wani SP, Singh HP, Sreedevi TK, Pathak P, Rego TJ, Shiferaw B and IyerShailaja Rama. 2003. Farmer-participatory integrated watershed management: Adarsha watershed, Kothapally India, An innovative and upscalable approach. A Case Study. Pages 123-147*in* Research towards integrated natural resources management: Examples of research problems, approaches and partnerships in action in the CGIAR (Harwood RR and Kassam

AH, eds.). Interim Science Council, Consultative Group on International Agricultural Research. Washington, DC, USA.

25. Wani SP, Sreedevi TK, Rockström J and Ramakrishna YS. 2009. Rainfed agriculture - Past trend and future prospects. Pages 1-35 *in*Rainfed agriculture: Unlocking the Potential. Comprehensive Assessment of Water Management in Agriculture Series (Wani SP, Rockström J and Oweis T, eds.). CAB International, Wallingford, UK.

26. Wani SP and Rockström J. 2011. Watershed development as a growth engine for sustainable development of dryland areas. Pages 35-52 *in* Integrated Watershed Management in Rainfed Agriculture (Wani Suhas P, Rockström J and Sahrawat KL, eds.). CRC Press, The Netherlands.

27. WMO. 2016. 2015 is hottest year on record. Press Release N° 2 dated 25 January 2016.https://www.wmo.int/media/content/2015-hottest-year-record[accessed: 08 February 2016].

Our Reminesances with Agricultural Extension

My Experiences with Agricultural Extension – Changing Lives and Livelihoods

Vijaya L. Mackrandilal
Formerly Senior Economist,
World Bank, Washington DC, USA

My first experience with agricultural extension happened without my awareness of it but it was a life changing experience for me, my family and countless other families in my home region and country.

I was born in the former British Guiana, a colony located in South America, but economically, politically and culturally we were part of the federation of the British West Indies. It is interesting to note that the only two surviving institutions of that federation are the University of the West Indies and the West Indies Cricket Team. As part of the federation, Guiana had a guaranteed market at preferential prices for its products. Rice, was a logical product for development, after sugar, given the abundant swamplands and rainfall . A rice research station was established at a place named Burma after the rice producing country of Burma, as it was then called. My family lived just a few miles away and were part of a development project launched to promote and support the high yielding products of the research station.

The Mahaicony-Abary Rice Development Scheme (MARDS), established during the second World War, was perhaps one of the earliest examples of what was later called Integrated Agricultural Development Projects by the World Bank . MARDS itself was part of a larger government sponsored rice development and export program with multiple economic, political and social objectives. MARDS provided a comprehensive package of supporting services . It started off with a research station and drainage and irrigation services which involved major infrastructural works and capital intensive operations and maintenance for the swamplands between two rivers, the Mahaicony and Abary. It then expanded into milling, credit, input supply, transportation and mechanized services.

Much has been written about the successes and failures of the program but rice remains a major export of Guyana (the new post independence name), the research station and MARDS continue to operate albeit with fluctuating levels of efficiency. The experience of the program offers some lessons which may be of relevance to agricultural programs today.

The first lessons are **organizational**. While political factors played a large part in the changing organizational arrangements, one feature was important for the sustainability of research and extension services over the years. The program by design or fortuitous events became **operationally self- funding and even profitable for several years** before it was undermined by hostile political factors. I do not have the details of the relationship between the research station and the services operations. In the minds of the local population, myself included, it was all one. The research findings were quickly scaled up on the fields surrounding the station and mill, the results were usually so impressive that adoption rates by farmers were between 80 and 100 % by the next season after each release of a new variety. Adoption rates in the rest of the country were somewhat slower but still impressive.

It is not clear exactly how the extension of research findings to farmers was carried out. There was no formal extension services in the early days . "Word of Mouth" seems to have been a powerful extension technique. The demonstration effect of the fields at Burma was also another powerful extension tool. Again by design or good fortune, the two main roads into my home region both passed by the research station and even city visitors would pause or stop to admire the fields of rice stretching for miles in all directions, beautifully green in the growing season, a rippling carpet of gold in the harvest season. Later, "Field or Demonstration days" were introduced mainly for mechanized services. The location of the rice mill right next to the research station also served as an extension tool. Farmers were able to observe and interact with the workers of the research station on a regular basis at no additional travelling or time costs.

This **close interaction** was another key organizational feature. The program evolved into one of the most integrated development projects I have ever encountered, while maintaining some public/private balance. The program was very similar to the cotton "baremes" of French West Africa, which were also highly successful until dismantled in the late 80s by the World Bank and its Benor style of national extension services which were later proven to be unsustainable within the socio economic conditions prevailing in West Africa. The integration of services and increased farmer interaction went hand in hand. The mill was first set up to process the paddy from the research station and demonstration fields. There were no roads for access to small farmer production along the two rivers, Mahaicony and Abary,

Those small farmers used small private mills along the rivers. Then the dams built for the drainage and irrigation scheme started being used as dry weather roads for increased farmer access to the research station but this was still limited since there were no mechanized means of road transportation while motorized barges on the rivers along with strong and predictable tides for row boats helped the small millers maintain their comparative advantage over the larger and more efficient mill at Burma which had no river access. This lasted only for a short while before the mechanization program integrated most farmers fully into the MARDS scheme. With the availability of tractors for both cultivation and transportation , an elaborate system of input distribution, credit and paddy collection was set up using farmers as sub-contractors working directly with MARDS.

The impressive yield improvements achieved by the program and the high adoption rates by farmers provided the foundation for a dramatic mechanization program. In less than twenty years there was a transformation of rice production from dependence on animal power and human labor to almost complete mechanization. Land availability was not a limiting factor since some of the original settler families had hundreds of acres of land unused or kept for grazing cattle. There were also more swamplands that could be re-claimed. At the peak of its success, the program was using small planes to sow seeds and spray the fields, a few large farmers were able to adopt even these practices through the rental program offered. The more basic aspects of the mechanization program – tractors, harvesters and supporting equipment – were supported by rental services from MARDS, other public support and very active private sector promotion through financing and hire-purchase arrangements. Other public support included concessions on taxes and custom duties on machinery, equipment and fuel as well as subsidies for the drainage and irrigation scheme.

Politics played a large part in the rise and fall of MARDS. The scheme was set up under the colonial administration to ensure rice supply during the Second World War when supplies from Asia were no longer reliable. Then from 1953 to 1964, the People's Progressive Party (PPP), actively supported MARDS and other rice development schemes partly because these farmers were an important part of its political base. The replacement of the PPP through the connivance of the British and American governments led to problems for the program and the rice industry as a whole.

The socio-economic benefits of the program are indisputable and persist to this day. MARDS had the most dramatic impact on incomes, education and quality of life that I have encountered, even after seventeen years in the World Bank . The higher yields, mechanization and expansion of holdings provided the basis for a complete socio-economic transformation not only in the immediate region but also in the rest of the country .

Higher incomes and mechanization facilitated a quantum leap in education. "Primary education became compulsory in 1876. Truancy, however, was common. The British planters and bureaucrats discouraged the education of the Indo-Guyanese indentured laborers. The government stated in 1904 that Indo-Guyanese should not be prosecuted if they objected on religious grounds to sending their daughters to school. Planters used this policy to discourage workers from sending their children to school. Not until 1933 was the Indo-Guyanese leadership successful in changing government policy."[1] But it was MARDS that transformed this policy into reality. Mechanization required literacy and training to repair and maintain machinery on farm, only MARDS had repair workshops. With higher incomes and lower labor needs, children could stay in school for much longer. In fact larger landowners kept only one son on the farm, other sons and girls were sent off for higher education. This set in place a trend that was copied elsewhere in the country, even where incomes were lower, and is continued even to this day. Of course, the downside is that education increases mobility and with erratic political and economic conditions, most of the Guyanese population now live outside the country.

Conclusions and Implications for Extension Services Elsewhere

Like all other successful programs I have encountered, such as the out-grower schemes of West and East Africa, there **was a cash crop with sufficient comparative advantage** to stimulate and sustain supporting credit, input supply and marketing services.

The research station was part of MARDS with sufficient land for demonstration effects as well as **income for sustainability**. Similar arrangements in the out-grower schemes and cotton programs of West Africa before they were dismantled by the World Bank and replaced with national research organizations with disastrous effects.

Demonstration effects , farmers like all consumers, need to be impressed, and there will always be leaders and followers in adoption, a fact that the World Bank chose to ignore in project design. All the successful research/extension programs I have encountered have impressive demonstration fields and large farmers with large fields.

The larger **and wealthier farmers** are better able to bear **the risks of early adoption** of new techniques, this may increase income disparities in the beginning, but over time these gaps closed in MARDS. Some of the farmer failures which I have encountered in other countries can be attributable to aggressive extension to farmers who were not sufficiently prepared for the risks.

Large, extended families played a part in the success of the program, in the beginning their labor availability was a factor, later they were better able to bear the risks of the inputs and mechanization. I have encountered similar examples in West Africa where larger family size and land holdings were criteria for inclusion in the out-grower schemes before the World Bank got involved and tried to improve income equality with inappropriate measures.

Close relationships between farmers and MARDS staff and scientists due to integration of all supporting services rather than the national or independent services recommended by donor agencies.

---

[1] http://www.photius.com/countries/guyana/society/guyana_society_education.html

Our Reminesances with Agricultural Extension

50 Years of my Journey with Agricultural Extension - Towards Changing Lives and Livelihoods

Prof. S.V. Reddy

My first exposure to Agricultural Extension concept and philosophy was during 1965 when I was final year student of BSc (Ag) at Bapatla College, Andhra Pradesh Agricultural University. The seeds of Agricultural Extension was sown in my mind by Dr. A. Adivi Reddy, then Professor and Head Extension division. In his teaching he used to give illustrations from epics and history of India to explain Philosophy of Agricultural Extension eg "Human beings have conquered the Himalayan mountains and hoisted a flag but they have not changed in their attitude to Life and service to humanity. Similarly a quote from Swami Vivekanda "Give me a few men and women who are pure and selfish and I shall shake the world" After quoting he used to relate it to Agricultural Extension. Similarly Dr. George Vasatha Rao used to teach Rural Sociology. This inspiration has dragged me to choose this profession for my specialisation where in I could understand the fight of rural people and development.

I have chosen this discipline during my master degree programme at Extension Education Institute, Hyderabad India. The teaching of Prof.Bhaskaram and Prof. C. Lakshmanna who were my friends, philosophers and guides has deeply impressed me to recognize the potential of the great discipline and plan for working with resource poor farmers and to research in to several aspects of Extension which has further motivated me to enter into Ph.D in a premier institute namely Indian Agricultural Research Institute (IARI). The teaching of my "Gurus" Dr. K.N. Singh, Dr. C.Prasad, Dr. S.N.Singh and Dr. B.N.Sahay has kindled new enthusiasm and contributed for widening my knowledge and skills. Late Dr. K.N. Singh used to say that "you cannot carry a atom bomb in a Bullock cart and try to win over the war" The extension methods , tools and techniques should also be equally sharp with the advancement of Agricultural technology. He added more efforts are needed for "Research in Extension" I came out of IARI with a flying colours getting 4.00 / 4.00 OGPA and a Gold Medal, which was awarded by Late Mrs. Indira Ghandi, the then Prime Minister of India in the presence of Father of Green Revolution, Dr. M.S. Swaminadha, the then Director of IARI.

I entered regular teaching position as an Assistant Professor (Educational Psychologist) at the same institute (EEI) where I did my MSc (Ag) in Extension Education Earlier I worked for an year at S.V.Agricultural Colloquies, Bapatla as Instructor and then went on leave to prosecute Ph.D. There was a big change from student to a teacher life. It was much more challenging because I was only Ph.D. then in the entire University in this discipline. Therefore I had a tremendous motivation to share new knowledge and skills with my students and colloques which has caused me good respect as a Professional. I had an opportunity to Guide 8 Ph.D Students and 10 MSc students as major advisor and several as minor advisor in a short period. EEI has also given me unique opportunity to train much of middle level extension workers from southern states, west Bengal and Orissa. It was a great exposure and learning experience. I was elevated as Professor and later as Head of EEI. During that tenure of 6 years I have introduced staff research programmes, Innovative Training Programmes and village adoption programme besides designing curriculum for several post graduate courses. I finally believed Agricultural Extension has to live in the villages and serve the rural poor. I had bit of ups and downs during this tenure because some of my colleagues could not see my meteotic raise in a very young age. With courage of conviction I had tied over the obstacles and emerged successfully. Hard work, tolerance and wisdom always pays.

During this time T&V system of Agricultural Extension was introduced National Wide and EEI has played a major role in conducting number of orientation and supportive Training Programmes under my leadership. For outstanding work done, I was chosen by Govt of India and World Bank to represent the country case study at Asian productivity council meeting at Changmai, Thailand (my first visit abroad) and given me opportunity to meet Daniel Benor, Architect of T&V system. He was impressed by my presentation and congratulated me saying “Young man representing a large country made intelligible impressions”. This appreciation has made my energy to rededicate and work furthermore for this discipline in general and farmers at large one fine morning, I got a call from Michel Baxtor, World Bank resident Mission, New Delhi asking me to apply for a Consultancy Position in World Bank programme as a Technical Advisor at Somalia, East Africa. This assignment has brought me changes in life by seeing the hunger and nutrition as well as despair among the people. My assignment has given me opportunity to coordinate among different donor agencies and convince them to divert some of the funding towards agricultural development. My knowledge of Agriculture Extension helped me to see farmers with empathy from psychological and sociological perspective to design programmes based on needs. Unfortunately I was evacuated amidst the civil war and was placed as consultant temporarily at New Delhi where in I was supervising IDA funding projects.

I was re-allocated to Uganda, East Africa as Consultant and Later as Agricultural Extension Services Advisor to the Ministry of Agriculture, Animal Industry and Fisheries (MAAIF). I had a long tenure of 7 years where in I was instrumental in designing Unified Agricultural Extension Programme under the able guidance of Dr. Daniel Benor, Dr. V.Venkateshan, Dr. Vijaya Macrandilal, Dr. Edquisimbing, Dr. Daniel Gustafson, and Dr

Rob Stowe. The present honable Minister for MAAIF Mr.Tress Buchayanandi was then the Director of Agriculture. This programme along with village level Participatory Approach, outsourcing, cost recovery concepts with involvement of state Governments has changed lives and Livelihoods of resource poor small scale farmers. Women empowerment and group dynamics was visible. During this time I had an opportunity to visit several African countries to advise and guide in Agricultural Extension. The first visit of Dr. James Wolfenson, the President, World Bank was arranged and he said this programme is a way forward for many developing countries. Similarly a great opportunity was to meet Dr. Normal Borlog, a Nobal Lamet in Agriculture. I had an opportunity to spend 10 days with him in the fields of Uganda meeting and interacting with farmers and farm women. A great learning experience which I cherished forever. I still remember Dr. Benor who is a champion of Sustainable Agriculture says "You can't wake up a lion and don't feed it". Agricultural Extension should have credibility otherwise farmers will revolt against us. I have learned a lot on Agricultural Extension Management with Dr.Benor. This opportunity has given me exposure to international extension. I can travel and see the Agricultural extension in action in several countries.

After end of tenure, I had several offers yet I decided to serve to the millions of resource poor farmers of India which has lead me to register a Professional Non Government Development Organization namely "Participatory Rural Development Initiative Society (PRDIS) during 1999. Dr.T. Goverdhan Reddy, Economist initially and Dr.S.J.Reddy Extensionist through out the life of the NGO supported the efforts. Dr.N.S.Reddy and Dr.M.Surya Mani, Consultants support is worth acknowledging. The main focus was on Agricultural Development and Natural Resource Management. The organization worked in several States of India including Chattisgarh and Jharkhand. Working with tribal people has again proved the power of Agricultural Extension. This has lead the organization to get a Presidential award by then Late A B J Abudul Kallam for promoting Sustainable Agricultural Development in tribal areas. In addition to capacity building, sustainable agricultural development a number of livelihood improvement activities were taken up benefiting millions of small scale resource poor farmers in rainfed areas. Timely councelling and guidance also helped farm families to eliminate the despair. Several extension approaches were implemented during this process. More than one hundred Extension Staff are working with the Organisation.

My family members have recently created a foundation by name "Sarvareddy Venkureddy Foundation for Development (SVFD)" dedicated for Agricultural Extension and welfare activities. My desire is that this discipline is grossly undermined. It is important that it is revamped and retooled. The lessons learnt from my 50 years of journey with Agricultural Extension is that with the dedicated Extension staff there is a big hope to change the lives and livelihoods of farm families. There is need for more investment for Agricultural Extension in Research, teaching, training, and field activities besides providing incentives and facilities for extension workers. This could be done through private public partnership. Let us all rededicate to eliminate hunger in the world with a goal of "Extension for All".

About the Authors

Daniel Gustafson, M.sc (Ag. Economics), Ph.D, Dy. Director General (Operations), FAO. As a US citizen started his career at the Inter-American Institute for Cooperation on Agriculture (IICA), working on a range of projects supporting Brazil's agricultural programmes, and rose to the position of Acting Country Representative for IICA. He served as FAO Country Representative in Kenya, Somalia, India and Bhutan. Subsequently, worked Director of FAO's Liaison Office for the US, Canada, and also FAO's Office of Support to Decentralization (OSD), based at FAO headquarters in Rome, Italy.
email: daniel.gustafson@fao.org

Dr. C. Prasad, M.Sc (Ag), Ph.D, Ex. Deputy Director General (Ag. Ext.), ICAR, New Delhi & served in different positions in India including Ex. Director, NAARM, Hyderabad and consultant FAO. Worked for different countries as Senior Advisor, FAO and done notable work for strengthening Agricultural Research, Education and Extension. He was honoured with number of awards and recognitions besides Life Time Achievement Award. He has over 250 papers to his credit on Agricultural Education and Extension including books, Bulletins, research papers and articles.
email: vardan_ngo@yahoo.com

Prof. R.K. Samanta, M.Sc (Ag), Ph.D, Consultant, Agri. Extn. Strategy Building, Rural Management and Human Resources Development, Formerly, Vice-chancellor, BCKV, West Bengal and Director,

MANAGE & NAARM, Hyderabad, India. He was expert member for different committee and served as Principal Investigator for different projects and also he has 22 books in the field of Agricultural Extension and Human Resource Development.
email: ranajitsamanta123@yahoo.com

Dr. M.N. Reddy, M.Sc (Ag), Ph.D former Director (Agril.Extn. &Commn.), National Institute of Agricultural Extension Management (MANAGE), GoI Hyderabad. He was Professor of Agricultural Extension in APAU for two decades. He served as Principal Coordinator for (DAESI) programme, Extension Reforms (ATMA), as International Consultant to Agricultural Sector Management Support Project, designed and launched Post Graduate Diploma in Agricultural Extension Management (PGDAEM) and also as National Facilitator for MANAGE-COVERDALE (U.K.) Program in the areas of Agricultural Extension Management. He was awarded Gold Medal in M.Sc.(Ag), and National Communication Research award for outstanding Research. *email: mnreddy2000@rediffmail.com*

Augustine Mujungu Nyamwegyendaho, is a Ugandan Specializing in Agricultural Extension and Training with over forty (40) years of experience. He was responsible for formulation and Implementation of Agricultural Policies and Programmes of the Ministry of Agriculture. As a consultant, he worked as training specialist on a World Bank funded Agricultural Extension Project in the Ministry of Agriculture responsible for Planning, Implementing, Monitoring and Evaluation of training activities. *email: augustine.nyamwe@gmail.com*

Dr.K.Narayana Gowda, Former Vice-Chancellor, UAS (B). He was recipient of Gold Medal for outstanding performance in Ph.D. Prof.Gowda served in 14 different positions in the University for more than four decades and he has published 74 papers in national and international journals and participated in 26 international seminars and conferences and authored seven books. He was Principal Investigator for six projects. He has visited ten different countries and participated in seminars and conferences besides established collaborative programmes. He received 16 awards; most important ones are Swami Sahajanand Extension Scientist Award of ICAR and Life Time achievement award of International Society of Extension Education (INSEE), Nagapur.

email: knarayanagowda@yahoo.co.in

Dr Bharat S. Sontakki, Ph.D (Agril. Extn), MBA (HRD) is currently working as Principal Scientist and Professor at the ICAR-National Academy of Agricultural Research Management, NAARM, Hyderabad. He specialized in management in agricultural extension and training and also educational technology. He has number of publications to his credit published in national and international journals. He has taught over 10 courses at masters and doctoral levels and rendered academic guidance to PG, Ph.D and MBA students.
email: bharatss@naarm.ernet.in

Dr.P.Gidda Reddy, M.Sc (Ag), Ph.D, former Director of Extension, ANGRAU. He worked in different capacities in the university. He has published 18 research articles in various National Journals and a book on "Farming Performance of Farm Women" and contributed an article on "Privatization of Extension" to the book "Privatization of Extension- Indian Experiences" published by MANAGE. He has rich experience of managing the out-reach programs. He has introduced innovative Flag Method in the field of Extension.
email: giddareddy53@yahoo.co.in

Dr. P. Chandra Shekara, M.Sc (Ag), Ph.D, **Director, (Agricultural Extension),** MANAGE, Hyderabad. 25 years of professional experience in Training, Research, Consultancy, Documentation and Dissemination in the field of Agricultural Extension Management. He has expertise in Agricultural Extension Management, Agri-Entrepreneurship Development, and Public-Private-Partnership. He is a recipient of Young Scientist Award-2007 and received Magnum National Honour Award -2011 and also Special Appreciation Award.
email: chandra@manage.gov.in

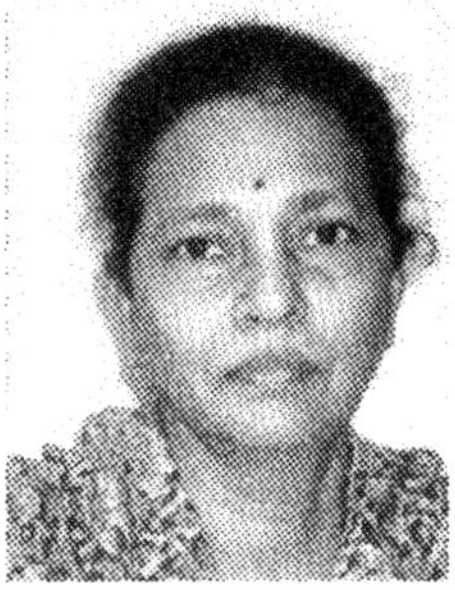

Prof. C. Beena, PhD in Psychology, Coordinator, Sahayam, Psychological Counselling Centre, Osmania University, Hyderabad. Beena Chintalapuri is committed to the field of Psychology which is evident from her research, training, teaching and Publications. She is associated with Human resources Development involved Organizations like Indian Space research Organization, Telangana Prisons and Teachers at different levels. She worked for 30 years in the University system. She held several prestigious positions as Registrar, Dean, Director, and Chairperson. She was recipient of Andhra Pradesh Best Teacher Award in the year 2008.

email:beena.chintalapuri@gmail.com

Dr. R. M. Prasad, M.Sc (Ag), Ph.D is a retired Associate Director of Extension of Kerala Agricultural University (KAU). He served as Advisor, Government of Meghalaya, Consultant (Agriculture) for the ADB and as Senior Fellow at NIRD, Hyderabad. Dr Prasad had served as Training Specialist in two European Union funded projects (KHDP and KMIP) in Kerala. He was also the National Facilitator of MANAGE, Hyderabad. Presently, he is the General Secretary of Farm Care Foundation (NGO) in Thrissur, Kerala.

email: drrmprasad@gmail.com

Dr. Richard W. Oliver, Founder & CEO, American Sentinel University, Aurora, Colorado. He served as a visiting professor of management at the Johnson Graduate School of Management at Cornell University and received the Outstanding Professor Award and a Dean's Teaching Award. Dr. Oliver had a 20-year career at Nortel Networks, where he was vice president of marketing, and he also worked for DuPont Co. Dr. Oliver is the author of seven books and more than 50 book chapters and journal articles about management, technology and education, and has lectured around the world on these topics.

email: rick.oliver@americansentinel.edu

Dr. V.P. Sharma, M.Sc (Ag), Ph.D working as Director, Information Technology, Documentation and Publications (ITDP) MANAGE, Hyderabad. Responsible for Planning, Implementing and Monitoring IT Resources and Services for the Institute including Hardware, Software and more importantly the human resources. Implemented Interactive Computer Video Project (ICVT). Resource person to various National and International organizations including FAO and Asian Productivity Organization, Tokyo, Japan and other National reputed institutions.

email: vpsharma@manage.gov.in

Dr. Shaik N.Meera, M.Sc (Ag), Ph.D, Principal Scientist, IIRR, Hyderabad. He made outstanding contributions in Indian rice sector with innovative extension methods and practical ICT approaches benefitting rice farmers and extension agencies. The Rice Knowledge Management Portal developed by him is acclaimed as one of the finest ICT applications in agriculture by Food and Agriculture Organization (FAO). He received several national and international awards such as Lal Bahadur Shastri Young Scientist Award of ICAR, INSEE Young Extension Professional and so on.

email: shaik_meera2005@yahoo.com

Dr.P.Venkataramaiah, MSc (Ag)., a Ph.D (Agri. Extn)., Dip.French & Russian, UAS Gold Medalist Former Prof. & Univ. Head of Extension, Former Principal, EEI, Hyderabad & Extn. Consultant: IOWA: USA Founder PPL: EEI. Prof. of Eminence for Far East countries & Taiwan under APC-Japan. World bank consultant (GOI). Trainer for ICISAT, MANAGE, NIRD, NPA, etc. Participated in UNESCO & SAARC seminars. Guided around 100 PG & Ph.D researchers. Authored books, manuals-RAWEP and architect of SES scale.

email: revathi.puppala@icloud.com

Dr. I. Sreenivasa Rao, M.Sc (Ag), Ph.D is the Professor and University Head, Department of Agricultural Extension, PJTSAU, Hyderabad. He got Doctoral degree with Gold Medal from ANGRAU. He has unique experience of more than 22 years in Teaching, Training and Extension as well as guiding Post graduate students. Dr.Rao has attended various Professional conferences to present research papers in USA, China and other countries. He has published nearly 40 research articles in reputed Indian and Foreign Journals.
email: illuris@gmail.com

Dr. S.Senthil Vinayagam, M.Sc (Ag), Ph.D has experience over 25 years in Agricultural Extension System. Currently, working as Professor and Principal Scientist (Agri. Extension) at ICAR- National Academy of Agricultural Research Management (NARRM), Hyderabad. He has Ph.D (Agri. Extension) and expertise in Entrepreneurship Development and Agri.- Business. He worked in various capacities in ICAR institutes and worked as Faculty Head. He has completed 14 research projects (4 Consultancy) and over 70 publications covering research papers, books, chapters, manuals and conference papers.
email: senthil@naarm.er.net.in

Prof. Devi Prasad Juvvadi, M.Sc (Ag), Ph.D is Director, Agriculture Management at Centre for Good Governance, Hyderabad. He has three decades of experience in agricultural research, teaching and consultancy. He worked as Assistant Professor at American University of Beirut for a decade and adjunct to US-Saudi Joint Commission on Economic Co-operation in Riyadh, Saudi Arabia. He also worked as Consultant to various international projects of UNDP, World Bank, DFID in Middle East, Africa and Europe.
email: deviprasad.j@cgg.gov.in

Dr.V. V. Sadamate, M.Sc (Ag), Ph.D Agricultural Extension Specialist & Former Adviser Agriculture, Planning Commission, GOI, New Delhi. He held different positions in Govt. of India, Ministry of Agriculture and Principal Consultant, FAO, New Delhi. He has number of publications and also guided PG students for Ph.D and M.Sc degrees. He was awarded with Feel Bright Fellowship in Post Doctoral Study in USA. Presently working as Agricultural Convergent Expert, MoRD, World Bank Project.
email:sadamatevv@gmail.com

Prof. Hemnath Rao H. M.B.A., Ph.D. Formerly Professor; Director, Centre for Poverty Studies and Rural Development; & Dean Administrative Staff College of India (ASCI) and currently Director, StratLead International. Designed and delivered numerous management capacity building programmes for senior civil servants and corporate executives and has led policy level studies for the Ministries of Agriculture, Rural Development, MSME, Water Resources, HRD and others in GoI besides consulting for ComSec, UNDP and World Bank.
email: hemnathrao@gmail.com

Dr.R.Ratnakar, M.Sc (Ag), Ph.D A professional in the field of Agricultural Extension and Administration with 33 years of experience and expertise in the areas, Agri and Rural development, Irrigation management, Training and Capacity building, M&E of development programs and projects, Research and Field Extension. Former Director, Southern Region, EEI, GoI. Presently the Team Leader, External Monitoring, AP, Telangana States Community Based Tank Management Project (World Bank). Consultant: Asian Development Bank and FAO projects. Recipient of International, National and State awards, cash prize and appreciation. As a visiting Professor he delivered special lectures in USA, Philippines, Hong Kong and so on.
email: ratnakar244@gmail.com

Dr. Geethakutty.P.S, M.Sc (Ag), Ph.D Professor and Project Co-ordinator, Centre for Gender Studies in Agriculture of the Kerala Agricultural University (KAU), Kerala, India and trained on gender and agriculture from IAC, Wagheningen, Netherlands and FAO, Rome. She had founded the Centre for Gender Concerns in Agriculture at KAU . She has been teaching and guiding students of Agriculture of KAU and imparting training to agricultural functionaries and farmers. She had served as Visiting Professor of the Leibniz University, Hanover in Germany, as Director of the Centre for Women and Gender Studies in the National Institute of Rural Development (NIRD),Hyderabad.

email: geethakutty@gmail.com

Dr AVR Kesava Rao, M.Sc (Ag), Ph.D is a Scientist (Agroclimatology) at the ICRISAT Development Centre, ICRISAT. He has about 30 years of experience in teaching, research and extension in Agroclimatology. He developed software for agroclimatic analysis. Trained on agrometeorology, remote sensing, GIS, climate science in India, Hadley Met Centre, UK and in crop modelling and precision farming at the University of Florida, USA. He guided a few M.Sc. and Ph.D. students in Agroclimatology. Visited several countries viz., USA, UK, Kenya etc. Has 52 research papers published in national and international journals, books and conference proceedings.

email: avrkrao@yahoo.com

~~U-197-35-0795-K~~

~~Rs 1995~~

~~11/11/18~~ (1)

(24) (24)

49